Discrete Hamiltonian Systems

Kluwer Texts in the Mathematical Sciences

VOLUME 16

A Graduate-Level Book Series

The titles published in this series are listed at the end of this volume.

Discrete Hamiltonian Systems

Difference Equations, Continued Fractions, and Riccati Equations

by

Calvin D. Ahlbrandt
University of Missouri

and

Allan C. Peterson
University of Nebraska

Springer-Science+Business Media, B.V.

A C.I.P. Catalogue record for this book is available from the Library of Congress.

DOI 10.1007/978-1-4757-2467-7

Printed on acid-free paper

Dedicated to our families

Evelyn, Robert, William, and Michael

and

Tina, Carla, David, and Carrie

CONTENTS

PREFACE

This book should be accessible to students who have had a first course in matrix theory. The existence and uniqueness theorem of Chapter 4 requires the implicit function theorem, but we give a self–contained constructive proof of that theorem. The reader willing to accept the implicit function theorem can read the book without an advanced calculus background. Chapter 8 uses the Moore–Penrose pseudo–inverse, but is accessible to students who have facility with matrices. Exercises are placed at those points in the text where they are relevant. For U.S. universities, we intend for the book to be used at the senior undergraduate level or beginning graduate level. Chapter 2, which is on continued fractions, is not essential to the material of the remaining chapters, but is intimately related to the remaining material. Continued fractions provide closed form representations of the extreme solutions of some discrete matrix Riccati equations. Continued fractions solution methods for Riccati difference equations provide an approach analogous to series solution methods for linear differential equations.

The book develops several topics which have not been available at this level. In particular, the material of the chapters on continued fractions (Chapter 2), symplectic systems (Chapter 3), and discrete variational theory (Chapter 4) summarize recent literature. Similarly, the material on transforming Riccati equations presented in Chapter 3 gives a self–contained unification of various forms of Riccati equations.

Motivation for our approach to difference equations came from the work of Harris, Vaughan, Hartman, Reid, Patula, Hooker, Erbe & Yan, and Bohner. Students usually encounter three term recurrences in the study of series solutions of second order linear differential equations. As with second order linear differential equations, the structure of the solution space of recurrence relations is in some ways best expressed in terms of the self-adjoint form of the difference equation. When one writes three term recurrence relations in self-adjoint form, the theory becomes more like differential equations than linear algebra. Then the powerful tools of discrete variational theory can be applied to the study of these difference equations. The present study is motivated by various linear fractional recurrence relations occuring in applications. These include matrix Riccati equations of classical Kalman filtering and the dual problem of the discrete regulator problem in optimal control theory. Vaughan's succinct 1970 paper in IEEE Transactions on Automatic Control revealed the associated symplectic structure of a corresponding linear system. We now explain the discrete Sturmian theory and variational origins of these discrete Hamiltonian systems. Our purpose here is to present recent unifications of those topics. This is not the final word on the subject, as closer connections between the subjects of optimal control, filtering theory, discrete variational theory, and symplectic continued fractions are being revealed in current research. In particular, we have not fully treated general discrete variational problems with constraints.

One central theme of the linear aspects of this book is the formulation of a discrete version of Reid's "Roundabout Theorem." Such theorems equate positive definiteness of a quadratic functional with existence of a nonsingular solution across the interval, which in turn is equivalent to the existence of a Hermitian solution of a discrete Riccati equation across the interval. In one sense, these results are Morse Index Theorems when the index is 0, but the Riccati equivalence is what has made Reid's Theorem applicable. Another way of viewing the Reid Roundabout Theorem is that it is a Jacobi condition equivalence to the second variation being positive definite. Consequently, strong sign conditions on the coefficients, which often occur in the applications, make the quadratic functional positive definite, provide disconjugacy of the linear system, and give existence of solutions of the matrix Riccati equation. When applied to half lines or to the whole real axis, disconjugacy implies existence of extreme solutions of matrix Riccati equations and representations of those extreme solutions by symplectic matrix continued fractions.

The book is somewhat modular in that we start anew in most chapters. We could have presented the theory in a general setting and then obtained the special cases relevant to applications as corollaries. Instead, we start with the scalar case in Chapter 1 where we introduce the topic of symplectic matrices.

Chapter 2 is a self-contained introduction to continued fractions. We first study scalar continued fractions associated with sequences of 2×2 matrices. We show how the theory differs if these matrices are companion matrices or if they are symplectic. Continued fractions are unified with the rest of the book via a "Pincherle" theorem which connects convergence of continued fractions with existence of recessive and dominant solutions of associated linear systems. It is interesting that the proof of equivalence of convergence of a classical continued fraction with convergence of an associated series, which dates back to Euler, is very closely related to the self-adjoint form of the associated linear recurrence relation. Thus for continued fractions, symplectic theory plays somewhat the same central role as the self–adjoint theory of three term recurrences. We show how classical continued fractions are related to our symplectic continued fractions and apply this relationship to Gautschi's Bessel function example. Then we present matrix continued fractions including symplectic continued fractions related to the discrete matrix Riccati equations of control and filtering. In an optional section we outline a theory of continued fractions in a normed ring.

Chapter 3 develops a theory of linear symplectic systems. Of special note is the reduction of order formula which allows discussion of recessive and dominant solutions. This chapter contains some general comments about matrix Riccati equations. We show how many of the discrete Riccati equations which arise in the applications are equivalent to those which are determined by symplectic linear systems and discrete variational theory.

Chapter 4 is on discrete variational theory. It can be read without completing the earlier material. We start the chapter with fixed stepsize of 1; but starting in Section 4.7, we allow variable step sizes. Chapter 4 is nonlinear in nature. In particular, we use a discretization of the minimal surface of revolution variational problem to show how spurious solutions can arise and how the implicit function theorem selects the correct solution. A general existence and uniqueness theorem for local solutions of discrete Hamiltonian systems is developed. We show that our discrete Hamiltonian systems provide symplectic flows, even in the nonlinear case. It has been observed that solutions of certain difference equations can cross asymptotes, whereas solutions of the corresponding continuous autonomous problem cannot. Also, solutions of difference equations can have global existence even though the corresponding solutions in the continous case have finite escape time. This computational anomaly is often a consequence of using fixed step size. However, the Existence and Uniqueness Theorem of Chapter 4 shows how the implicit function theorem dictates that variable step size may be required in order to avoid spurious solutions. Furthermore, from a given point in phase space, if one lets the step size h go to zero, one can see that discrete linear Hamiltonian solutions give estimates for the derivatives of the continuous solution through the base point. This allows transition from discrete Riccati equations with variable step size h to the usual continuous Riccati equations. This is not intuitive because the right–hand side of a discrete Riccati equation is linear fractional, whereas the right–hand side of a continuous Riccati equation is a quadratic matrix polynomial.

Chapter 5 presents a Sturmian theory for symmetric three term recurrences. This development is largely self–contained. There is some repetition of earlier material, but, historically, this development came before the more general theory of Chapter 3. The reverse reduction of order theorem presented there is new. This culminated in the general unification of reduction of order formulas for systems which is presented in Exercise 3.34 on pages 111–112. That exercise also presents a novel summation operator S which does not require the index shifting in the usual fundamental theorem for difference operators. That summation operator S has additivity properties much like the integral operator property $\int_a^b = \int_a^c + \int_c^b$.

Chapter 6 discusses discrete Riccati equations associated with the symmetric three term recurrence relations of Chapter 5.

Chapter 7 constructs a Green's function for nonhomogeneous three term equations. Chapter 8 gives some disconjugacy and disfocality theorems for bounded intervals. These are patterned after integral tests for nonoscillation of differential equations.

Chapter 9 is an introduction to the Bohner theory. This subject was motivated by the question of existence of a Reid Roundabout Theorem for discrete linear Hamiltonian systems which include even–order scalar self–adjoint difference equations as a special case. A complication occurs because the B matrix for that example is $n \times n$ of rank 1 and the problem is not equivalent to a three term recurrence. A satisfactory resolution of that problem has recently been presented by Martin Bohner.

We would like to express our appreciation to those who helped us in editing and proofreading. Our current students, Doug Anderson, Don Steiger, Tammy Voepel, and John Weatherwax, commented on several drafts of this book. We also thank our co–authors for participating in the development of this subject over the years. In particular, various aspects of this work would not have been possible without the help of Carmen Chicone, Steve Clark, Michael Heifetz, John Hooker, Bill Patula, Tim Peil, and Jerry Ridenhour.

The connection with continued fractions was made possible by guidance from Jerry Lange, Hans–J. Runckel, Paul Levrie, and Milos Znojil.

Finally, we thank Martin Bohner at Universität Hohenheim in Stuttgart for his reading of the whole manuscript, for resolving LaTeX difficulties, and for his incredible work in extending the theory to the singular case. Only a sample of Bohner's work is included here, but Chapter 9 provides an introduction to the "Bohner Theory" cited in the bibliography.

1

SECOND ORDER SCALAR DIFFERENCE EQUATIONS

1.1 DIFFERENCE EQUATIONS AND RECURRENCE RELATIONS

We shall start our study with second order linear difference equations. We will show how they can be written as equivalent first order systems which have a particular form called symplectic. Later chapters will show how these symplectic systems contain discrete linear Hamiltonian systems. We will also use the linear theory in order to motivate the symplectic structure of general nonlinear discrete Hamiltonian systems. There are interconnections between these subjects and the topics of discrete variational theory, discrete matrix Riccati equations, and what we call symplectic continued fractions. But we will start with some discussion about the simplest scalar problems.

In this chapter we will be concerned with the scalar second order formally self–adjoint difference equation

$$Lu(t) \equiv \Delta[p(t)\Delta u(t-1)] + q(t)u(t) = 0 \tag{1.1}$$

where t takes on values in the discrete interval $[a+1, b+1] \equiv \{a+1, \ldots, b+1\}$, where $a \leq b$ are integers, and Δ is the forward difference operator defined by $\Delta u(t) = u(t+1) - u(t)$. We assume $p(t)$ is nonzero and real valued for each t in $[a+1, b+2]$. We also assume that $q(t)$ is real valued on $[a+1, b+1]$.

Expanding out the terms in $Lu(t)$ it is easy to see that

$$Lu(t) = p(t+1)u(t+1) + c(t)u(t) + p(t)u(t-1) \tag{1.2}$$

where

$$c(t) = q(t) - p(t) - p(t+1), \qquad t \in [a+1, b+1]. \tag{1.3}$$

Theorem 1.1 *The initial value problem (abbreviated by* **IVP***)*

$$\begin{aligned} Lu(t) &= h(t), \quad t \in [a+1, b+1], \\ u(t_0) &= u_0, \qquad u(t_0+1) = u_1, \end{aligned}$$

where $t_0 \in [a, b+1]$, u_0, u_1 are complex constants, and $h(t)$ is defined on the interval $[a+1, b+1]$, has a unique solution $u(t)$ which is defined on $[a, b+2]$.

Proof: Using (1.2) we can write the nonhomogeneous difference equation $Lu(t) = h(t)$ in the form of a nonhomogeneous three term recurrence relation

$$p(t+1)u(t+1) + c(t)u(t) + p(t)u(t-1) = h(t).$$

The conclusion of the theorem follows from the fact that this last equation can be uniquely solved for $u(t+1)$ from known values of $u(t)$ and $u(t-1)$ since $p(t+1) \neq 0$. It also can be uniquely solved for $u(t-1)$ from known values of $u(t)$ and $u(t+1)$ since $p(t) \neq 0$. □

Now consider a three term homogeneous recurrence relation

$$p_2(t)u(t+1) + p_1(t)u(t) + p_0(t)u(t-1) = 0, \tag{1.4}$$

for $t \in [a+1, b+1]$, where the coefficient functions are defined on $[a+1, b+1]$ and we assume

$$p_0(t)p_2(t) \neq 0, \qquad \text{for each } t \text{ in } [a+1, b+1]. \tag{1.5}$$

It follows from (1.2) and (1.3) that the self–adjoint equation (1.1) can be written in the form (1.4). We now would like to show that (1.4) can be transformed to an equivalent self–adjoint difference equation. To see this multiply both sides of (1.4) by a function $\alpha(t) \neq 0$ to be determined later. Then (1.4) has the same solutions as

$$\alpha(t)p_2(t)u(t+1) + \alpha(t)p_1(t)u(t) + \alpha(t)p_0(t)u(t-1) = 0. \tag{1.6}$$

Comparing this with (1.1) and (1.2) we see that we want

$$\begin{aligned} \alpha(t+1)p_0(t+1) &= \alpha(t)p_2(t), \text{ i.e.,} \\ \alpha(t+1) &= \frac{p_2(t)}{p_0(t+1)}\alpha(t). \end{aligned}$$

Take

$$\alpha(t) = A \prod_{s=a+1}^{t-1} \frac{p_2(s)}{p_0(s+1)}, \qquad t \in [a+1, b+2]$$

where, for convenience, we have set $p_0(b+2) \equiv p_0(b+1)$ and $\alpha(a+1) = A$ for some nonzero constant A. Note that in the above formula for $\alpha(t)$ wuth $t = a+1$, i.e., for $\alpha(a+1)$ we are using the usual *product convention* that the product from a lower index of $a+1$ to an upper index of a is defined to be 1.

Choose $p(t)$ as

$$p(t) = \alpha(t)p_0(t), \qquad t \in [a+1, b+2]$$

and note that $p(t) \neq 0$ for each t in $[a+1, b+2]$. Equation (1.6) now takes the form

$$p(t+1)u(t+1) + \alpha(t)p_1(t)u(t) + p(t)u(t-1) = 0,$$

i.e.,

$$p(t+1)\Delta u(t) + p(t+1)u(t) + \alpha(t)p_1(t)u(t) - p(t)\Delta u(t-1) + p(t)u(t) = 0,$$

which becomes

$$\Delta[p(t)\Delta u(t-1)] + q(t)u(t) = 0,$$

where

$$q(t) = p(t) + p(t+1) + \alpha(t)p_1(t).$$

Thus our theory also applies to nonsymmetric three term recurrences because they can always be transformed into formally self–adjoint equations.

Example 1.2 *Write the Fibonacci recurrence relation*

$$u(t+1) = u(t) + u(t-1), \qquad t \geq 1,$$

in self–adjoint form.

Proceed as above by writing

$$u(t+1) - u(t) - u(t-1) = 0$$

as

$$\alpha(t)u(t+1) - \alpha(t)u(t) - \alpha(t)u(t-1) = 0$$

where we want

$$-\alpha(t+1) = \alpha(t).$$

Choose $\alpha(t) = (-1)^t$ *for*

$$(-1)^t u(t+1) - (-1)^t u(t) - (-1)^t u(t-1) = 0,$$

i.e,

$$(-1)^t \Delta u(t) - (-1)^{t-1} \Delta u(t-1) + (-1)^{t-1} u(t) = 0,$$

and

$$\Delta[(-1)^{t-1} \Delta u(t-1)] + (-1)^{t-1} u(t) = 0.$$

Multiply this last equation by -1 *to see that the Fibonacci recurrence relation is equivalent to the self–adjoint equation*

$$\Delta[(-1)^t \Delta u(t-1)] + (-1)^t u(t) = 0.$$

1.2 SECOND ORDER SCALAR EQUATIONS AS SYMPLECTIC SYSTEMS

We now wish to show that second order self–adjoint difference equations can be written as equivalent first order symplectic systems. A complex 2×2 matrix M is said to be *symplectic* provided

$$M^* J M = J$$

where

$$J = \begin{bmatrix} 0 & 1 \\ -1 & 0 \end{bmatrix}$$

and M^* denotes the conjugate transpose of M. Note $J^2 = -I$, where I is the 2×2 identity matrix.

Exercise 1.3 *Show that a* 2×2 *real matrix* M

$$M = \begin{bmatrix} e & f \\ g & h \end{bmatrix}$$

is symplectic if and only if $\det M = 1$.

Theorem 1.4 *The set of all* 2×2 *symplectic matrices forms a group under matrix multiplication.*

Proof: Since

$$I^*JI = J$$

the identity matrix is symplectic. Now assume M is symplectic. Then

$$M^*JM = J$$

which implies that

$$|\det M|^2 = 1.$$

Hence M is nonsingular and M^{-1} exists. Since $J^2 = -I$, we obtain

$$J^{-1} = -J = J^*.$$

The assumption that M is symplectic, i.e.,

$$M^*JM = J$$

implies

$$(J^{-1}M^*J)M = I$$

and hence M has a left inverse (thus an inverse) given by

$$M^{-1} = J^{-1}M^*J = J^*M^*J. \tag{1.7}$$

In order to show that the inverse of a symplectic matrix is symplectic, use

$$(M^{-1})^*JM^{-1} = (J^*MJ)JM^{-1} = J^*M(-I)M^{-1} = -J^* = J.$$

Hence M^{-1} is a symplectic matrix. The associative law of multiplication holds because it holds for matrix multiplication. Finally assume M and N are symplectic and consider

$$(MN)^*J(MN) = N^*M^*JMN = N^*JN = J.$$

Therefore MN is symplectic and we have proven that the set of symplectic 2×2 matrices is a group under multiplication. □

Note that the form of M^{-1} given in (1.7) implies that for M labeled as

$$M = \begin{bmatrix} e & f \\ g & h \end{bmatrix} \qquad \text{we have} \qquad M^{-1} = \begin{bmatrix} \bar{h} & -\bar{f} \\ -\bar{g} & \bar{e} \end{bmatrix}. \tag{1.8}$$

This is reminiscent of the rule for inverting 2×2 matrices with determinant one. Indeed it reduces to that for 2×2 symplectic matrices with real entries. However, for the example of the diagonal matrix D with i and i on the diagonal, this rule is not the same as the usual rule, since $\det D = -1$ and furthermore the usual rule does not have complex conjugates on the entries. This rule for finding the inverse will extend to the $2n \times 2n$ symplectic case, (see Chapter 3), where the $n \times n$ block entries on the main diagonal are interchanged, the signs are changed on the off diagonal blocks and conjugate transposes are taken of all entries.

We now show how to write the self–adjoint difference equation $Lu(t) = 0$ as an equivalent symplectic system. Let $u(t)$ be a solution of $Lu(t) = 0$ and set

$$\begin{aligned} y(t) &= u(t-1), \qquad & t \in [a+1, b+3] \\ z(t) &= p(t)\Delta u(t-1), \qquad & t \in [a+1, b+2]. \end{aligned}$$

Then for $t \in [a+1, b+2]$,

$$\Delta y(t) = \Delta u(t-1) = \frac{1}{p(t)} z(t)$$

implies

$$y(t+1) = y(t) + \frac{1}{p(t)} z(t). \tag{1.9}$$

Also, for $t \in [a+1, b+1]$,

$$\begin{aligned} \Delta z(t) &= \Delta[p(t)\Delta u(t-1)] = -q(t)u(t) \\ &= -q(t)y(t+1) \\ &= -q(t)y(t) - \frac{q(t)}{p(t)} z(t). \end{aligned}$$

Hence, for $t \in [a+1, b+1]$,

$$z(t+1) = -q(t)y(t) + \left[1 - \frac{q(t)}{p(t)}\right] z(t). \tag{1.10}$$

Define a vector $x(t)$ for $t \in [a+1, b+2]$ by

$$x(t) = \begin{bmatrix} y(t) \\ z(t) \end{bmatrix}.$$

Then by equations (1.9) and (1.10)

$$x(t+1) = M(t)x(t), \qquad t \in [a+1, b+1] \tag{1.11}$$

where

$$M(t) = \begin{bmatrix} 1 & \frac{1}{p(t)} \\ -q(t) & 1 - \frac{q(t)}{p(t)} \end{bmatrix}, \tag{1.12}$$

for $t \in [a+1, b+1]$. Since

$$M^*(t)JM(t) = \begin{bmatrix} 1 & -q(t) \\ \frac{1}{p(t)} & 1-\frac{q(t)}{p(t)} \end{bmatrix} \begin{bmatrix} -q(t) & 1-\frac{q(t)}{p(t)} \\ -1 & -\frac{1}{p(t)} \end{bmatrix} = J,$$

we conclude that $M(t)$ is a symplectic matrix for each $t \in [a+1, b+1]$. Because of this property of the coefficient matrix $M(t)$, we say the system (1.11) is a *symplectic system.*

Conversely, if

$$x(t) = \begin{bmatrix} y(t) \\ z(t) \end{bmatrix}$$

is a solution of the symplectic system (1.11) on $[a+1, b+2]$, then $u(t) = y(t+1)$ is a solution of the scalar equation $Lu(t) = 0$ on $[a, b+2]$ if we define $u(b+2)$ appropriately.

Exercise 1.5 *Use Exercise* 1.3 *to show that* $M(t)$ *of equation* (1.12) *is symplectic for each* $t \in [a+1, b+1]$.

We now show an alternate way of writing $Lu(t) = 0$ as an equivalent symplectic system. Let $u(t)$ be a solution of $Lu(t) = 0$ and this time set

$$\begin{aligned} y(t) &= u(t), & t \in [a, b+2] \\ z(t) &= p(t)\Delta u(t-1), & t \in [a+1, b+2]. \end{aligned}$$

Then for $t \in [a+1, b+1]$

$$\Delta z(t) = \Delta[p(t)\Delta u(t-1)] = -q(t)u(t).$$

Hence

$$z(t+1) = -q(t)y(t) + z(t) \tag{1.13}$$

for $t \in [a+1, b+1]$. On the other hand, for $t \in [a, b+1]$

$$\begin{aligned} z(t+1) &= p(t+1)\Delta u(t) \\ &= p(t+1)y(t+1) - p(t+1)y(t). \end{aligned}$$

Thus, for $t \in [a, b+1]$,

$$y(t+1) = y(t) + \frac{1}{p(t+1)} z(t+1).$$

Using (1.13) we get that

$$y(t+1) = \left[1 - \frac{q(t)}{p(t+1)}\right] y(t) + \frac{1}{p(t+1)} z(t) \tag{1.14}$$

for $t \in [a+1, b+1]$. From (1.13) and (1.14) we have that

$$x(t+1) = M_1(t)x(t), \qquad t \in [a+1, b+1]$$

where

$$x(t) = \begin{bmatrix} y(t) \\ z(t) \end{bmatrix}, \qquad M_1(t) = \begin{bmatrix} 1 - \frac{q(t)}{p(t+1)} & \frac{1}{p(t+1)} \\ -q(t) & 1 \end{bmatrix}. \tag{1.15}$$

Note that for $t \in [a+1, b+1]$

$$\begin{aligned} M_1^*(t)JM_1(t) &= \begin{bmatrix} 1 - \frac{q(t)}{p(t+1)} & -q(t) \\ \frac{1}{p(t+1)} & 1 \end{bmatrix} \begin{bmatrix} -q(t) & 1 \\ -1 + \frac{q(t)}{p(t+1)} & \frac{-1}{p(t+1)} \end{bmatrix} \\ &= \begin{bmatrix} 0 & 1 \\ -1 & 0 \end{bmatrix} = J. \end{aligned}$$

So $M_1(t)$ is symplectic on $[a+1, b+1]$.

Example 1.6 *Let $u(t)$ and $v(t)$ be solutions of $Lu(t) = 0$. Then*

$$X(t) = \begin{bmatrix} u(t-1) & v(t-1) \\ p(t)\Delta u(t-1) & p(t)\Delta v(t-1) \end{bmatrix}$$

and

$$X_1(t) = \begin{bmatrix} u(t) & v(t) \\ p(t)\Delta u(t-1) & p(t)\Delta v(t-1) \end{bmatrix}$$

have constant rank on $[a+1, b+2]$.

Proof: Since $u(t)$ and $v(t)$ are solutions of $Lu(t) = 0$ we have

$$X(t+1) = M(t)X(t) \quad \text{and} \quad X_1(t+1) = M_1(t)X_1(t)$$

for M and M_1 of (1.12) and (1.15), respectively. Since $|\det M(t)| = 1 \neq 0$, $M(t)$ is nonsingular and hence $X(t)$ has constant rank on $[a+1, b+2]$. Similarly $X_1(t)$ has constant rank on $[a+1, b+2]$. □

We introduced symplectic systems here to motivate what we will do in Chapter 3. In Chapter 3 we will study $2n$–dimensional symplectic systems which include system (1.1) as a special case.

For the rest of this chapter we study equation (1.1).

1.3 WRONSKIANS OF SOLUTIONS

If $u(t)$ and $v(t)$ are defined on $[a, b+2]$ we define the *Wronskian* (or *Casoratian*) of u, v by

$$w[u(t), v(t)] = \begin{vmatrix} u(t) & v(t) \\ u(t+1) & v(t+1) \end{vmatrix} = \begin{vmatrix} u(t) & v(t) \\ \Delta u(t) & \Delta v(t) \end{vmatrix} \tag{1.16}$$

for $t \in [a, b+1]$.

Theorem 1.7 *If $u(t)$, $v(t)$ are complex valued functions on $[a, b+2]$, then*

$$\overline{v(t)}Lu(t) - u(t)\overline{Lv(t)} = \Delta\{v(t); u(t)\}$$

for $t \in [a+1, b+1]$, where the Lagrange bracket $\{v(t); u(t)\}$ is defined by

$$\{v(t); u(t)\} = p(t)w[\overline{v(t-1)}, u(t-1)]$$

for $t \in [a+1, b+1]$.

Proof: For $t \in [a+1, b+1]$, consider

$$\begin{aligned}
\overline{v(t)}Lu(t) &= \overline{v(t)}\Delta[p(t)\Delta u(t-1)] + \overline{v(t)}q(t)u(t) \\
&= \Delta[\overline{v(t-1)}p(t)\Delta u(t-1)] - \Delta\overline{v(t-1)}p(t)\Delta u(t-1) + \overline{v(t)}q(t)u(t) \\
&= \Delta\{\overline{v(t-1)}p(t)\Delta u(t-1) - u(t-1)p(t)\Delta\overline{v(t-1)}\} \\
&\quad + u(t)\Delta[p(t)\Delta\overline{v(t-1)}] + u(t)q(t)\overline{v(t)} \\
&= \Delta\{p(t)w[\overline{v(t-1)}, u(t-1)]\} + u(t)\overline{Lv(t)}
\end{aligned}$$

which gives us the desired result. □

Exercise 1.8 *Derive Theorem 1.7 by calculating $\Delta\{v(t); u(t)\}$.*

As a corollary of Theorem 1.7 we get the following result.

Corollary 1.9 (Liouville's Formula) *If $u(t)$ and $v(t)$ are solutions of the equation $Lu(t) = 0$, then*

$$\{u(t); v(t)\} = \text{ constant}, \qquad t \in [a+1, b+1].$$

In particular

$$w[\overline{u(t)}, v(t)] = C/p(t), \qquad t \in [a, b+1]$$

where C is a constant.

1.4 PREPARED SOLUTIONS

It follows from Corollary 1.9 that if $u(t)$ is a solution of $Lu(t) = 0$, then

$$\{u(t); u(t)\} = C, \qquad t \in [a, b+1],$$

where C is a constant. If this constant C is zero then we say $u(t)$ is a *prepared solution* of $Lu(t) = 0$. Hence if $u(t)$ is a prepared solution of $Lu(t) = 0$, then

$$p(t)w[\overline{u(t-1)}, u(t-1)] = 0, \qquad t \in [a, b+1]$$

and therefore

$$p(t)\overline{u(t-1)}\Delta u(t-1) = p(t)u(t-1)\Delta\overline{u(t-1)}, \qquad t \in [a, b+1].$$

Consequently,

$$u(t-1)\overline{u(t)} = \overline{u(t-1)}u(t)$$

for $t \in [a, b+1]$. Actually $u(t)$ is a prepared solution of $Lu(t) = 0$ iff ($\equiv$ *if and only if*) $u(t-1)\overline{u(t)}$ is real valued on $[a+1, b+2]$.

Theorem 1.10 *The function $u(t)$ is a prepared solution of $Lu(t) = 0$ if and only if $u(t) = Cv(t)$ where C is complex and $v(t)$ is a real solution of $Lu(t) = 0$.*

Proof: First assume $u(t) = Cv(t)$ where $v(t)$ is a real solution of $Lu(t) = 0$. Then $u(t)$ is a solution of $Lu(t) = 0$ with

$$u(t-1)\overline{u(t)} = |C|^2 v(t-1)v(t)$$

real valued on $[a+1, b+2]$. Hence $u(t)$ is a prepared solution of $Lu(t) = 0$. Conversely, assume $u(t)$ is a nontrivial prepared solution of $Lu(t) = 0$. Because of the linearity of L, it suffices to show that two consecutive values of $u(t)$ lie on the same straight line through the origin. Let $t_0 \in [a+1, b+2]$ such that

$$u(t_0 - 1) \neq 0, \qquad u(t_0) \neq 0.$$

Then we can write $u(t_0 - 1) = re^{i\theta}$ and $u(t_0) = se^{i\phi}$ where $r > 0$ and $s > 0$. Since

$$u(t_0 - 1)\overline{u(t_0)} = rse^{i(\theta - \phi)}$$

is real, $\theta - \phi = n\pi$ for some integer n. It follows that the line through $u(t_0 - 1)$ and $u(t_0)$ in the complex plane passes through the origin. Now assume $t_1 \in [a+1, b+1]$ and $u(t_1) = 0$. It follows that $u(t_1 - 1) \neq 0$ and $u(t_1 + 1) \neq 0$. Using (1.2) and $u(t_1) = 0$ we get that

$$p(t_1 + 1)u(t_1 + 1) + p(t_1)u(t_1 - 1) = 0.$$

Hence,

$$u(t_1 + 1) = -\frac{p(t_1)}{p(t_1 + 1)}u(t_1 - 1).$$

It follows that $u(t_1 - 1)$ and $u(t_1 + 1)$ lie on the same line passing through the origin in the complex plane. Thus, if $u(t)$ is a prepared solution of $Lu(t) = 0$, then the points $u(t)$, $a \leq t \leq b+2$, all lie on the same line passing through the origin in the complex plane. It follows that

$$u(t) = Cv(t)$$

where C is complex and $v(t)$ is a real valued solution of $Lu(t) = 0$. □

1.5 GENERALIZED ZEROS OF SOLUTIONS

We now make a definition of *generalized zeros* . A nontrivial prepared solution $u(t)$ of $Lu(t) = 0$ has a *generalized zero at a* if and only if $u(a) = 0$. We say $u(t)$ has a *generalized zero at t_0* $\in [a+1, b+2]$ provided $u(t_0 - 1) \neq 0$ and

$$p(t_0)u(t_0 - 1)\overline{u(t_0)} \leq 0.$$

If $p(t) > 0$ and we consider only real valued solutions, then this definition agrees with that given by Hartman [76].

Example 1.11 *The Fibonacci difference equation*

$$\Delta[(-1)^t \Delta u(t-1)] + (-1)^t u(t) = 0$$

has solutions

$$u(t) = \lambda_1^t, \qquad v(t) = \lambda_2^t,$$

where

$$\lambda_1 = \frac{1-\sqrt{5}}{2} < 0, \qquad \lambda_2 = \frac{1+\sqrt{5}}{2} > 0.$$

Since

$$\begin{aligned} p(t)u(t-1)u(t) &= (-1)^t \lambda_1^{2t-1} \\ p(t)v(t-1)v(t) &= (-1)^t \lambda_2^{2t-1} \end{aligned}$$

$u(t)$ has a generalized zero at each even integer and $v(t)$ has a generalized zero at each odd integer. Note that by Hartman's definition $v(t)$ has no generalized zeros and $u(t)$ has a generalized zero at every integer. Consequently, our terminology gives the interlacing property of generalized zeros of linearly independent solutions whereas Hartman's definition does not.

Exercise 1.12 *Find two linearly independent solutions of the difference equation*

$$\Delta[(-1)^t \Delta u(t-1)] + 2(-1)^t u(t) = 0$$

and determine where their generalized zeros are.

1.6 DISCONJUGACY AND THE REID ROUNDABOUT THEOREM

We say that $Lu(t) = 0$ is *disconjugate on* $[a, b+2]$ provided there is no nontrivial prepared solution of $Lu(t) = 0$ with two generalized zeros on $[a, b+2]$. By Theorem 1.10 if $u(t)$ is a nontrivial prepared solution of $Lu(t) = 0$, then $u(t) = Cv(t)$ where $C \neq 0$ and $v(t)$ is a real solution of $Lu(t) = 0$. It is easy to see that $u(t)$ and $v(t)$ have the same generalized zeros. Hence $Lu(t) = 0$ is disconjugate on $[a, b+2]$ iff there is no real solution with two generalized zeros on $[a, b+2]$.

Exercise 1.13 *Show that if $p(t) > 0$ on $[a+1, b+2]$, then if $u(t)$ is a nontrivial prepared solution of $Lu(t) = 0$ with $u(t_0) = 0$ where $a+1 \leq t_0 \leq b+1$, then*

$u(t_0-1)$ and $u(t_0+1)$ lie on the opposite sides of a line with respect to the origin. Because of this we say a nontrivial prepared solution can not have a "double" generalized zero.

Theorem 1.14 (Reduction of Order) *If $v(t)$ is a nonzero solution of $Lu(t)=0$ for $t \geq t_0$, then any solution $u(t)$ for $t \geq t_0$ is of the form*

$$u(t) = \overline{v(t)}\left[A + B\sum_{s=t_0}^{t-1} \frac{1}{p(s+1)\overline{v(s)v(s+1)}}\right]$$

where

$$A = \frac{u(t_0)}{\overline{v(t_0)}}, \qquad B = \{v(t); u(t)\}.$$

Proof: Assume $v(t)$ is a nonzero solution of $Lu(t)=0$ for $t \geq t_0$ and $u(t)$ is a second solution. By Liouville's formula

$$\begin{aligned}
\Delta\left\{\frac{u(t)}{\overline{v(t)}}\right\} &= \frac{w[\overline{v(t)}, u(t)]}{\overline{v(t)v(t+1)}} \\
&= \frac{\{u(t+1); v(t+1)\ \}}{p(t+1)\overline{v(t)v(t+1)}} \\
&= \frac{B}{p(t+1)\overline{v(t)v(t+1)}}
\end{aligned}$$

where

$$B = \{v(t); u(t)\}.$$

Summing from t_0 to $t-1$ we obtain

$$\frac{u(t)}{\overline{v(t)}} - \frac{u(t_0)}{\overline{v(t_0)}} = B\sum_{s=t_0}^{t-1} \frac{1}{p(s+1)\overline{v(s)v(s+1)}}.$$

It follows that

$$u(t) = \overline{v(t)}\left[A + B\sum_{s=t_0}^{t-1} \frac{1}{p(s+1)\overline{v(s)v(s+1)}}\right],$$

where

$$A = \frac{u(t_0)}{\overline{v(t_0)}}.$$

□

Corollary 1.15 *If in Theorem 1.14 we also assume $v(t)$ is a prepared solution, then*

$$u(t) = v(t)\{A + BS_v(t_0;t)\}, \qquad t \geq t_0$$

where

$$S_v(t_0;t_0) = 0, \qquad S_v(t_0;t) = \sum_{s=t_0}^{t-1} \frac{1}{p(s+1)v(s)\overline{v(s+1)}}, \qquad t > t_0,$$

and

$$A = \frac{u(t_0)}{v(t_0)}, \qquad B = \{v(t); u(t)\}.$$

Proof: By Theorem 1.14

$$u(t) = \overline{v(t)}\left\{A_0 + B\sum_{s=t_0}^{t-1} \frac{1}{p(s+1)\overline{v(s)v(s+1)}}\right\}$$

where

$$A_0 = \frac{u(t_0)}{\overline{v(t_0)}}, \qquad B = \{v(t); u(t)\}.$$

By Theorem 1.10 $v(t) = cv_0(t)$ where $c \neq 0$ is a complex number and $v_0(t)$ is a real solution of $Lu(t) = 0$. Note that

$$\begin{aligned}
\overline{v(t)}A_0 &= \overline{v(t)}\frac{u(t_0)}{\overline{v(t_0)}} \\
&= \bar{c}v_0(t)\frac{u(t_0)}{\bar{c}v_0(t_0)} \\
&= cv_0(t)\frac{u(t_0)}{cv_0(t_0)} \\
&= v(t)\frac{u(t_0)}{v(t_0)} = v(t)A.
\end{aligned}$$

Also

$$\begin{aligned}
\overline{v(t)}\sum_{s=t_0}^{t-1} \frac{1}{p(s+1)\overline{v(s)v(s+1)}} &= \bar{c}v_0(t)\sum_{s=t_0}^{t-1} \frac{1}{p(s+1)\bar{c}v_0(s)\overline{v(s+1)}} \\
&= cv_0(t)\sum_{s=t_0}^{t-1} \frac{1}{p(s+1)cv_0(s)\overline{v(s+1)}} \\
&= v(t)\sum_{s=t_0}^{t-1} \frac{1}{p(s+1)v(t)\overline{v(s+1)}}.
\end{aligned}$$

The result follows easily from this observation. □

Exercise 1.16 *Show directly that if $v(t)$ is a nonzero prepared solution of $Lu(t) = 0$ for $t \geq t_0$, then*

$$u(t) = v(t) \sum_{s=t_0}^{t-1} \frac{1}{p(s+1)v(s)\overline{v(s+1)}},$$

is also a prepared solution of $Lu(t) = 0$.

Theorem 1.17 *If $Lu(t) = 0$ is disconjugate on $[a, b+2]$, then the boundary value problem*

$$Lu(t) = h(t), \qquad t \in [a+1, b+1],$$
$$u(a) = y_1, \ u(b+2) = y_2,$$

where $h(t)$ is a given function defined on $[a+1, b+1]$ and y_1, y_2 are given constants, has a unique solution.

Exercise 1.18 *Prove Theorem* 1.17.

We now wish to connect the topic of disconjugacy with the existence of a Green's function.

Suppose that $g(t, s)$ is a function which satisfies the following conditions:

(a) $g(t, s)$ is defined for $a \leq t \leq b+2$, $a+1 \leq s \leq b+1$;

(b) $Lg(t, s) = \delta_{ts}$, $a+1 \leq t, s \leq b+1$ where δ_{ts} is the Kronecker delta function ($\delta_{ts} = 0$ for $t \neq s$ and $= 1$ for $t = s$);

(c) $g(a, s) = 0 = g(b+2, s)$, $a+1 \leq s \leq b+1$.

Assume $g(t, s)$ satisfies **(a)**–**(c)** and set

$$u(t) = \sum_{s=a+1}^{b+1} g(t, s)h(s), \qquad t \in [a, b+2].$$

We now show that $u(t)$ is a solution of the boundary value problem (abbreviated by **BVP**)

$$Lu(t) = h(t)$$
$$u(a) = 0, \ u(b+2) = 0.$$

By **(c)**

$$\begin{aligned} u(a) &= \sum_{s=a+1}^{b+1} g(a,s)h(s) = 0 \\ u(b+2) &= \sum_{s=a+1}^{b+1} g(b+2,s)h(s) = 0. \end{aligned}$$

Next consider

$$\begin{aligned} Lu(t) &= \sum_{s=a+1}^{b+1} Lg(t,s)h(s) \\ &= \sum_{s=a+1}^{b+1} \delta_{ts}h(s) \\ &= h(t). \end{aligned}$$

Before we show that if $Lu(t) = 0$ is disconjugate on $[a, b+2]$, there is a function satisfying **(a)**–**(c)**; we would like to define the *Cauchy function* for $Lu(t) = 0$.

We define the *Cauchy function* $u(t,s)$ for $Lu(t) = 0$ to be, for each fixed s in $[a, b+1]$, the solution of the initial value problem (abbreviated by **IVP**)

$$\begin{aligned} Lu(t) &= 0 \\ u(s) &= 0, \qquad u(s+1) = \frac{1}{p(s+1)}. \end{aligned}$$

Example 1.19 *Find the Cauchy function for*

$$\Delta[p(t)\Delta u(t-1)] = 0 \quad \textit{for } t \geq s.$$

Since $\Delta[p(t)\Delta u(t-1,s)] = 0$ *we have* $p(t)\Delta u(t-1,s) = A$ *and*

$$p(s+1)\Delta u(s,s) = A = p(s+1)\frac{1}{p(s+1)}.$$

Hence $A = 1$ *and*

$$\begin{aligned} \Delta u(t,s) &= \frac{1}{p(t+1)}, \\ u(t,s) - u(s,s) &= \sum_{\tau=s}^{t-1} \frac{1}{p(\tau+1)}. \end{aligned}$$

Therefore the Cauchy function for $\Delta[p(t)\Delta u(t-1)] = 0$ *is given by*

$$u(t,s) = \sum_{\tau=s+1}^{t} \frac{1}{p(\tau)}, \qquad t \geq s.$$

Exercise 1.20 *Find the Cauchy function for*

$$\Delta^2 u(t-1) = 0$$

for $t \geq s$.

Exercise 1.21 *Show that if* $u(t)$ *and* $v(t)$ *are real valued linearly independent solutions of* $Lu(t) = 0$, *then the Cauchy function* $u(t,s)$ *is given by*

$$u(t,s) = \frac{\begin{vmatrix} u(s) & v(s) \\ u(t) & v(t) \end{vmatrix}}{p(s+1)\begin{vmatrix} u(s) & v(s) \\ u(s+1) & v(s+1) \end{vmatrix}} = \frac{\begin{vmatrix} u(s) & v(s) \\ u(t) & v(t) \end{vmatrix}}{\{u(s); v(s)\}}.$$

Exercise 1.22 *Find the Cauchy function for*

$$\Delta\left[\left(\frac{1}{6}\right)^t \Delta u(t-1)\right] + 2\left(\frac{1}{6}\right)^{t+1} u(t) = 0.$$

for $t \geq s$.

Now assume $Lu(t) = 0$ is disconjugate on $[a, b+2]$. It follows that the Cauchy function $u(t,s)$ satisfies $u(b+2,a) \neq 0$. Hence we can define $g(t,s)$ on $[a, b+2] \times [a+1, b+1]$ by

$$g(t,s) = \begin{cases} -\dfrac{u(t,a)u(b+2,s)}{u(b+2,a)}, & t \leq s \\ u(t,s) - \dfrac{u(t,a)u(b+2,s)}{u(b+2,a)}, & s \leq t. \end{cases} \tag{1.17}$$

We now show that $g(t,s)$ satisfies **(a)**–**(c)**. Note that

$$g(a,s) = -\frac{u(a,a)u(b+2,s)}{u(b+2,a)} = 0$$

and

$$g(b+2,s) = u(b+2,s) - \frac{u(b+2,a)u(b+2,s)}{u(b+2,a)} = 0$$

for $a+1 \leq s \leq b+1$. Hence **(c)** holds. We now show that (b) holds. Assume $t \leq s-1$, then

$$Lg(t,s) = -\frac{u(b+2,s)}{u(b+2,a)}Lu(t,a) = 0 = \delta_{ts}.$$

Now assume $t \geq s+1$. Then

$$Lg(t,s) = Lu(t,s) - \frac{u(b+2,s)}{u(b+2,a)}Lu(t,a) = 0 = \delta_{ts}.$$

Finally using (1.2),

$$\begin{aligned} Lg(s,s) &= p(s+1)g(s+1,s) + c(s)g(s,s) + p(s)g(s-1,s) \\ &= p(s+1)u(s+1,s) - \frac{u(b+2,s)}{u(b+2,a)}Lu(s,s) = 1 = \delta_{ss}. \end{aligned}$$

It follows that $g(t,s)$ satisfies **(a)**–**(c)**.

Theorem 1.23 *Assume $Lu(t) = 0$ is disconjugate on $[a, b+2]$. Then there is a unique function $g(t,s)$ satisfying* **(a)**–**(c)** *given by* (1.17), *called the Green's function for the boundary value problem (abbreviated by* **BVP**)

$$\begin{aligned} &Lu(t) = 0 \\ &u(a) = 0 = u(b+2). \end{aligned}$$

The unique solution of

$$\begin{aligned} &Lu(t) = h(t), \qquad t \in [a+1, b+1] \\ &u(a) = 0 = u(b+2) \end{aligned}$$

is given by

$$u(t) = \sum_{s=a+1}^{b+1} g(t,s)h(s).$$

If $p(t) > 0$ on $[a+1, b+2]$, then $g(t,s) < 0$ on the square $a+1 \leq t,\ s \leq b+1$.

Proof: Assume $g(t,s)$ given by (1.17) and $h(t,s)$ satisfy **(a)**–**(c)**. Fix s in the interval $[a+1,b+1]$ and set

$$u(t) = g(t,s) - h(t,s).$$

Then

$$\begin{aligned} Lu(t) &= Lg(t,s) - Lh(t,s) \\ &= \delta_{ts} - \delta_{ts} \\ &= 0, \qquad t \in [a+1,b+1]. \end{aligned}$$

Furthermore

$$\begin{aligned} u(a) &= g(a,s) - h(a,s) = 0, \\ u(b+2) &= g(b+2,s) - h(b+2,s) = 0. \end{aligned}$$

Hence $u(t)$ solves the **BVP**

$$\begin{aligned} &Lu(t) = 0 \\ &u(a) = 0 = u(b+2). \end{aligned}$$

It follows from Theorem 1.17 that $u(t) \equiv 0$. Hence

$$g(t,s) = h(t,s), \qquad a \le t \le b+2.$$

It follows that $g(t,s) = h(t,s)$ for $a \le t,\ s \le b+2$. Now assume $p(t) > 0$ on $[a+1,b+2]$. Then the Cauchy function $u(t,s)$ satisfies

$$u(t,s) > 0$$

for $a \le s < t \le b+2$. Hence for $a+1 \le t \le s \le b+1$

$$g(t,s) = -\frac{u(t,a)u(b+2,s)}{u(b+2,a)} < 0.$$

Now consider for fixed $s \in [a+1,b+1]$

$$g(t,s) = u(t,s) - \frac{u(t,a)u(b+2,s)}{u(b+2,a)}.$$

Note that $g(t,s)$ is a solution of $Lu(t) = 0$ on $[s,b+2]$ with $g(s,s) < 0$ and $g(b+2,s) = 0$. It follows from the disconjugacy assumption that

$$g(t,s) < 0 \text{ on } [s,b+1].$$

Since s was arbitrary in $[a+1,b+1]$ we have that

$$g(t,s) < 0$$

for $a+1 \le s \le t \le b+1$. The other parts of this theorem were proved previously. □

Corollary 1.24 *If $Lu(t) = 0$ is disconjugate on $[a, b+2]$ then the unique solution of the* **BVP**

$$\begin{aligned} Lu(t) &= h(t), \qquad t \in [a+1, b+1] \\ u(a) &= A, \qquad u(b+2) = B \end{aligned}$$

is given by

$$u(t) = v(t) + \sum_{s=a+1}^{b+1} g(t,s)h(s)$$

$t \in [a, b+2]$, *where $v(t)$ solves the* **BVP**

$$\begin{aligned} Lv(t) &= 0 \\ v(a) &= A, \qquad v(b+2) = B. \end{aligned}$$

Exercise 1.25 *Prove Corollary* 1.24.

Example 1.26 *Find the Green's function for the* **BVP**

$$\begin{aligned} &\Delta[p(t)\Delta u(t-1)] = 0 \\ &u(a) = 0 = u(b+2). \end{aligned}$$

By Example 1.19 *the Cauchy function for $\Delta[p(t)\Delta u(t-1)] = 0$ is given by*

$$u(t,s) = \sum_{\tau=s+1}^{t} \frac{1}{p(\tau)}.$$

Hence by (1.17) *we have that for $t \le s$*

$$g(t,s) = -\frac{\sum_{\tau=a+1}^{t} p^{-1}(\tau) \sum_{\tau=s+1}^{b+2} p^{-1}(\tau)}{\sum_{\tau=a+1}^{b+2} p^{-1}(\tau)}.$$

For $t \ge s$, we have by (1.17) *that*

$$\begin{aligned} g(t,s) &= \sum_{\tau=s+1}^{t} p^{-1}(\tau) - \frac{\sum_{\tau=a+1}^{t} p^{-1}(\tau) \sum_{\tau=s+1}^{b+2} p^{-1}(\tau)}{\sum_{\tau=a+1}^{b+2} p^{-1}(\tau)} \\ &= \frac{\sum_{\tau=s+1}^{t} p^{-1}(\tau) \left[\sum_{\tau=a+1}^{t} p^{-1}(\tau) + \sum_{\tau=t+1}^{b+2} p^{-1}(\tau)\right]}{\sum_{\tau=a+1}^{b+2} p^{-1}(\tau)} \end{aligned}$$

$$
\begin{aligned}
&-\frac{\sum_{\tau=a+1}^{t} p^{-1}(\tau) \sum_{\tau=s+1}^{b+2} p^{-1}(\tau)}{\sum_{\tau=a+1}^{b+2} p^{-1}(\tau)} \\
&= \frac{-\sum_{\tau=a+1}^{t} p^{-1}(\tau) \sum_{\tau=t+1}^{b+2} p^{-1}(\tau) + \sum_{\tau=s+1}^{t} p^{-1}(\tau) \sum_{\tau=t+1}^{b+2} p^{-1}(\tau)}{\sum_{\tau=a+1}^{b+2} p^{-1}(\tau)} \\
&= -\frac{\sum_{\tau=a+1}^{s} p^{-1}(\tau) \sum_{\tau=t+1}^{b+2} p^{-1}(\tau)}{\sum_{\tau=a+1}^{b+2} p^{-1}(\tau)}.
\end{aligned}
$$

Hence

$$
g(t,s) = \begin{cases} -\dfrac{\sum_{\tau=a+1}^{t} p^{-1}(\tau) \sum_{\tau=s+1}^{b+2} p^{-1}(\tau)}{\sum_{\tau=a+1}^{b+2} p^{-1}(\tau)}, & t \le s \\ -\dfrac{\sum_{\tau=a+1}^{s} p^{-1}(\tau) \sum_{\tau=t+1}^{b+2} p^{-1}(\tau)}{\sum_{\tau=a+1}^{b+2} p^{-1}(\tau)}, & s \le t. \end{cases} \tag{1.18}
$$

We will use the following lemma in the proof of the next theorem.

Lemma 1.27 *If $M > 0$ and $N > 0$, then*

$$
\frac{4}{M+N} \le \frac{1}{M} + \frac{1}{N}.
$$

Proof: Note the following equivalent statements:

$$
\begin{aligned}
(M-N)^2 &\ge 0 \\
M^2 - 2MN + N^2 &\ge 0 \\
M^2 + 2MN + N^2 &\ge 4MN \\
(M+N)^2 &\ge 4MN \\
\frac{4}{M+N} &\le \frac{M+N}{MN} \\
\frac{4}{M+N} &\le \frac{1}{M} + \frac{1}{N}.
\end{aligned}
$$

□

Exercise 1.28 *Show that if $p(t) > 0$ on $[a, b+1]$ and $q(t) \le 0$ on $[a+1, b+1]$, then $Lu(t) = 0$ is disconjugate on $[a, b+2]$.*

Theorem 1.29 *Assume $p(t) > 0$ on $[a+1, b+2]$ and $q(t) \geq 0$ on $[a+1, b+1]$. If*

$$D \equiv 4\left\{\sum_{\tau=a+1}^{b+2} p^{-1}(\tau)\right\}^{-1} - \sum_{\tau=a+1}^{b+1} q(\tau) > 0,$$

then $Lu(t) = 0$ is disconjugate on $[a, b+2]$.

Proof: Let $u(t)$ be the solution of the **IVP**

$$\begin{aligned} &Lu(t) = 0 \\ &u(a) = 0, \qquad u(a+1) = 1. \end{aligned}$$

It suffices to show that $u(t) > 0$ on $[a+1, b+2]$. Assume not, then there is a $c \in [a+2, b+2]$ such that

$$\begin{aligned} u(t) &> 0 \text{ on } [a, c-1] \\ u(c) &\leq 0. \end{aligned}$$

Since

$$4\left\{\sum_{\tau=a+1}^{c} p^{-1}(\tau)\right\}^{-1} - \sum_{\tau=a+1}^{c-1} q(\tau) \geq 4\left\{\sum_{\tau=a+1}^{b+2} p^{-1}(\tau)\right\}^{-1} - \sum_{\tau=a+1}^{b+1} q(\tau) = D,$$

we can assume without loss of generality that $c = b+2$. Let $B = u(b+2)$, then $B \leq 0$ and $u(t)$ solves the **BVP**

$$\begin{aligned} &\Delta[p(t)\Delta u(t-1)] = -q(t)u(t) \\ &u(a) = 0, \qquad u(b+2) = B. \end{aligned}$$

Hence, by Corollary 1.24 with $Lu(t) = \Delta[p(t)\Delta u(t-1)] = 0$ (show that this simple difference equation is disconjugate), we get that

$$u(t) = v(t) + \sum_{s=a+1}^{b+1} g(t,s)[-q(s)u(s)], \qquad t \in [a, b+2],$$

where $g(t, s)$ is given by (1.18) and $v(t)$ is the unique solution of the **BVP**

$$Lv(t) = 0, \qquad v(a) = 0, \qquad v(b+2) = B.$$

Since $g(t, s) \leq 0$,

$$u(t) = v(t) + \sum_{s=a+1}^{b+1} |g(t,s)|q(s)u(s).$$

Since $\Delta[p(t)\Delta v(t-1)] = 0$ is disconjugate on $[a, b+2]$ and $v(t)$ satisfies $v(a) = 0$ and $v(b+2) = B \le 0$, it follows that

$$v(t) \le 0 \text{ on } [a, b+2].$$

Thus

$$u(t) \le \sum_{s=a+1}^{b+1} |g(t,s)|q(s)u(s).$$

Pick $t_0 \in [a+1, b+1]$ such that

$$u(t_0) = \max_{a+1 \le t \le b+1} u(t) > 0.$$

Then

$$\begin{aligned} u(t_0) &\le \sum_{s=a+1}^{b+1} \{|g(t_0,s)|q(s)u(s)\} \\ &\le u(t_0) \sum_{s=a+1}^{b+1} \{|g(t_0,s)|q(s)\}. \end{aligned}$$

Hence

$$1 \le |g(t_0,t_0)| \sum_{s=a+1}^{b+1} q(s).$$

Using (1.18) we get that

$$1 \le \frac{\sum_{\tau=a+1}^{t_0} p^{-1}(\tau) \sum_{\tau=t_0+1}^{b+2} p^{-1}(\tau)}{\sum_{\tau=a+1}^{b+2} p^{-1}(\tau)} \sum_{s=a+1}^{b+1} q(s).$$

Taking $M = \sum_{\tau=a+1}^{t_0} p^{-1}(\tau)$, $N = \sum_{\tau=t_0+1}^{b+2} p^{-1}(\tau)$, in the above inequality yields

$$1 \le \frac{MN}{M+N} \sum_{s=a+1}^{b+1} q(s).$$

By Lemma 1.27 we obtain

$$\begin{aligned} 4 &\le MN\left[\frac{1}{M} + \frac{1}{N}\right] \sum_{s=a+1}^{b+1} q(s) \\ &\le [N+M] \sum_{s=a+1}^{b+1} q(s) \\ &= \sum_{\tau=a+1}^{b+2} p^{-1}(\tau) \sum_{s=a+1}^{b+1} q(s). \end{aligned}$$

Hence

$$D = 4\left\{\sum_{\tau=a+1}^{b+2} p^{-1}(\tau)\right\}^{-1} - \sum_{s=a+1}^{b+1} q(s) \le 0$$

which is a contradiction. □

Example 1.30 *Show for the difference equation*

$$\Delta^2 u(t-1) + 2u(t) = 0$$

that $D = 0$ and this equation is not disconjugate on $[0,2]$.

Here

$$\begin{aligned} D &= 4\left\{\sum_{\tau=1}^{2} p^{-1}(\tau)\right\}^{-1} - \sum_{\tau=1}^{1} q(\tau) \\ &= 4 \cdot \frac{1}{2} - 2 = 0. \end{aligned}$$

Let $u(t)$ be the solution of the **IVP** *$\Delta^2 u(t-1) + 2u(t) = 0$, $u(0) = 0$, $u(1) = 1$. It follows that $u(t)$ is real valued, hence prepared, and also has $u(2) = 0$. Therefore, $\Delta^2 u(t-1) + 2u(t) = 0$ is not disconjugate on $[0,2]$.*

We say $Lu(t) = 0$ is *disfocal* on $[a, b+2]$ provided no nontrivial prepared solution $u(t)$ of $Lu(t) = 0$ has a generalized zero followed by a generalized zero of $\Delta u(t)$ on $[a, b+2]$.

Theorem 1.31 *If $Lu(t) = 0$ is disfocal on $[a, b+2]$, then the* **BVP**

$$\begin{aligned} Lu(t) &= h(t), & t &\in [a+1, b+1] \\ u(a) &= u_0, & \Delta u(b+1) &= u_1 \end{aligned}$$

has a unique solution.

Exercise 1.32 *Prove Theorem* 1.31.

Assume $k(t,s)$ is a function satisfying the following properties

(a') $k(t,s)$ is defined for $a \le t \le b+2, \quad a+1 \le s \le b+1$;

(b') $Lk(t,s) = \delta_{ts}, \qquad a+1 \le t, s \le b+1;$

(c') $k(a,s) = 0 = \Delta k(b+1,s), \qquad a+1 \le s \le b+1.$

Set

$$u(t) = \sum_{s=a+1}^{b+1} k(t,s)h(s).$$

Then, using **(b')**,

$$\begin{aligned} Lu(t) &= \sum_{s=a+1}^{b+1} Lk(t,s)h(s) = \sum_{s=a+1}^{b+1} \delta_{ts}h(s) \\ &= h(t), \qquad \text{for } t \in [a+1,b+1]. \end{aligned}$$

Further by **(c')**,

$$\begin{aligned} u(a) &= \sum_{s=a+1}^{b+1} k(a,s)h(s) = 0 \\ \Delta u(b+1) &= \sum_{s=a+1}^{b+1} \Delta k(b+1,s)h(s) = 0. \end{aligned}$$

Thus

$$u(t) = \sum_{s=a+1}^{b+1} k(t,s)h(s)$$

solves the **BVP**

$$\begin{aligned} &Lu(t) = h(t), \qquad t \in [a+1,b+1] \\ &u(a) = 0 = \Delta u(b+1). \end{aligned}$$

Assume $Lu(t) = 0$ is disfocal on $[a, b+2]$. Then the Cauchy function $u(t,s)$ for $Lu(t) = 0$ satisfies $\Delta u(t,s) \neq 0$ for $s+1 \le t \le b+1$. Then we can define for $a \le t \le b+2$, $a+1 \le s \le b+1$.

$$k(t,s) = \begin{cases} -\dfrac{u(t,a)\Delta u(b+1,s)}{\Delta u(b+1,a)}, & t \le s \\ u(t,s) - \dfrac{u(t,a)\Delta u(b+1,s)}{\Delta u(b+1,a)}, & s \le t. \end{cases} \tag{1.19}$$

We now show that $k(t,s)$ satisfies **(a')**–**(c')**. Clearly **(a')** holds. Next

$$k(a,s) = -\frac{u(a,a)\Delta u(b+1,s)}{\Delta u(b+1,a)} = 0$$

and

$$\Delta k(b+1,s) = \Delta u(b+1,s) - \frac{\Delta u(b+1,a)\Delta u(b+1,s)}{\Delta u(b+1,a)} = 0,$$

for $a+1 \le s \le b+1$. Hence **(c')** holds. For $t < s$

$$Lk(t,s) = -\frac{\Delta u(b+1,s)}{\Delta u(b+1,a)} Lu(t,a) = 0 = \delta_{ts}.$$

For $t > s$

$$Lk(t,s) = Lu(t,s) - \frac{\Delta u(b+1,s)}{\Delta u(b+1,a)} Lu(t,a) = 0 = \delta_{ts}.$$

Finally assume $t = s$. Then by (1.2)

$$\begin{aligned} Lk(t,s) &= p(t+1)k(t+1,t) + c(t)k(t,t) + p(t)k(t-1,t) \\ &= -\frac{\Delta u(b+1,s)}{\Delta u(b+1,a)} Lu(t,a) + p(t+1)u(t+1,t) \\ &= 1 \ = \ \delta_{tt}. \end{aligned}$$

Theorem 1.33 *Assume $Lu(t) = 0$ is disfocal on $[a, b+2]$. Then there is a unique function $k(t,s)$ satisfying* **(a')**–**(c')** *called the Green's function for the* **BVP**

$$\begin{aligned} &Lu(t) = 0 \\ &u(a) = 0, \qquad \Delta u(b+1) = 0. \end{aligned}$$

The unique solution of

$$\begin{aligned} &Lu(t) = h(t) \\ &u(a) = u_0, \qquad \Delta u(b+1) = u_1 \end{aligned}$$

is given by

$$u(t) = v(t) + \sum_{s=a+1}^{b+1} k(t,s)h(s),$$

$t \in [a, b+2]$, where $v(t)$ is the unique solution of the **BVP**

$$\begin{aligned} &Lv(t) = 0 \\ &v(a) = u_0, \ \Delta v(b+1) = u_1. \end{aligned}$$

If $p(t) > 0$ on $[a+1, b+2]$, then $k(t,s) < 0$ for $a+1 \le t \le b+1$, $a+1 \le s \le b+1$.

Example 1.34 *Find the Green's function for the focal* **BVP**

$$\Delta[p(t)\Delta u(t-1)] = 0$$
$$u(a) = 0, \qquad \Delta u(b+1) = 0.$$

From Example 1.19 *the Cauchy function for* $\Delta[p(t)\Delta u(t-1)] = 0$ *is given by*

$$u(t,s) = \sum_{\tau=s+1}^{t} p^{-1}(\tau), \qquad t \geq s.$$

It then follows from (1.19) *that*

$$k(t,s) = \begin{cases} -\sum_{\tau=a+1}^{t} p^{-1}(\tau), & t \leq s \\ -\sum_{\tau=a+1}^{s} p^{-1}(\tau), & s \leq t. \end{cases} \tag{1.20}$$

Theorem 1.35 *Assume* $p(t) > 0$ *on* $[a+1, b+2]$ *and* $q(t) \geq 0$ *on* $[a+1, b+1]$. *If*

$$F \equiv \left\{ \sum_{t=a+1}^{b+1} \frac{1}{p(t)} \right\}^{-1} - \sum_{t=a+1}^{b+1} q(t) > 0,$$

then $Lu(t) = 0$ *is disfocal on* $[a, b+2]$.

Proof: Assume $Lu(t) = 0$ is not disfocal on $[a, b+2]$. Then there is a nontrivial real solution $u(t)$ of $Lu(t) = 0$ and integers t_1, t_2, $a \leq t_1 < t_2 \leq b+1$ such that

$$\begin{aligned} u(t_1) &\leq 0 \\ u(t) &> 0 \text{ on } [t_1+1, t_2] \\ \Delta u(t) &> 0 \text{ on } [t_1, t_2-1] \\ \Delta u(t_2) &\leq 0. \end{aligned}$$

Since

$$\left\{ \sum_{t=t_1+1}^{t_2} \frac{1}{p(t)} \right\}^{-1} - \sum_{t=t_1+1}^{t_2} q(t) \geq \left\{ \sum_{t=a+1}^{b+1} \frac{1}{p(t)} \right\}^{-1} - \sum_{t=a+1}^{b+1} q(t) = F > 0,$$

we can assume without loss of generality that $t_1 = a$ and $t_2 = b+1$. Let $A = u(a) \leq 0$ and $B = \Delta u(b+1) \leq 0$. Then $u(t)$ solves the **BVP**

$$\Delta[p(t)\Delta u(t-1)] = -q(t)u(t)$$
$$u(a) = A, \quad \Delta u(b+1) = B.$$

By using Theorem 1.33 with $Lu(t) = \Delta[p(t)\Delta u(t-1)] = 0$, (show that this equation is disfocal),

$$u(t) = v(t) + \sum_{s=a+1}^{b+1} k(t,s)[-q(s)u(s)]$$

where $k(t,s)$ is given by (1.20) and $v(t)$ is the unique solution of the **BVP**

$$\begin{aligned} &\Delta[p(t)\Delta v(t-1)] = 0 \\ &v(a) = A, \qquad \Delta v(b+1) = B. \end{aligned}$$

Since $\Delta[p(t)\Delta v(t-1)] = 0$ is disfocal and $A \leq 0$, $B \leq 0$ we have

$$v(t) \leq 0 \text{ on } [a, b+2].$$

Consequently,

$$u(t) \leq \sum_{s=a+1}^{b+1} |k(t,s)|q(s)u(s).$$

Hence

$$\begin{aligned} u(b+1) &\leq \sum_{s=a+1}^{b+1} |k(b+1,s)|q(s)u(s) \\ &\leq u(b+1) \sum_{s=a+1}^{b+1} |k(b+1,s)|q(s). \end{aligned}$$

It follows that

$$\begin{aligned} 1 &\leq |k(b+1,b+1)| \sum_{s=a+1}^{b+1} q(s) \\ &\leq \sum_{t=a+1}^{b+1} \frac{1}{p(t)} \sum_{s=a+1}^{b+1} q(s). \end{aligned}$$

Thus

$$F = \left\{ \sum_{t=a+1}^{b+1} \frac{1}{p(t)} \right\}^{-1} - \sum_{s=a+1}^{b+1} q(s) \leq 0$$

which is a contradiction. □

Example 1.36 *Show that the difference equation*

$$\Delta^2 u(t-1) + q(t)u(t) = 0, \qquad t \in [a+1, b+1]$$

where

$$q(t) = \begin{cases} 0, & a \le t \le b \\ \frac{1}{b-a+1}, & t = b+1 \end{cases}$$

is not disfocal on $[a, b+2]$ *and* $F = 0$.

First

$$\begin{aligned} F &= \left\{ \sum_{t=a+1}^{b+1} \frac{1}{p(t)} \right\}^{-1} - \sum_{t=a+1}^{b+1} q(t) \\ &= \{b-a+1\}^{-1} - \frac{1}{b-a+1} = 0. \end{aligned}$$

Now let $u(t)$ *be the solution of the difference equation in this example satisfying* $u(a) = 0,\ u(a+1) = 1$. *It follows that* $\Delta u(b+1) = 0$. *Hence this difference equation is not disfocal on* $[a, b+2]$.

Define a set of functions A called *admissible variations* by

$$A = \{\eta \mid \eta : [a, b+2] \to \mathrm{C} \text{ such that } \eta(a) = \eta(b+2) = 0\}.$$

Then define J on A by

$$J\eta = \sum_{t=a+1}^{b+2} p(t)|\Delta\eta(t-1)|^2 - \sum_{t=a+1}^{b+1} q(t)|\eta(t)|^2.$$

In many of the calculations below it is convenient to extend the domain of definitions of p, q and η by $p(t) = p(a+1)$, $t \le a$, $p(t) = p(b+2)$, $t > b+2$, $q(t) = q(a+1)$, $t \le a$, $q(t) = q(b+1)$, $t \ge b+2$, $\eta(t) = 0$, $t < a$ and $t > b+2$. An immediate consequence of this is that (since $\eta(b+2) = 0$) we can write

$$J\eta = \sum_{t=a+1}^{b+2} \{p(t)|\Delta\eta(t-1)|^2 - q(t)|\eta(t)|^2\}.$$

Lemma 1.37 *If* $\eta \in A$, *then*

$$J\eta = -\sum_{t=a+1}^{b+2} \overline{\eta(t)} L\eta(t).$$

Proof: Let $\eta \in A$ and consider

$$J\eta = \sum_{t=a+1}^{b+2} \{p(t)\overline{\Delta\eta(t-1)}\Delta\eta(t-1) - q(t)\overline{\eta(t)}\eta(t)\}.$$

Using summation by parts on the first term we obtain

$$\begin{aligned} J\eta &= [p(t)\Delta\eta(t-1)\overline{\eta(t-1)}]_{t=a+1}^{b+3} - \sum_{t=a+1}^{b+2} \overline{\eta(t)}[\Delta[p(t)\Delta\eta(t-1)] + q(t)\eta(t)] \\ &= -\sum_{t=a+1}^{b+2} \overline{\eta(t)} L\eta(t). \end{aligned}$$

□

Lemma 1.38 *Assume* $u(t)$ *is a prepared solution of* $Lu(t) = 0$ *and* t_1, t_2 *satisfy* $a+1 \leq t_1 < t_2 \leq b+2$. *If we define*

$$\eta(t) = \begin{cases} 0, & a \leq t \leq t_1 - 1 \\ u(t), & t_1 \leq t \leq t_2 - 1 \\ 0, & t_2 \leq t \leq b+2 \end{cases}$$

then

$$J\eta = p(t_1)\overline{u(t_1-1)}u(t_1) + p(t_2)\overline{u(t_2-1)}u(t_2).$$

Proof: Since $\eta(a) = \eta(b+2) = 0$, $\eta \in A$ and hence by Lemma 1.37

$$\begin{aligned} J\eta &= -\sum_{t=a+1}^{b+2} \overline{\eta(t)} L\eta(t) \\ &= -\sum_{t=t_1}^{t_2-1} \overline{u(t)} L\eta(t). \end{aligned}$$

First assume $t_1 = t_2 - 1$. Then use (1.2) and (1.3) to expand L and solve for $c(t_1)u(t_1)$ to obtain

$$\begin{aligned} J\eta &= -\overline{u(t_1)}L\eta(t_1) = -\overline{u(t_1)}[c(t_1)u(t_1)] \\ &= \overline{u(t_1)}[p(t_1+1)u(t_1+1) + p(t_1)u(t_1-1)] \\ &= p(t_1)u(t_1-1)\overline{u(t_1)} + p(t_2)u(t_2)\overline{u(t_2-1)} \\ &= p(t_1)\overline{u(t_1-1)}u(t_1) + p(t_2)u(t_2)\overline{u(t_2-1)} \end{aligned}$$

because $u(t)$ is a prepared solution. Now assume $t_1 < t_2 - 1$. Then

$$\begin{aligned} J\eta &= -\overline{u(t_1)}L\eta(t_1) - \sum_{t=t_1+1}^{t_2-2} \overline{u(t)}Lu(t) - \overline{u(t_2-1)}L\eta(t_2-1) \\ &= -\overline{u(t_1)}L\eta(t_1) - \overline{u(t_2-1)}L\eta(t_2-1) \\ &= \overline{u(t_1)}p(t_1)u(t_1-1) + \overline{u(t_2-1)}p(t_2)u(t_2) \\ &= p(t_1)\overline{u(t_1-1)}u(t_1) + p(t_2)\overline{u(t_2-1)}u(t_2). \end{aligned}$$

□

We say J is *positive definite* on A provided $J\eta \geq 0$ for all $\eta \in A$ and $J\eta = 0$ if and only if $\eta = 0$. Assume $z(t)$ is a function defined on $[a, b+2]$ such that $z(t) + p(t) \neq 0$ on $[a+1, b+2]$. For all such functions z we define the *Riccati operator* R by

$$Rz(t) = \Delta z(t) + q(t) + \frac{z^2(t)}{z(t)+p(t)}$$

for $t \in [a+1, b+1]$. The equation

$$Rz(t) = 0 \tag{1.21}$$

is called the Riccati difference equation. If $z(t)$ is defined on $[a+1, b+2]$, $z(t) + p(t) \neq 0$ on $[a+1, b+1]$ and

$$Rz(t) = 0$$

for each $t \in [a+1, b+1]$, then we say $z(t)$ is a solution of the Riccati equation (1.21) on $[a+1, b+2]$. In our next theorem we will use the following two lemmas concerning the Riccati equation.

Lemma 1.39 *Assume $u(t)$ is real valued, defined on $[a, b+2]$, and satisfies $p(t)u(t-1)u(t) > 0$ on $[a+1, b+2]$. If we make the Riccati substitution*

$$z(t) = \frac{p(t)\Delta u(t-1)}{u(t-1)}$$

for $t \in [a, b+2]$, then $z(t) + p(t) > 0$ on $[a+1, b+2]$ and

$$Lu(t) = u(t)Rz(t)$$

for $t \in [a+1, b+1]$.

Proof: First note that for $t \in [a, b+1]$

$$\begin{aligned} z(t) &= \frac{p(t)[u(t) - u(t-1)]}{u(t-1)} \\ &= p(t)\frac{u(t)}{u(t-1)} - p(t). \end{aligned}$$

Hence

$$z(t) + p(t) = \frac{p(t)u(t)}{u(t-1)} > 0$$

on $[a+1, b+2]$. Furthermore,

$$\frac{u(t-1)}{p(t)} = \frac{u(t)}{z(t) + p(t)} \tag{1.22}$$

for $t \in [a+1, b+2]$. Consider

$$\begin{aligned} u(t)\Delta z(t) &= \frac{u(t-1)\Delta[p(t)\Delta u(t-1)] - p(t)[\Delta u(t-1)]^2}{u(t-1)} \\ &= \Delta[p(t)\Delta u(t-1)] - \left[\frac{p(t)\Delta u(t-1)}{u(t-1)}\right]^2 \frac{u(t-1)}{p(t)}. \end{aligned}$$

Using (1.22) we have that

$$u(t)\Delta z(t) = \Delta[p(t)\Delta u(t-1)] - z^2(t)\frac{u(t)}{z(t) + p(t)}.$$

Adding $q(t)u(t)$ to both sides it follows that

$$u(t)Rz(t) = Lu(t)$$

for $t \in [a+1, b+1]$. □

Lemma 1.40 *Assume $z(t)$ is a real valued solution of the Riccati equation (1.21) on $[a+1, b+3]$ with $z(t) + p(t) > 0$ on $[a+1, b+2]$. If $u(t)$ is defined on $[a, b+2]$, then*

$$Ju = \{z(t)|u(t-1)|^2\}_{a+1}^{b+3} + \sum_{t=a+1}^{b+2} |F(t)|^2$$

where

$$F(t) = \sqrt{z(t) + p(t)}\ \Delta u(t-1) - \frac{z(t)u(t)}{\sqrt{z(t) + p(t)}}$$

for $t \in [a+1, b+2]$.

Proof: Consider, for $t \in [a+1, b+2]$,

$$
\begin{aligned}
\Delta\{z(t)|u(t-1)|^2\} &= \Delta\{z(t)\overline{u(t-1)}u(t-1)\} \\
&= \Delta z(t)\overline{u(t)}u(t) + z(t)\Delta\overline{u(t-1)}u(t) + z(t)\overline{u(t-1)}\Delta u(t-1) \\
&= \left[-q(t) - \frac{z^2(t)}{z(t)+p(t)}\right]\overline{u(t)}u(t) \\
&\quad + z(t)\Delta\overline{u(t-1)}u(t) + z(t)\overline{u(t-1)}\Delta u(t-1) \\
&= \{p(t)|\Delta u(t-1)|^2 - q(t)|u(t)|^2\} \\
&\quad - p(t)\Delta\overline{u(t-1)}\Delta u(t-1) + z(t)\Delta\overline{u(t-1)}u(t) \\
&\quad + z(t)\overline{u(t-1)}\Delta u(t-1) - \frac{z^2(t)}{z(t)+p(t)}\overline{u(t)}u(t) \\
&= \{p(t)|\Delta u(t-1)|^2 - q(t)|u(t)|^2\} \\
&\quad - [z(t)+p(t)]\overline{\Delta u(t-1)}\Delta u(t-1) + z(t)\overline{\Delta u(t-1)}\Delta u(t-1) \\
&\quad + z(t)\overline{\Delta u(t-1)}u(t) + z(t)\overline{u(t-1)}\Delta u(t-1) \\
&\quad - \frac{z^2(t)}{z(t)+p(t)}\overline{u(t)}u(t) \\
&= \{p(t)|\Delta u(t-1)|^2 - q(t)|u(t)|^2\} \\
&\quad - [z(t)+p(t)]\overline{\Delta u(t-1)}\Delta u(t-1) + z(t)\overline{\Delta u(t-1)}u(t) \\
&\quad + z(t)\overline{u(t)}\Delta u(t-1) - \frac{z^2(t)}{z(t)+p(t)}\overline{u(t)}u(t) \\
&= \{p(t)|\Delta u(t-1)|^2 - q(t)|u(t)|^2\} \\
&\quad - \left\{\sqrt{z(t)+p(t)}\;\overline{\Delta u(t-1)} - \frac{z(t)\overline{u(t)}}{\sqrt{z(t)+p(t)}}\right\}.
\end{aligned}
$$

$$\begin{aligned}&\left\{\sqrt{z(t)+p(t)}\ \Delta u(t-1)-\frac{z(t)u(t)}{\sqrt{z(t)+p(t)}}\right\}\\ =\ &\{p(t)|\Delta u(t-1)|^2-q(t)|u(t)|^2\}-|F(t)|^2.\end{aligned}$$

Summing both sides from $a+1$ to $b+2$ we get the desired result. □

Theorem 1.41 (Reid Roundabout Theorem) *The following conditions are equivalent:*

(a) *$Lu(t)=0$ is disconjugate on $[a,b+2]$.*

(b) *$Lu(t)=0$ has a real solution satisfying $p(t)u(t-1)u(t)>0$ on $[a+1,b+2]$.*

(c) *The Riccati equation*

$$Rz(t)=\Delta z(t)+q(t)+\frac{z^2(t)}{z(t)+p(t)}=0$$

has a solution $z(t)$ on $[a+1,b+2]$ satisfying $z(t)+p(t)>0$ on $[a+1,b+2]$.

(d) *J is positive definite on A.*

Proof: To see that **(a)** implies **(b)** assume $Lu(t)=0$ is disconjugate on $[a,b+2]$. Let $v(t)$ be a solution satisfying the initial conditions

$$\begin{aligned}v(a)&=0\\ v(a+1)&=1.\end{aligned}$$

It follows that

$$p(t)v(t-1)v(t)>0$$

on $[a+2,b+2]$. Let $u_\varepsilon(t)$ be the solution satisfying the initial conditions

$$\begin{aligned}u_\varepsilon(a)&=\varepsilon p(a+1)\\ u_\varepsilon(a+1)&=1.\end{aligned}$$

Then

$$\lim_{\varepsilon\to 0}u_\varepsilon(t)=v(t)$$

for $t\in[a,b+2]$. For $\varepsilon_0>0$, sufficiently small, $u(t)\equiv u_{\varepsilon_0}(t)$ satisfies

$$p(t)u(t-1)u(t)>0$$

on $[a+2, b+2]$. Furthermore,

$$p(a+1)u(a)u(a+1) = \varepsilon_0 p^2(a+1) > 0.$$

Hence $p(t)u(t-1)u(t) > 0$ on $[a+1, b+2]$ and **(b)** holds.

To see that **(b)** implies **(c)** assume $u(t)$ is a real solution of $Lu(t) = 0$ with $p(t)u(t-1)u(t) > 0$ on $[a+1, b+2]$. Make the Riccati substitution

$$z(t) = \frac{p(t)\Delta u(t-1)}{u(t)}$$

for $t \in [a+1, b+2]$. By Lemma 1.39, $z(t) + p(t) > 0$ on $[a+1, b+2]$ and

$$Lu(t) = u(t)Rz(t)$$

for $t \in [a+1, b+1]$. It follows that $Rz(t) = 0$ for $t \in [a+1, b+1]$ and **(c)** holds.

To see that **(c)** implies **(d)** assume the Riccati equation $Rz(t) = 0$ has a solution $z(t)$ on $[a+1, b+2]$ with $z(t) + p(t) > 0$ on $[a+1, b+2]$. Define

$$z(b+3) = z(b+2) - q(b+2) - \frac{z^2(b+2)}{z(b+2) + p(b+2)}.$$

Then this $z(t)$ is a solution of $Rz(t) = 0$ on $[a+1, b+3]$ and we can apply Lemma 1.40. Let $\eta \in A$, then by Lemma 1.40

$$J\eta = \{z(t)|\eta(t-1)|^2\}_{a+1}^{b+3} + \sum_{t=a+1}^{b+2} |F(t)|^2$$

where

$$F(t) = \sqrt{z(t) + p(t)}\ \Delta\eta(t-1) - \frac{z(t)\eta(t)}{\sqrt{z(t) + q(t)}}$$

for $t \in [a+1, b+2]$. Since $\eta(a) = \eta(b+2) = 0$,

$$J\eta = \sum_{t=a+1}^{b+2} |F(t)|^2 \geq 0.$$

Clearly $J\eta = 0$ when $\eta = 0$. To show that J is positive definite on A it remains to show that $J\eta = 0$ implies $\eta = 0$. So assume $\eta \in A$ and $J\eta = 0$. Then from above

$$J\eta = \sum_{t=a+1}^{b+2} |F(t)|^2 = 0.$$

It follows that $F(t) = 0$ for $a + 1 \le t \le b + 2$. Hence

$$\sqrt{z(t)+p(t)}\ \Delta\eta(t-1) = \frac{z(t)\eta(t)}{\sqrt{z(t)+p(t)}}$$

for $t \in [a+1, b+2]$. Solving for $\eta(t)$ we get that

$$\eta(t) = \frac{z(t)+p(t)}{p(t)}\eta(t-1)$$

for $t \in [a+1, b+2]$. Since $\eta(a) = 0$ we get that $\eta(t) \equiv 0$ on $[a, b+2]$ and **(d)** holds.

Finally to see that **(d)** implies **(a)**, assume J is positive definite on A. Assume $Lu(t) = 0$ is not disconjugate on $[a, b+2]$. Then there is a real solution $u(t)$ of $Lu(t) = 0$ such that

$$\begin{aligned} p(t_1)u(t_1-1)u(t_1) &\le 0, \qquad u(t_1) \ne 0 \\ p(t_2)u(t_2-1)u(t_2) &\le 0, \qquad u(t_2-1) \ne 0 \end{aligned}$$

where

$$a + 1 \le t_1 < t_2 \le b + 2.$$

Define

$$\eta(t) = \begin{cases} 0, & a \le t \le t_1 - 1 \\ u(t), & t_1 \le t \le t_2 - 1 \\ 0, & t_2 \le t \le b+2. \end{cases}$$

Since $\eta(a) = 0 = \eta(b+2)$ and $\eta(t_1) \ne 0$, $\eta \in A - \{0\}$. Hence $J\eta > 0$. But by Lemma 1.38

$$J\eta = p(t_1)u(t_1-1)u(t_1) + p(t_2)u(t_2-1)u(t_2) \le 0$$

which is a contradiction. Hence **(a)** holds. □

1.7 DISFOCALITY

Let A_f denote the set

$$A_f = \{\eta \mid \eta : [a, b+2] \to \mathrm{R} \quad \text{such that} \quad \eta(a) = 0,\ \Delta\eta(b+1) = 0\}$$

and define $J_f : A_f \to \mathrm{R}$ by

$$J_f\eta = \sum_{t=a+1}^{b+1} \{p(t)[\Delta\eta(t-1)]^2 - q(t)\eta^2(t)\}.$$

We say $Lu(t) = 0$ is *C-disfocal on* $[a, b+2]$ provided there does not exist a nontrivial prepared solution $u(t)$ of $Lu(t) = 0$ with $\Delta u(b+1) = 0$ which has a generalized zero in $[a, b+2]$. The C in C–disfocal is used to refer to W.A. Coppel. See Theorem 8 on page 11 in Coppel [44] for the definition in the continuous case.

For extensions of many of the results in this section and the next section to the vector case, see Peil and Peterson [113].

Example 1.42 *The difference equation*

$$\Delta^2 u(t-1) + \frac{138 - 119t}{10} u(t) = 0$$

is C-disfocal on $[0, 3]$ *but is not C-disfocal on* $[0, 2]$.

Exercise 1.43 *Verify the above example.*

Theorem 1.44 *Assume* $p(t) > 0$ *on* $[a+1, b+2]$. *Then the following are equivalent:*

(a) $Lu(t) = 0$ *is C-disfocal on* $[a, b+2]$.

(b) $Lu(t) = 0$ *has a positive solution* $u(t)$ *on* $[a, b+2]$ *with* $\Delta u(b+1) = 0$.

(c) *The Riccati equation* $Rz(t) = 0$ *has a solution on* $[a+1, b+2]$ *which satisfies* $z(t) + p(t) > 0$ *on* $[a+1, b+2]$ *and* $z(b+2) = 0$.

(d) J_f *is positive definite on* A_f.

Proof: **(a) ⇒ (b):** Assume $Lu(t) = 0$ is C-disfocal on $[a, b+2]$. Let $u(t)$ be the solution of the **IVP**

$$\begin{aligned} Lu(t) &= 0 \\ u(b+1) &= 1 \\ \Delta u(b+1) &= 0. \end{aligned}$$

Then it follows that $u(t) > 0$ on $[a, b+2]$.

(b) $\Rightarrow$ (c): Assume $u(t)$ is a positive solution of $Lu(t) = 0$ on $[a, b+2]$ with $\Delta u(b+1) = 0$. Set

$$z(t) = \frac{p(t)\Delta u(t-1)}{u(t-1)}$$

for $t \in [a+1, b+2]$. Then as before $z(t) + p(t) > 0$ on $[a+1, b+2]$ and $z(t)$ is a solution of the Riccati equation $Rz(t) = 0$ on $[a+1, b+2]$. Also $z(b+2) = 0$.

(c) $\Rightarrow$ (d): Assume $z(t)$ is a solution of the Riccati equation $Rz(t) = 0$ on $[a+1, b+2]$ with $z(t) + p(t) > 0$ on $[a+1, b+2]$ and $z(b+2) = 0$. By the proof of Lemma 1.40, for $\eta \in A_f$,

$$\Delta[z(t)\eta^2(t-1)] = \{p(t)[\Delta\eta(t-1)]^2 - q(t)\eta^2(t)\} - F^2(t)$$

where

$$F(t) = \sqrt{z(t)+p(t)}\Delta\eta(t-1) - \frac{z(t)\eta(t)}{\sqrt{z(t)+p(t)}}.$$

Summing from $a+1$ to $b+1$ we obtain

$$[z(t)\eta^2(t-1)]_{a+1}^{b+2} = J_f\eta - \sum_{t=a+1}^{b+1} F^2(t).$$

Since $\eta(a) = 0$ and $z(b+2) = 0$ we get that

$$J_f\eta = \sum_{t=a+1}^{b+1} F^2(t).$$

Hence $J_f\eta \geq 0$ for all $\eta \in A_f$. Clearly $J_f 0 = 0$. Finally assume $\eta \in A_f$ and $J_f\eta = 0$. Then for $t \in [a+1, b+1]$,

$$F(t) = 0.$$

Hence

$$\sqrt{z(t)+p(t)}\Delta\eta(t-1) = \frac{z(t)\eta(t)}{\sqrt{z(t)+p(t)}}$$

which implies the sequence of statements

$$\begin{aligned} [z(t)+p(t)][\eta(t)-\eta(t-1)] &= z(t)\eta(t), \\ p(t)\eta(t) &= [z(t)+p(t)]\eta(t-1), \quad \text{and} \\ \eta(t) &= \frac{z(t)+p(t)}{p(t)}\eta(t-1) \end{aligned}$$

for $t \in [a+1, b+1]$. Since $\eta(a) = 0$ we have that $\eta(t) = 0$ for $t \in [a, b+1]$. Since $\Delta\eta(b+1) = 0$, $\eta(t) \equiv 0$ on $[a, b+2]$. Hence J_f is positive definite on A_f.

(d) ⇒ (a): Assume J_f is positive definite on A_f. Assume $Lu(t) = 0$ is not C-disfocal on $[a, b+2]$. Then there is a nontrivial real solution $u(t)$ and an integer t_1 such that $a+1 \leq t_1 \leq b+1$ with

$$u(t_1 - 1)u(t_1) \leq 0, \qquad u(t_1) \neq 0, \qquad \Delta u(b+1) = 0.$$

Define η by

$$\eta(t) = \begin{cases} 0, & a \leq t \leq t_1 - 1 \\ u(t), & t_1 \leq t \leq b+2. \end{cases}$$

Since $\eta(a) = 0$, $\Delta\eta(b+1) = 0$, $\eta(t_1) = u(t_1) \neq 0$ we have that $\eta \in A_f - \{0\}$. Hence $J_f\eta > 0$. But

$$J_f\eta = \sum_{t=a+1}^{b+1} \{p(t)[\Delta\eta(t-1)]^2 - q(t)\eta^2(t)\}.$$

Using summation by parts on the first term under the sum we obtain

$$J_f\eta = \{\eta(t-1)p(t)\Delta\eta(t-1)\}\Big|_{a+1}^{b+2} - \sum_{t=a+1}^{b+1} \eta(t)L\eta(t).$$

It follows that

$$\begin{aligned} J_f\eta &= -\sum_{t=t_1}^{b+1} u(t)L\eta(t) \\ &= -u(t_1)L\eta(t_1) \\ &= -u(t_1)[p(t_1+1)\eta(t_1+1) + c(t_1)\eta(t_1) + p(t_1)\eta(t_1-1)] \\ &= p(t_1)u(t_1-1)u(t_1) \\ &\leq 0, \end{aligned}$$

which is a contradiction. □

1.8 COMPARISON THEOREMS FOR DISFOCALITY

We will consider the two equations

$$L_iu(t) = \Delta[p_i(t)\Delta u(t-1)] + q_i(t)u(t) = 0,$$

$i = 1, 2$, where $p_i(t) > 0$ on $[a+1, b+2]$ and $q_i(t)$ is defined on $[a+1, b+1]$, $i = 1, 2$. For $i = 1, 2$, define $J_f^i : A_f \to \mathrm{R}$ by

$$J_f^i \eta = \sum_{t=a+1}^{b+1} \{p_i(t)[\Delta\eta(t-1)]^2 - q_i(t)\eta^2(t)\}.$$

Corollary 1.45 *If $L_1 u(t) = 0$ is C-disfocal on $[a, b+2]$ and*

$$\begin{aligned} p_2(t) &\geq p_1(t) > 0 && \text{on } [a+1, b+2] \\ q_2(t) &\leq q_1(t) && \text{on } [a+1, b+1], \end{aligned}$$

then $L_2 y(t) = 0$ is C-disfocal on $[a, b+2]$.

Proof: Since $L_1 u(t) = 0$ is C-disfocal on $[a, b+2]$ we have by Theorem 1.44 that J_f^1 is positive definite on A_f. For $\eta \in A_f$ consider

$$\begin{aligned} J_f^2 \eta &= \sum_{t=a+1}^{b+1} \{p_2(t)[\Delta\eta(t-1)]^2 - q_2(t)\eta^2(t)\} \\ &\geq \sum_{t=a+1}^{b+1} \{p_2(t)[\Delta\eta(t-1)]^2 - q_1(t)\eta^2(t)\} \\ &= J_f^1 \eta. \end{aligned}$$

It follows that J_f^2 is positive definite on A_f and hence by Theorem 1.44, $L_2 u(t) = 0$ is C-disfocal on $[a, b+2]$. □

Corollary 1.46 *Assume $L_i u(t) = 0$, $i = 1, 2$ is C-disfocal on $[a, b+2]$. Assume $p_i(t) > 0$ on $[a+1, b+2]$, $i = 1, 2$. Set*

$$\begin{aligned} p(t) &= \lambda_1 p_1(t) + \lambda_2 p_2(t), && t \in [a+1, b+2], \\ q(t) &= \lambda_1 q_1(t) + \lambda_2 q_2(t), && t \in [a+1, b+1], \end{aligned}$$

where $\lambda_i \geq 0$, $i = 1, 2$, $\lambda_1 + \lambda_2 > 0$. Then $Lu(t) = 0$ is C-disfocal on $[a, b+2]$.

Proof: The result follows easily from Theorem 1.44 using the fact that

$$J_f = \lambda_1 J_f^1 + \lambda_2 J_f^2.$$

□

Corollary 1.47 *If $Lu(t) = 0$ is C-disfocal on $[a, b+2]$, then*

$$p(a+1) > \sum_{t=a+1}^{b+1} q(t).$$

Proof: By Theorem 1.44, J_f is positive definite on A_f. Define

$$\eta(t) = \begin{cases} 0, & t = a \\ 1, & a+1 \leq t \leq b+2. \end{cases}$$

Since $\eta \in A_f - \{0\}$,

$$J_f \eta > 0.$$

Hence

$$\sum_{t=a+1}^{b+1} \{p(t)[\Delta\eta(t-1)]^2 - q(t)\eta^2(t)\} > 0.$$

Simplifying we get that

$$p(a+1) - \sum_{t=a+1}^{b+1} q(t) > 0.$$

□

Corollary 1.48 *If $Lu(t) = 0$ is C-disfocal on $[a, b+2]$, then*

$$p(t) + p(t+1) > q(t)$$

for $t \in [a+1, b]$ and

$$p(b+1) > q(b+1).$$

Proof: By Theorem 1.44, J_f is positive definite on A_f. Fix $t_0 \in [a+1, b]$ and set

$$\eta(t) = \begin{cases} 1, & t = t_0 \\ 0, & \text{otherwise.} \end{cases}$$

Then $\eta \in A_f - \{0\}$, hence

$$J_f \eta > 0.$$

Thus

$$\sum_{t=a+1}^{b+1} \{p(t)[\Delta\eta(t-1)]^2 - q(t)\eta^2(t)\} > 0.$$

Simplifying,

$$p(t_0) + p(t_0 + 1) - q(t_0) > 0.$$

Since $t_0 \in [a+1, b]$ is arbitrary, we have that

$$p(t) + p(t+1) > q(t)$$

for $t \in [a+1, b]$. Next define

$$\eta_1(t) = \begin{cases} 0, & t \in [a, b] \\ 1, & t = b+1, \ b+2. \end{cases}$$

Then $\eta_1 \in A_f - \{0\}$, so again

$$J_f \eta_1 > 0.$$

Thus

$$\sum_{t=a+1}^{b+1} \{p(t)[\Delta \eta_1(t-1)]^2 - q(t)\eta_1^2(t)\} > 0.$$

Simplifying, we get that

$$p(b+1) - q(b+1) > 0.$$

□

Corollary 1.49 *The self-adjoint equation $Lu(t) = 0$ is C-disfocal on $[a, b+2]$ if and only if the symmetric tridiagonal matrix*

$$S = \begin{bmatrix} -c(a+1) & p(a+2) & 0 & \cdots & 0 \\ p(a+2) & -c(a+2) & & \ddots & \\ 0 & & \ddots & & 0 \\ \cdots & \cdots & \cdots & -c(b+1) & p(b+1) \\ 0 & \cdots & 0 & p(b+1) & p(b+2) \end{bmatrix}$$

is positive definite.

Proof: For $\eta \in A_f$, consider

$$J_f \eta = \sum_{t=a+1}^{b+1} \{p(t)[\Delta \eta(t-1)]^2 - q(t)\eta^2(t)\}$$

$$\begin{aligned}
&= \sum_{t=a+1}^{b+1} \{p(t)\eta^2(t) - 2p(t)\eta(t)\eta(t-1) + p(t)\eta^2(t-1) - q(t)\eta^2(t)\} \\
&= \sum_{t=a+1}^{b+1} \{[p(t) + p(t+1) - q(t)]\eta^2(t) - 2p(t)\eta(t)\eta(t-1)\} \\
&\qquad - p(b+2)\eta^2(b+1) \\
&= -\sum_{t=a+1}^{b+1} \{c(t)\eta^2(t) + 2p(t)\eta(t)\eta(t-1)\} - p(b+2)\eta^2(b+1) \\
&= \alpha^T S\alpha
\end{aligned}$$

where

$$\alpha^T = [\eta(a+1), \eta(a+2), \ldots, \eta(b+1)].$$

It follows that J_f is positive definite on A_f if and only if S is a positive definite matrix. Hence the result follows from Theorem 1.44. □

1.9 NOTES

For second order linear homogeneous equations, with nonvanishing lead coefficient, there exists an integrating factor which allows the equation to be put into self–adjoint form. Our Example 1.2, where we transform the the Fibonacci recurrence in self-adjoint form is different from the continuous case in that the "integrating factor" is not of fixed sign. Thus the usual assumption of positive $p(t)$ in second order differential equations

$$(p(t)y')' + q(t)y = 0$$

is not natural for difference equations. From a variational viewpoint, if the equation is a Jacobi equation, then the coefficient $p(t)$ being of one sign is the Legendre condition. However, for Jacobi difference equations, the Legendre conditions requires that $c(t)$ of equation (1.3) be of one sign. Thus the discrete calculus of variations provides an explanation of this distinction between the continuous and the discrete cases. Many authors have studied difference equations with the preconception that a sign condition on $p(t)$ was somehow natural. This led to a difficulty in providing Sturmian theory for examples such as the Fibonacci recurrence, where one solution seemed to be oscillatory while another was oscillatory. The definition we use of generalized zeros brings these exceptional cases into the general theory. The evolution of this thinking can be seen in our papers.

The term "prepared" was first used by Phillip Hartman in his study of second order linear matrix differential equations. He was thinking of the solution being prepared for the application of the reduction of order theorem in order to generate "principal solutions" (which we have called recessive). The concept of "conjugate solutions" for linear Hamiltonian systems was used by Radon. Reid, by allowing complex solutions, introduced the term "conjoined" because conjugate already had a variational meaning in "conjugate points" and his definition of conjoined pairs of solutions was a more general concept than Radon's conjugate pairs. Thus for an individual solution to be conjoined with itself, Reid said "self-conjoined". Coppel used the term "isotropic" for the same idea. One can observe that when something has many different names in the literature, it probably is important.

2

CONTINUED FRACTIONS

2.1 THE SCALAR CASE

Riccati equations of the form developed in the previous chapter are closely related to continued fractions. With the connections between these subjects as our goal, we make a fresh start in neutral notation. Recall some basic facts about linear fractional (Möbius) transformations. Suppose that A is a nonsingular 2×2 matrix

$$A = \begin{bmatrix} a & b \\ c & d \end{bmatrix} \tag{2.1}$$

with real or complex entries. The associated linear fractional transformation is formally defined by

$$T_A(z) = \frac{az + b}{cz + d}.$$

With matrix continued fractions in mind, we write this as

$$T_A(z) = (az + b)(cz + d)^{-1}, \tag{2.2}$$

where z is a real or complex variable. We say that z is in the domain of T_A if $cz + d \neq 0$ and we say that $z = \infty$ is in the domain of T_A if $c \neq 0$ and make the definition $T_A(\infty) = ac^{-1}$. If σ is a nonzero scalar, then A may be rescaled by multiplying by σ because

$$T_{\sigma A}(z) \equiv T_A(z). \tag{2.3}$$

Hence A can be rescaled without changing the associated linear fractional transformation.

Suppose that m is a fixed real number. Let A_k be a family of nonsingular 2×2 matrices, $k = m, m+1, \ldots$, and let $T_{A_k}(z)$ be defined by

$$T_{A_k}(z) = \frac{a_k z + b_k}{c_k z + d_k}$$

which we write as

$$T_{A_k}(z) = (a_k z + b_k)(c_k z + d_k)^{-1}. \tag{2.4}$$

We now compare the *functional composite*

$$T_{A_m} \circ T_{A_{m+1}} \circ \cdots \circ T_{A_k}(z) \tag{2.5}$$

with the *formal composite*

$$T_{A_m A_{m+1} \cdots A_k}(z). \tag{2.6}$$

Although we used the motivation of $T_A(z)$ for z a real or complex number, we now regard z as playing the role of a *generator* of an asssociated continued fraction. It is often the case that the entries in the matrix A might involve a real variable x or a complex variable z or other parameters and they should not be confused with the generator. For those cases, it might be useful to replace the generator z with some other symbol, such as w. In the matrix case we will denote the generator by $\mathcal{Z}$.

Theorem 2.1 *The formal composite* (2.6) *is an extension of the functional composite* (2.5) *in the sense that when z is in the domain of the functional composite, then z is in the domain of the formal composite and their functional values agree. Furthermore, if ∞ is in the domain of the functional composite, then ∞ is in the domain of the formal composite and*

$$T_{A_m} \circ T_{A_{m+1}} \circ \cdots \circ T_{A_k}(\infty) = T_{A_m A_{m+1} \cdots A_k}(\infty). \tag{2.7}$$

Proof: We introduce a bivariate Möbius function defined for ordered pairs $\vec{z} = (z_1, z_2)$ by

$$T_A(\vec{z}) = (az_1 + bz_2)(cz_1 + dz_2)^{-1}. \tag{2.8}$$

For $j = m, \ldots, k$ define numerators $\mathcal{N}$ and denominators $\mathcal{D}$ by

$$\begin{bmatrix} \mathcal{N}_{j,k}(\vec{z}) \\ \mathcal{D}_{j,k}(\vec{z}) \end{bmatrix} = A_j A_{j+1} \cdots A_k \begin{bmatrix} z_1 \\ z_2 \end{bmatrix}. \tag{2.9}$$

Note that for z real or complex and $T_j \equiv T_{A_j}$ we have

$$(T_j \circ \cdots \circ T_k)(z) = (\mathcal{N}_{j,k}(\vec{z}))(\mathcal{D}_{j,k}(\vec{z}))^{-1} \text{ for } \vec{z} = (z, 1) \tag{2.10}$$

and

$$(T_j \circ \cdots \circ T_k)(\infty) = (\mathcal{N}_{j,k}(\infty))(\mathcal{D}_{j,k}(\infty))^{-1}, \text{ for } \vec{z} = (1, 0). \tag{2.11}$$

If z is in the domain of the functional composite, then z is in the domain of each formal composite for $j = k, k-1, \ldots, m$ and we have $\mathcal{D}_{j,k}(\vec{z}) \neq 0$ for each j. Thus $\mathcal{D}_{m,k}(\vec{z}) \neq 0$. Thus from (2.9) $\vec{z}$ is in the domain of the formal composite and (2.9), (2.10), and (2.11) imply that when the functional composite is defined, then so is the formal composite and they have the same value. The choice of $j = m$ gives the theorem. □

Let m be fixed. We base our definition of continued fractions on the formal composites evaluated at $z = \infty$ for $k = m, m+1, \ldots$. Introduce the notation

$$\begin{bmatrix} p_k & q_k \\ r_k & s_k \end{bmatrix} = A_m A_{m+1} \cdots A_k, \tag{2.12}$$

for $k = m, m+1, \ldots$. Define the sequence of *approximants* as

$$T_{A_m A_{m+1} \cdots A_k}(\infty) = (p_k)(r_k)^{-1}, \text{ for } k = m, m+1, \ldots. \tag{2.13}$$

The factorization for $k > m$ of

$$A_m A_{m+1} \cdots A_k = (A_m A_{m+1} \cdots A_{k-1}) A_k$$

gives

$$\begin{bmatrix} p_k & q_k \\ r_k & s_k \end{bmatrix} = \begin{bmatrix} p_{k-1} & q_{k-1} \\ r_{k-1} & s_{k-1} \end{bmatrix} A_k \tag{2.14}$$

which also holds for $k = m$ if we make the definitions

$$p_{m-1} = 1, \ q_{m-1} = 0, \ r_{m-1} = 0, \ s_{m-1} = 1. \tag{2.15}$$

Then we have the conclusion that the functions $[p \quad q]$ and $[r \quad s]$ are solutions of the same system

$$[u_k \quad v_k] = [u_{k-1} \quad v_{k-1}] A_k \tag{2.16}$$

with different initial conditions.

The continued fraction

$$\{T_{A_m A_{m+1} \cdots A_k}(\infty)\} = \left\{\frac{p_k}{r_k}\right\}, \quad k = m, m+1, \ldots \tag{2.17}$$

is said to *converge* if the partial denominators r_k are nonzero for large k and the sequence $\{p_k r_k^{-1}\}$ has a finite limit.

We now make an observation about the corresponding functional composites.

Theorem 2.2 *Suppose that the formal composite continued fraction* (2.17) *converges to a value* Ω_m. *If there exists a value* $p \geq m$ *such that for* $k \geq p$, *no zeros occur in denominators for the functional composites*

$$T_{A_m} \circ T_{A_{m+1}} \circ \cdots \circ T_{A_k}(\infty), \tag{2.18}$$

then the sequence of functional composites also converges to Ω_m.

This theorem says that if we have convergence of the formal composite continued fraction and if from some point on, we can compute the functional composites, then we can use the resulting limit as the limit of the formal composites, i.e., under these circumstances, *if you can compute it, you can reliably use it.*

Proof: If no zero divisors occur for $k \geq p$, then ∞ is in the domain of each of these functional composites. Since the formal composite is an extension of the functional composite, convergence of the formal composites implies convergence of the functional composites and they must converge to the same limit, namely Ω_m. □

Since we are accustomed to having the variables on the right in linear systems, the *key idea* for connecting the topic of continued fractions with the topic of linear recurrences is to take the transpose of (2.14), i.e.,

$$\begin{bmatrix} p_k & r_k \\ q_k & s_k \end{bmatrix} = A_k^T \begin{bmatrix} p_{k-1} & r_{k-1} \\ q_{k-1} & s_{k-1} \end{bmatrix}. \tag{2.19}$$

Set $M(k) \equiv A_k^T$ and use new labels for the entries of $M(k)$ denoted by

$$M(k) \equiv \begin{bmatrix} e_k & f_k \\ g_k & h_k \end{bmatrix} = A_k^T. \tag{2.20}$$

Then each of the column vector functions

$$x_1(k) \equiv \begin{bmatrix} y_1(k) \\ z_1(k) \end{bmatrix} = \begin{bmatrix} p_{k-1} \\ q_{k-1} \end{bmatrix} \quad \text{and} \quad x_2(k) \equiv \begin{bmatrix} y_2(k) \\ z_2(k) \end{bmatrix} = \begin{bmatrix} r_{k-1} \\ s_{k-1} \end{bmatrix} \tag{2.21}$$

are solutions of the linear system

$$x(k+1) = \begin{bmatrix} y(k+1) \\ z(k+1) \end{bmatrix} = M(k) \begin{bmatrix} y(k) \\ z(k) \end{bmatrix} = M(k)x(k) \tag{2.22}$$

and the initial conditions obtained from (2.15) for fixed m become

$$y_1(m) = 1, \; z_1(m) = 0, \qquad y_2(m) = 0, \; z_2(m) = 1. \tag{2.23}$$

Now convergence of the continued fraction (2.17) to Ω_m is equivalent to $y_2(k)$ being nonzero for large k and existence of the limit

$$\lim_{k\to\infty} y_1(k)\,(y_2(k))^{-1} = \Omega_m. \tag{2.24}$$

□

Exercise 2.3 *Use* (2.17) *and the above formulas to find the limit of the continued fraction corresponding to the sequence of constant matrices*

$$A_k = M^T(k) \quad \textit{for} \quad M(k) = \begin{bmatrix} 0 & 1 \\ -6 & 5 \end{bmatrix}.$$

A useful identity is obtained by taking the determinant of the matrix equation

$$\begin{bmatrix} y_1(t+1) & y_2(t+1) \\ z_1(t+1) & z_2(t+1) \end{bmatrix} = M(t) \begin{bmatrix} y_1(t) & y_2(t) \\ z_1(t) & z_2(t) \end{bmatrix}. \tag{2.25}$$

Let us introduce the notation

$$\{x_1(t); x_2(t)\} = \det \begin{bmatrix} y_1(t) & y_2(t) \\ z_1(t) & z_2(t) \end{bmatrix} = \begin{vmatrix} y_1(t) & y_2(t) \\ z_1(t) & z_2(t) \end{vmatrix} = y_1(t)z_2(t) - z_1(t)y_2(t). \tag{2.26}$$

Then we have the following result.

Proposition 2.4 (Determinant Formula) *Let $x_1(t)$ and $x_2(t)$ be the solutions of system* (2.22) *determined by the respective initial conditions* (2.23). *Suppose $\mu(t) = \det M(t)$. Then the initial value problem* **IVP**

$$\{x_1(t+1); x_2(t+1)\} = \mu(t)\,\{x_1(t); x_2(t)\}, \qquad \{x_1(m); x_2(m)\} = 1 \tag{2.27}$$

implies the determinant formula

$$y_1(t)z_2(t) - z_1(t)y_2(t) = \{x_1(t); x_2(t)\} = \begin{cases} 1, & \textit{for } t = m, \\ \mu(t-1)\cdots\mu(m), & \textit{for } t > m. \end{cases} \tag{2.28}$$

Note that this determinant formula may be written in the form

$$y_1(t)z_2(t) - z_1(t)y_2(t) = \prod_{s=m}^{t-1} \mu(s), \tag{2.29}$$

for $t \geq m$, if we use the convention that $\prod_{s=m}^{m-1} \mu(s) = 1$.

The next two theorems are based on ideas of Pincherle and Gautschi; namely, the equivalence of convergence of continued fractions with the existence of a recessive solution $y_0(k), z_0(k)$ with $y_0(m) \neq 0$.

Theorem 2.5 (Genesis of a Recessive Solution) *Assume that the A_k are nonsingular and the continued fraction* (2.17) *converges to a finite limit Ω_m. For $M(k) \equiv A_k^T$, let y_0, z_0 be the solution of the linear system* (2.22) *defined by*

$$\begin{aligned} y_0(k) &= y_1(k) - y_2(k)\Omega_m \\ z_0(k) &= z_1(k) - z_2(k)\Omega_m. \end{aligned} \tag{2.30}$$

Then

1. $\begin{bmatrix} y_0(k) \\ z_0(k) \end{bmatrix} \neq \begin{bmatrix} 0 \\ 0 \end{bmatrix}$ *for each k;*

2. *If* $\begin{bmatrix} y(k) \\ z(k) \end{bmatrix}$ *is any solution of* (2.22) *such that*

$$\begin{bmatrix} y_0(k) & y(k) \\ z_0(k) & z(k) \end{bmatrix} \tag{2.31}$$

is nonsingular, then $y(k)$ is nonzero for large k and

$$\frac{y_0(k)}{y(k)} \to 0 \text{ as } k \to \infty. \tag{2.32}$$

These properties of y_0 will become our defining properties of a recessive solution.

Proof: Since $y_0(m) = 1$, condition 1 is a consequence of the nonsingularity of $M(k) = A_k^T$. From the definition of convergence $y_2(k) \equiv r_{k-1}$ is nonzero for large k. Hence

$$\frac{y_0(k)}{y_2(k)} = \frac{y_1(k)}{y_2(k)} - \Omega_m \to \Omega_m - \Omega_m = 0. \tag{2.33}$$

Again, since the $M(k)$ are nonsingular the matrix in (2.31) is of constant rank, so nonsingularity at one k implies nonsingularity for all k. Since these solutions y_1, z_1 and y_2, z_2 constitute a basis for the solution space of (2.22) there exist constants C_1 and C_2 such that

$$\begin{aligned} y(k) &= y_1(k)C_1 + y_2(k)C_2 \\ z(k) &= z_1(k)C_1 + z_2(k)C_2 \end{aligned} \tag{2.34}$$

namely, $C_1 = y(m)$, $C_2 = z(m)$. Nonsingularity of the matrix in (2.31) for $k = m$ makes

$$\det \begin{bmatrix} 1 & C_1 \\ -\Omega_m & C_2 \end{bmatrix} = C_2 + \Omega_m C_1 \neq 0. \tag{2.35}$$

But

$$\frac{y(k)}{y_2(k)} = \frac{y_1(k)}{y_2(k)} C_1 + \frac{y_2(k)}{y_2(k)} C_2 \rightarrow \Omega_m C_1 + C_2 \neq 0 \tag{2.36}$$

makes $y(k)$ nonzero for large k. We complete the proof by showing

$$\frac{y_0(k)}{y(k)} = \frac{y_1(k) - y_2(k)\Omega_m}{y_1(k)C_1 + y_2(k)C_2} = \frac{y_1(k)/y_2(k) - \Omega_m}{(y_1(k)/y_2(k))C_1 + C_2} \rightarrow \frac{0}{\Omega_m C_1 + C_2} = 0. \tag{2.37}$$

□

We say that a solution $x_0 = \begin{bmatrix} y_0 \\ z_0 \end{bmatrix}$ is a *recessive solution* of

$$x(k+1) = M(k)x(k) \tag{2.38}$$

if properties 1 and 2 of Theorem 2.5 hold.

Theorem 2.6 (Pincherle & Gautschi) *Assume that all A_k are nonsingular and $M(k) \equiv A_k^T$. A necessary and sufficient condition for convergence of the continued fraction* (2.17) *is that there exists a recessive solution at ∞ of the form $x_0 = \begin{bmatrix} y_0 \\ z_0 \end{bmatrix}$ of* (2.38) *with $y_0(m) \neq 0$. Furthermore, if the continued fraction converges, then it converges to*

$$\Omega_m = -\frac{z_0(m)}{y_0(m)}. \tag{2.39}$$

Proof: Necessity of the existence of such a recessive solution was shown in Theorem 2.5. In order to prove that the criterion is sufficient for convergence, assume that x_0 is a recessive solution with $y_0(m) \neq 0$. For x_1 and x_2 the solutions of (2.38) defined by the initial conditions (2.23) there exist constants C_1, C_2 such that $x_0 = x_1C_1 + x_2C_2$. Thus $C_1 = y_0(m) \neq 0$, $C_2 = z_0(m)$, and $y_0(k) = y_1(k)C_1 + y_2(k)C_2$. Now

$$\begin{bmatrix} y_0(m) & y_2(m) \\ z_0(m) & z_2(m) \end{bmatrix} = \begin{bmatrix} C_1 & 0 \\ C_2 & 1 \end{bmatrix}$$

is nonsingular since $C_1 \neq 0$. Thus x_0 and x_2 are linearly independent. Since x_0 is recessive we must have $y_2(k) \neq 0$ for large k and

$$\frac{y_0(k)}{y_2(k)} = \frac{y_1(k)C_1 + y_2(k)C_2}{y_2(k)} = \frac{y_1(k)}{y_2(k)}C_1 + C_2 \to 0$$

which implies that

$$\frac{y_1(k)}{y_2(k)} \to -\frac{C_2}{C_1}.$$

Hence, the continued fraction converges to

$$\Omega_m = -C_2/C_1 = -z_0(m)/y_0(m). \tag{2.40}$$

□

The connection between continued fractions and Riccati equations is made by setting $w(k) = z(k)/y(k)$. Then if $y(k)$ and $y(k+1)$ are nonzero, we formally use (2.20), (2.22) to obtain

$$w(k+1) = \frac{g_k y(k) + h_k z(k)}{e_k y(k) + f_k z(k)}$$

and

$$w(k+1) = \frac{g_k + h_k w(k)}{e_k + f_k w(k)} \tag{2.41}$$

which is a form of a discrete Riccati equation. In particular, if there exists a recessive solution with $y_0(m) \neq 0$, then

$$-w_0(m) = \lim_{k\to\infty} \{T_{A_m A_{m+1} \cdots A_k}(\infty)\}. \tag{2.42}$$

Exercise 2.7 *Suppose all the $a_k = 0$ and all b_k and c_k are nonzero. Show that*

$$T_{A_m A_{m+1} \cdots A_k A_{k+1}}(\infty) = T_{A_m A_{m+1} \cdots A_k}(0).$$

2.2 SCALAR SYMPLECTIC CONTINUED FRACTIONS

For 2×2 real matrices M, we saw in Exercise 1.3 on page 4 of Chapter 1 that being symplectic is the same as having determinant 1. Consider 2×2 symplectic matrices $M(t)$ of the form of those in equation (1.12) on page 6 of Chapter 1 written in the present notation with $e = 1$ and $f \neq 0$. That is, we now consider determinant 1 coefficient matrices $M(t)$ of the special form

$$M(t) = \begin{bmatrix} 1 & f(t) \\ g(t) & 1 + g(t)f(t) \end{bmatrix}, \tag{2.43}$$

whose transpose is given by

$$A(t) = M^T(t) = \begin{bmatrix} 1 & g(t) \\ f(t) & 1 + f(t)g(t) \end{bmatrix}. \tag{2.44}$$

A continued fraction determined by a sequence of symplectic matrices will be called a *symplectic continued fraction.*

The linear system

$$\begin{bmatrix} y(t+1) \\ z(t+1) \end{bmatrix} = \begin{bmatrix} 1 & f(t) \\ g(t) & 1 + g(t)f(t) \end{bmatrix} \begin{bmatrix} y(t) \\ z(t) \end{bmatrix} \tag{2.45}$$

is of the form of a discrete linear Hamiltonian system (see Chapter 4)

$$\begin{aligned} \Delta y(t) &= f(t)z(t) \\ \Delta z(t) &= g(t)y(t+1) \end{aligned} \tag{2.46}$$

with discrete momentum variable $z(t) = f^{-1}(t)\Delta y(t)$. Eliminate z to obtain the corresponding second order self–adjoint difference equation (a discrete Jacobi equation)

$$\Delta \left[f^{-1}(t)\Delta y(t)\right] = g(t)y(t+1). \tag{2.47}$$

Note that the above matrix A has

$$T_A(z) = \frac{z+g}{fz+1+fg} = \frac{z+g}{f(z+g)+1} = \frac{1}{f + \cfrac{1}{g+z}} \equiv \frac{1}{f} \underset{+}{} \frac{1}{g+z} \tag{2.48}$$

where the + in the denominator means that the latter fraction is in the denominator of the first. For m fixed, we now give the associated functional composite approximants for the symplectic continued fraction

$$\{T_{A_m A_{m+1} \cdots A_k}(\infty)\}, \quad k = m, m+1, \ldots. \tag{2.49}$$

The first two functional approximants are given by

$$\frac{1}{f(m)} \quad \text{and} \quad \frac{1}{f(m)}\,+\,\frac{1}{g(m)}\,+\,\frac{1}{f(m+1)} \tag{2.50}$$

and the general functional composite approximant for $k > m$ is of the form

$$\frac{1}{f(m)}\,+\,\frac{1}{g(m)}\,+\,\frac{1}{f(m+1)}\,+\cdots+\,\frac{1}{g(k)}\,+\,\frac{1}{f(k+1)}. \tag{2.51}$$

Use $z(t) = f^{-1}(t)(y(t+1) - y(t))$ in the determinant formula (2.28) on page 49 and use the fact that M has determinant 1, to see that

$$y_1(t)f^{-1}(t)(y_2(t+1) - y_2(t)) - y_2(t)f^{-1}(t)(y_1(t+1) - y_1(t)) = 1; \tag{2.52}$$

which, after cancellation of two terms and multiplication by $-f(t)$, can be rewritten as

$$\Delta\left[\frac{y_1(t)}{y_2(t)}\right] = \frac{y_1(t+1)}{y_2(t+1)} - \frac{y_1(t)}{y_2(t)} = \frac{-f(t)}{y_2(t)y_2(t+1)} \tag{2.53}$$

for values of t such that $y_2(t)$ and $y_2(t+1)$ are nonzero. We are now prepared to establish equivalence between convergence of symplectic continued fractions and convergence of an associated series.

Theorem 2.8 (Series Equivalence for the Symplectic Case) *A symplectic continued fraction determined by* (2.17) *on page 47, for $A = M^T$ with the above matrix $M(t)$, converges if and only if the solution $y_2(t)$ of* (2.22), (2.23) *is nonzero for large t and for k taken sufficiently large, the series*

$$\sum_{t=k}^{\infty} \frac{f(t)}{y_2(t)y_2(t+1)} \tag{2.54}$$

converges.

Proof: Replace t in (2.53) by τ and sum from k to t to obtain

$$\frac{y_1(t+1)}{y_2(t+1)} - \frac{y_1(k)}{y_2(k)} = -\sum_{\tau=k}^{t} \frac{f(\tau)}{y_2(\tau)y_2(\tau+1)}. \tag{2.55}$$

Hence the continued fraction $\{y_1(t)y_2^{-1}(t)\}$ converges if and only if there exists a value k such that $y_2(\tau)$ is nonzero for $\tau \geq k$ and the series (2.54) converges.□

The solution y_2 is said to be *dominant* if $y_2(t)$ is nonzero for large t and the series (2.54) converges. Thus the solution $y_2(t)$ is a dominant solution of equation (2.47) if and only if the continued fraction converges.

2.3 COMPANION MATRICES

If we expand out the linear second order difference equation

$$\Delta\left[f^{-1}(t)\Delta y(t)\right] = g(t)y(t+1) \tag{2.56}$$

we obtain a linear three term recurrence relation

$$y(t+2) = a(t)y(t) + b(t)y(t+1), \tag{2.57}$$

where $a(t)$ is nonzero and

$$a(t) \;=\; -f(t+1)f^{-1}(t) \tag{2.58}$$
$$b(t) \;=\; 1 + f(t+1)f^{-1}(t) + f(t+1)g(t). \tag{2.59}$$

(Caution, here a and b are not the entries of A at the start of this chapter.) Conversely, if $a(t)$ is nonzero and $f(t)$, $g(t)$ are solutions of

$$f(t+1) \;=\; -a(t)f(t), \quad \text{with} \quad f(m) \neq 0, \tag{2.60}$$
$$g(t) \;=\; f^{-1}(t+1)[b(t)-1] - f^{-1}(t), \tag{2.61}$$

then the recurrence relation (2.57) is equivalent to the difference equation (2.47). The choice of $M(t)$ of Section 1, equation (2.20), as the companion matrix $M(t) = C(t)$ for equation (2.57) is of the form

$$C(t) = \begin{bmatrix} 0 & 1 \\ a(t) & b(t) \end{bmatrix}, \quad \text{with transpose} \quad B(t) = C^T(t) = \begin{bmatrix} 0 & a(t) \\ 1 & b(t) \end{bmatrix}, \tag{2.62}$$

both of which have determinant of $-a(t)$. Furthermore,

$$T_B(z) = \frac{a}{z+b} = \frac{a}{b+z}. \tag{2.63}$$

Since,

$$T_{B(m)}(z) = \frac{a(m)}{b(m)+z} = \frac{0\cdot z + a(m)}{1\cdot z + b(m)} \quad \Rightarrow \quad T_{B(m)}(\infty) = 0$$

and

$$T_{B(m)} \circ T_{B(m+1)}(\infty) = \frac{a(m)}{b(m)+0} = \frac{a(m)}{b(m)},$$

the first three functional composites evaluated at $z = \infty$ are

$$0, \quad \frac{a(m)}{b(m)}, \quad \text{and} \quad \frac{a(m)}{b(m)} + \frac{a(m+1)}{b(m+1)}. \tag{2.64}$$

Except for the first case, the general functional composite approximant is of the form

$$\frac{a(m)}{b(m)} + \frac{a(m+1)}{b(m+1)} + \cdots + \frac{a(k)}{b(k)}, \tag{2.65}$$

for $k = m, m+1, \ldots$. However, the variable z in the initial conditions (2.23) is different for the cases of symplectic M than in the case of the companion matrix C. For the companion matrix case $z(m) = y(m+1)$. We have no recourse other than to introduce separate names for what we have been calling $y_1(t)$ and $y_2(t)$ in the two cases.

For the symplectic case, let $u_1(t)$ and $u_2(t)$ be the solutions of the linear second order equation (2.56) which satisfy the initial conditions

$$u_1(m) = 1,\ u_1(m+1) = 1, \qquad u_2(m) = 0,\ u_2(m+1) = f(m). \tag{2.66}$$

Then the *symplectic approximants* are $u_1(k)[u_2(k)]^{-1}$ and convergence means $u_2(k) \neq 0$ for large k and there exists an $\Omega(m)$ such that

$$u_1(k)[u_2(k)]^{-1} \to \Omega(m). \tag{2.67}$$

For the rest of this section, we will return to the companion case. Let $v_1(t)$ and $v_2(t)$ be the solutions of the recurrence relation (2.57) which satisfy the initial conditions

$$v_1(m) = 1,\ v_1(m+1) = 0, \qquad v_2(m) = 0,\ v_2(m+1) = 1. \tag{2.68}$$

Then the *companion approximants* are $v_1(k)[v_2(k)]^{-1}$ and convergence means $v_2(k) \neq 0$ for large k and there exists a $\Gamma(m)$ such that

$$v_1(k)[v_2(k)]^{-1} \to \Gamma(m). \tag{2.69}$$

The *negative* of the determinant of both sides of the equation

$$[x_1(t+1) \quad x_2(t+1)] = C(t)\,[x_1(t) \quad x_2(t)]$$

becomes

$$v_1(t+2)v_2(t+1) - v_1(t+1)v_2(t+2) = -a(t)(v_1(t+1)v_2(t) - v_1(t)v_2(t+1)). \tag{2.70}$$

Iterate this back to $t = m$ for the result

$$v_1(t+1)v_2(t) - v_1(t)v_2(t+1) = \rho(t) \equiv \begin{cases} -1, & t = m \\ -(-1)^{t-m}a(t-1)\cdots a(m), & t > m \end{cases} \tag{2.71}$$

since in the case $t > m$, we are multiplying (-1) by the $[(t-1)-m]+1 = t-m$ factors $-a(t-1), -a(t-2), \ldots, -a(m)$. After dividing this last expression by $v_2(t)v_2(t+1)$, the identity

$$\Delta\left[\frac{v_1(t)}{v_2(t)}\right] = \frac{v_1(t+1)}{v_2(t+1)} - \frac{v_1(t)}{v_2(t)} = \frac{\rho(t)}{v_2(t)v_2(t+1)} \tag{2.72}$$

leads us to the following equivalence with convergence of infinite series.

Theorem 2.9 (Series Equivalence for Companion Matrices) *The companion matrix continued fraction for $M(t) = C(t)$ and for $B(t) = M^T(t)$ converges if and only if the solution $v_2(t)$ of equation (2.57) determined by the initial conditions in (2.68) is nonzero for large t and for k taken sufficiently large, the series*

$$\sum_{t=k}^{\infty} \frac{\rho(t)}{v_2(t)v_2(t+1)} \tag{2.73}$$

converges.

Proof: See the proof of Theorem 2.8. □

Observe that the sequence $\rho(t)$ satisfies the initial value problem

$$\rho(t+1) = -a(t)\rho(t), \qquad \rho(m) = -1. \tag{2.74}$$

If we set $f(t) = \rho(t)$, and define $g(t)$ by equation (2.61) on page 55, we observe that the companion matrix continued fraction series equivalence is somehow trying to make the problem self–adjoint.

Exercise 2.10 (Transformations of Companion Matrix C. F.s) *Let $c(t)$ be a sequence of nonzero numbers with $c(m-1) \equiv 1$. Suppose $\alpha(t)$, $t \geq m$, is determined by the* **IVP** *$\alpha(m) = 1$ and $\alpha(t+1) = c(t-1)\alpha(t)$. Multiply the three term recurrence*

$$y(t+2) = a(t)y(t) + b(t)y(t+1),$$

by $\alpha(t+2)$, introduce $\hat{y}(t) \equiv \alpha(t)y(t)$ and deduce the equation

$$\hat{y}(t+2) = \hat{a}(t)\hat{y}(t) + \hat{b}(t)\hat{y}(t+1), \tag{2.75}$$

for

$$\hat{a}(t) = a(t)c(t)c(t-1), \qquad \hat{b}(t) = b(t)c(t). \tag{2.76}$$

Show that $\hat{y}(m) = y(m)$, $\hat{y}(m+1) = y(m+1)$ and the continued fractions determined by $\hat{C}(t)$ and by $C(t)$ have the same approximants. We write the corresponding functional composite approximants as

$$\frac{a(m)c(m)}{b(m)c(m)} + \frac{a(m+1)c(m)c(m+1)}{b(m+1)c(m+1)} + \cdots + \frac{a(k)c(k-1)c(k)}{b(k)c(k)}$$
$$= \frac{a(m)}{b(m)} + \frac{a(m+1)}{b(m+1)} + \cdots + \frac{a(k)}{b(k)}. \tag{2.77}$$

2.4 SYMPLECTIC C.F.S VERSUS COMPANION MATRIX C.F.S

We now show that in many cases one can convert a symplectic continued fraction **(SCF)** to a companion matrix continued fraction **(CCF)** and conversely.

Theorem 2.11 (SCF Approximants versus CCF Approximants) *Let $f(t)$ and $a(t)$ be nonzero for $t \geq m$ and assume $a(t)$, $b(t)$, $f(t)$, and $g(t)$ are related by equations* (2.58), (2.59). *Then*

$$\frac{1}{f(m)} + \frac{1}{g(m)} + \frac{1}{f(m+1)} + \cdots + \frac{1}{g(k)} + \frac{1}{f(k+1)}$$
$$= \frac{1}{f(m)}\left[1 + \frac{a(m)}{b(m)} + \frac{a(m+1)}{b(m+1)} + \cdots + \frac{a(k)}{b(k)}\right]. \tag{2.78}$$

If there exists a solution $y_0(t)$ which is recessive at ∞ and has $y_0(m) \neq 0$, then the **SCF** *converges to*

$$\Omega(m) = \frac{1}{f(m)}\left[1 - \frac{y_0(m+1)}{y_0(m)}\right] \tag{2.79}$$

and the **CCF** *converges to*

$$\Gamma(m) = -\frac{y_0(m+1)}{y_0(m)}. \tag{2.80}$$

Futhermore, the **SCF** *converges if and only if the* **CCF** *converges.*

Proof: In the first case the above equality is interpreted as being

$$\frac{1}{f(m)} = \frac{1}{f(m)}$$

with the understanding that the first approximant for the **CCF** is 0. The second functional composite approximant of the **SCF** is

$$\begin{aligned}
&\frac{1}{f(m)} + \frac{1}{g(m)} + \frac{1}{f(m+1)} \\
&\quad = \frac{1}{f(m)} + \frac{f(m+1)}{f(m+1)g(m)+1} \\
&\quad = \frac{1}{f(m)} + \frac{f(m+1)}{b(m) - f(m+1)f^{-1}(m)} \\
&\quad = \frac{b(m) - f(m+1)f^{-1}(m)}{f(m)b(m) - f(m)f(m+1)f^{-1}(m) + f(m+1)} \\
&\quad = \frac{b(m)+a(m)}{f(m)b(m)} \\
&\quad = \frac{1}{f(m)}\left[1 + \frac{a(m)}{b(m)}\right].
\end{aligned}$$

The induction step makes use of formula (2.78) with m and k shifted one to the right. Namely, we assume

$$\begin{aligned}
&\frac{1}{f(m+1)} + \cdots + \frac{1}{g(k+1)} + \frac{1}{f(k+2)} \\
&\quad = \frac{1}{f(m+1)}\left[1 + \frac{a(m+1)}{b(m+1)} + \frac{a(m+2)}{b(m+2)} + \cdots + \frac{a(k+1)}{b(k+1)}\right].
\end{aligned}$$

Set

$$r = \frac{a(m+1)}{b(m+1)} + \frac{a(m+2)}{b(m+2)} + \cdots + \frac{a(k+1)}{b(k+1)}.$$

Then use the induction hypothesis to rewrite the left side of formula (2.78) with k replaced by $k+1$ as

$$\begin{aligned}
&\frac{1}{f(m)} + \frac{1}{g(m)} + \frac{1+r}{f(m+1)} \\
&\quad = \frac{1}{f(m)} + \frac{f(m+1)}{f(m+1)g(m)+1+r} \\
&\quad = \frac{1}{f(m)} + \frac{f(m+1)}{b(m) - f(m+1)f^{-1}(m) + r} \\
&\quad = \frac{b(m)+a(m)+r}{f(m)[b(m) - f(m+1)f^{-1}(m) + r] + f(m+1)} \\
&\quad = \frac{b(m)+r+a(m)}{f(m)[b(m)+r]} \\
&\quad = \frac{1}{f(m)}\left[1 + \frac{a(m)}{b(m)+r}\right]
\end{aligned}$$

which completes the induction step. The remaining statements follow from the fact that the discrete Jacobi equation (2.56) on page 55 and the three term recurrence (2.57) have the same solutions. Therefore a solution $y_0(t)$ is a recessive solution with $y_0(m)$ nonzero for one equation if and only if it is for the other. Thus the **SCF** converges if and only if the **CCF** converges. Equations (2.79) and (2.80) are consequences of the different forms that $z_0(m)$ has in the expression $-z_0(m)y_0^{-1}(m)$ of Theorem 2.6 on page 51 for the two cases of **SCF** and **CCF**. Namely, for the symplectic case, $\Omega(m) = -z_0(m)/y_0(m)$ and $z_0(m) = f^{-1}(m)(y_0(m+1) - y_0(m))$ while for the companion case, $z_0(m) = y_0(m+1)$ and $\Gamma(m) = -y_0(m+1)/y_0(m)$. □

2.5 RATIOS OF BESSEL FUNCTIONS

We now give an example which compares symplectic continued fractions with companion matrix continued fractions.

Exercise 2.12 *Suppose that m is a fixed real number. The Bessel functions $J_k(x)$ for fixed $x > 0$ and variable $k = m-1, m, \ldots$ are known to be the recessive solution of a symmetric recurrence relation [67]*

$$-y_{k+1} - y_{k-1} + \frac{2k}{x}y_k = 0 \tag{2.81}$$

which is equivalent to

$$-\Delta^2(y_{k-1}) + \left[\frac{2k}{x} - 2\right]y_k = 0. \tag{2.82}$$

Choose $f(t) \equiv 1$, $g(t) = \left[\frac{2(t+1)}{x} - 2\right]$. Verify that the associated symplectic continued fraction functional composite is given by

$$\begin{aligned}\Omega(m) &= \lim_{k\to\infty} \frac{1}{1} + \frac{1}{g(m)} + \frac{1}{1} + \cdots + \frac{1}{g(k)} + \frac{1}{1} \\ &= \lim_{k\to\infty}\left\{1 + \frac{1}{\frac{2(m+1)}{x}} + \frac{1}{\frac{2(m+2)}{x}} + \cdots + \frac{1}{\frac{2(k+1)}{x}}\right\}\end{aligned} \tag{2.83}$$

if the inverses exist for large k. Show that this implies (replace m by $m-1$)

$$\begin{aligned}\frac{J_m(x)}{J_{m-1}(x)} &= 1 - \Omega(m-1) = \lim_{k\to\infty}\left[\frac{1}{\frac{2m}{x}} - \frac{1}{\frac{2(m+1)}{x}} - \cdots - \frac{1}{\frac{2(k+1)}{x}}\right] \\ &= \lim_{k\to\infty}\left[\frac{x}{2m} - \frac{x^2}{2(m+1)} - \cdots - \frac{x^2}{2(k+1)}\right]\end{aligned} \tag{2.84}$$

at any $x > 0$ *where* $J_{m-1}(x) \neq 0$.

Instead of computing the ratios of Bessel functions by the approximants for **CCF**, one can use the ratio $v_1(k)/v_2(k)$ of equation (2.70) and let k iterate from $m+1$. If one does this for fixed m and x on an interval and plots the solutions, the poles will become spaced by π as x goes to ∞ by Sturmian theory [77, page 336]. Since the sequence of Bessel functions $J_{m-1}(x)$, $J_m(x)$, $J_{m+1}(x), \ldots$ is a recessive solution, the recurrence relation can not be used to compute a sequence of values of $J_n(x)$ from known values of $J_{m-1}(x)$ and $J_m(x)$ [67]. It is known [141, page 241] that for $m = 1/2$ and $x > 0$,

$$\frac{J_{1/2}(x)}{J_{-1/2}(x)} = \frac{\sqrt{\frac{2}{\pi x}}\sin x}{\sqrt{\frac{2}{\pi x}}\cos x} = \tan x,$$

if x is not an odd multiple of $\pi/2$. For x replaced by z and $m = 1/2$, the continued fraction (2.84) is known to represent $\tan z$ everywhere except at the poles. Note that in that case, the values of $2k$ in the denominators are $1, 3, 5, \ldots$. For some history of the use of this continued fraction in establishing that π is irrational, see Simmons [141, page 264].

Exercise 2.13 *Find a symplectic continued fraction from a companion matrix continued fraction for the Fibonacci recurrence relation*

$$y_{n+2} = y_n + y_{n+1}. \tag{2.85}$$

2.6 MATRIX CONTINUED FRACTIONS

The objective of this section is to provide a matrix continued fraction representation of the "minimal solution", denoted by $W_-(m)$, of a matrix Riccati equation. Then we will present a reverse continued fraction representation of the "maximal solution", denoted by $W_+(m)$, of the same Riccati equation. These minimal and maximal solutions are associated with the recessive solutions at $+\infty$ and $-\infty$, respectively. It turns out that under strong sign conditions on the matrix coefficients, these continued fractions converge because of "disconjugacy" on $(-\infty, \infty)$. For standard discrete regulator or discrete Kalman filtering, these sign conditions hold and the associated continued fractions converge to the desired values.

For $n \times n$ matrices A, B, C, D with real or complex entries and

$$\mathcal{A} = \begin{bmatrix} A & B \\ C & D \end{bmatrix} \tag{2.86}$$

define a matrix Möbius transformation by

$$T_{\mathcal{A}}\mathcal{Z} = (A\mathcal{Z} + B)(C\mathcal{Z} + D)^{-1} \tag{2.87}$$

and formally define $T_{\mathcal{A}}(\infty) = AC^{-1}$.

For m fixed and $k = m, m+1, \ldots$ introduce the notation

$$\begin{bmatrix} P(k) & Q(k) \\ R(k) & S(k) \end{bmatrix} = \mathcal{A}(m)\mathcal{A}(m+1)\cdots\mathcal{A}(k) = \begin{bmatrix} P(k-1) & Q(k-1) \\ R(k-1) & S(k-1) \end{bmatrix} \mathcal{A}(k)$$

and define the sequence of approximants as (we use $\mathcal{A}(m)$ to avoid subscripts on subscripts)

$$T_{\mathcal{A}(m)\mathcal{A}(m+1)\cdots\mathcal{A}(k)}(\infty) = P(k)R^{-1}(k) \tag{2.88}$$

for $k = m,\ m+1, \ldots$. This sequence of approximants is called a *matrix continued fraction* which we abbreviate by **MCF**. Note that Schelling [138] uses the German label **MKB** for Matrizenkettenbrüche formulated in a different manner.

Let $M = \mathcal{A}^*$, where $*$ may denote transpose or conjugate transpose, be labeled by

$$M \equiv \begin{bmatrix} E & F \\ G & H \end{bmatrix}, \tag{2.89}$$

where the block entries are $n \times n$. Again, for m fixed and $M(k)$ a sequence of such nonsingular $2n \times 2n$ matrices $M(k)$, $k = m,\ m+1,\ \ldots$, let

$$X_1(t) = \begin{bmatrix} Y_1(t) \\ Z_1(t) \end{bmatrix}, \qquad X_2(t) = \begin{bmatrix} Y_2(t) \\ Z_2(t) \end{bmatrix},$$

be the $2n \times n$ solutions of

$$X(t+1) = M(t)X(t) \tag{2.90}$$

with initial conditions

$$X_1(m) = \begin{bmatrix} I \\ 0 \end{bmatrix}, \qquad X_2(m) = \begin{bmatrix} 0 \\ I \end{bmatrix}. \tag{2.91}$$

Slight modifications in the scalar case presented above show that convergence of this **MCF** to a matrix $\Omega(m)$ is equivalent to nonsingularity of $Y_2(t)$ for large t and convergence of $Y_2^{-1}(t)Y_1(t)$ to $\Omega^*(m)$. A $2n \times n$ solution $X_0(t)$ of $X(t+1) = M(t)X(t)$ is said to be *recessive* at ∞ if

1. *$X_0(t)$ has full column rank of n*

2. *If $X = \begin{bmatrix} Y \\ Z \end{bmatrix}$ is a solution such that $[X_0 \quad X]$ is nonsingular,*

then $Y(k)$ is nonsingular for large k and

$$Y^{-1}(k)Y_0(k) \to 0 \qquad as \quad k \to \infty.$$

We now give a matrix version of Pincherle's theorem.

Theorem 2.14 (Pincherle's Theorem for MCFs) *Assume that m is fixed and $\mathcal{A}(k)$ is nonsingular for $k = m,\ m+1,\ \ldots$. A necessary and sufficient condition for convergence of the* **MCF** *with approximants* (2.88) *is that there exists a $2n \times n$ recessive solution X_0 of* (2.90) *with $Y_0(m)$ nonsingular. Furthermore, if the continued fraction converges to $\Omega(m)$, then $\Omega^*(m) = -Z_0(m)Y_0^{-1}(m)$.*

Proof: Modify the proofs of Theorems 2.5 and 2.6 or see [9]. □

Now if $M(t)$ is symplectic, then we can say more. Let us assume that $*$ denotes conjugate transpose. Then Theorem 3.51 of Chapter 3, page 125, which we call the *connection theorem,* equates convergence of the above continued fraction, dominance of the above solution $X_2(t)$ at ∞ and existence of a solution $X_0(t)$ which is recessive at ∞ and has $Y_0(m)$ nonsingular. Furthermore, if the continued fraction converges to $\Omega(m)$, then $\Omega(m)$ is Hermitian and

$$\Omega(m) = -Z_0(m)Y_0^{-1}(m). \tag{2.92}$$

Furthermore, if we have disconjugacy on $[m-1, \infty)$, then $W_0(t)$ corresponds to the "minimal solution" of the associated discrete Riccati equation.

The special case of the **MCF** associated with the symplectic matrix

$$M(t) = \begin{bmatrix} I & F(t) \\ G(t) & I + G(t)F(t) \end{bmatrix} \tag{2.93}$$

with Hermitian $F(t)$ and $G(t)$ has first two functional approximants of

$$\frac{I}{F(m)}, \qquad \frac{I}{F(m)} + \frac{I}{G(m)} + \frac{I}{F(m+1)}, \tag{2.94}$$

and the general approximant for $k > m$ is

$$\frac{I}{F(m)} + \frac{I}{G(m)} + \frac{I}{F(m+1)} + \cdots + \frac{I}{G(k)} + \frac{I}{F(k+1)}. \tag{2.95}$$

If this sequence of functional approximants converges, then it converges to $-W_-(m)$, i.e., the negative of the "minimal solution" $W_-(t)$ which is the Riccati solution corresponding to the recessive solution at $+\infty$. Under the assumption that the $F(k)$ and $G(k)$ are positive definite, then we do have convergence for each m and $W_-(t)$ is the unique eventually minimal solution of the associated matrix Riccati equation [5, 7].

We wish to show that the reverse continued fraction may be defined in a manner such that when the sequence of functional composite approximants

$$G(m-1) + \frac{I}{F(m-1)}, \qquad G(m-1) + \frac{I}{F(m-1)} + \frac{I}{G(m-2)} + \frac{I}{F(m-2)}, \tag{2.96}$$

and the general case for $k < m$

$$G(m-1) + \frac{I}{F(m-1)} + \frac{I}{G(m-2)} + \frac{I}{F(m-2)} + \cdots + \frac{I}{G(k)} + \frac{I}{F(k)} \tag{2.97}$$

converges to a limit as $k \to -\infty$, then it converges to $W_+(m) = Z_+(m)Y_+^{-1}(m)$, where $X_+(t)$ is the recessive solution at $-\infty$.

In order to define the *reverse continued fraction*, let M be a symplectic matrix of the form (2.89). Usually we assume that E is nonsingular. Set $\mathcal{M} = M^{*-1}$. Then from the form of the inverse of a symplectic matrix given in Chapter 3,

$$\mathcal{M} \equiv M^{*-1} = \begin{bmatrix} H^* & -F^* \\ -G^* & E^* \end{bmatrix}^* = \begin{bmatrix} H & -G \\ -F & E \end{bmatrix} \quad \text{and set} \quad A = \begin{bmatrix} H & G \\ F & E \end{bmatrix}. \tag{2.98}$$

Then

$$\mathcal{W} = T_{\mathcal{M}}(\mathcal{Z}) \quad \text{iff} \quad -\mathcal{W} = T_A(-\mathcal{Z})$$

and

$$\mathcal{W} = T_{\mathcal{M}}(\infty) \quad \text{iff} \quad -\mathcal{W} = T_A(-\infty) = T_A(\infty).$$

A reverse continued fraction for fixed m and $k < m$ is defined by

$$T_{\mathcal{M}(m-1)\mathcal{M}(m-2)\cdots\mathcal{M}(k)}(\infty) = \frac{P_k}{R_k} \tag{2.99}$$

where

$$\begin{bmatrix} P_k & Q_k \\ R_k & S_k \end{bmatrix} \equiv \mathcal{M}(m-1)\mathcal{M}(m-2)\cdots\mathcal{M}(k)$$

and

$$\begin{bmatrix} P_{k-1} & Q_{k-1} \\ R_{k-1} & S_{k-1} \end{bmatrix} = \begin{bmatrix} P_k & Q_k \\ R_k & S_k \end{bmatrix} \mathcal{M}(k-1) = \begin{bmatrix} P_k & Q_k \\ R_k & S_k \end{bmatrix} M^{*-1}(k-1).$$

This also holds for $k = m$ if we assume that $P_m = I$, $Q_m = 0$, $R_m = 0$, and $S_m = I$. Note the difference between these conditions and those for the forward continued fraction in (2.15) on page 47. Postmultiply by $M^*(k-1)$ and then take the $*$ of both sides for

$$\begin{bmatrix} P_k^* & R_k^* \\ Q_k^* & S_k^* \end{bmatrix} = M(k-1) \begin{bmatrix} P_{k-1}^* & R_{k-1}^* \\ Q_{k-1}^* & S_{k-1}^* \end{bmatrix}.$$

Let $X_1(m)$ and $X_2(m)$ be the solutions with initial conditions (2.91). Then $P_k = Y_1^*(k)$ and $R_k = Y_2^*(k)$.

A left version of Pincherle's Theorem would say that the reverse continued fraction (2.99) converges as $k \to -\infty$ to a limit of $\Gamma(m)$ if and only if there exists a solution $Y_+(t)$, $Z_+(t)$ which is recessive at $-\infty$ and has $Y_+(m)$ nonsingular. Furthermore, the functional composite approximants for the A of (2.98) are

$$-P_k R_k^{-1} = T_{A(m-1)} \circ T_{A(m-2)} \circ \cdots \circ T_{A(k)}(\infty). \tag{2.100}$$

Exercise 2.15 (A Möbius Function Identity) *Suppose that nonsingular matrices M and A (not necessarily symplectic) are related by*

$$\mathcal{M} = \begin{bmatrix} H & -G \\ -F & E \end{bmatrix}, \qquad A = \begin{bmatrix} H & G \\ F & E \end{bmatrix}. \tag{2.101}$$

Consider a sequence of such matrices $\mathcal{M}(t)$, $A(t)$ for $t = m-1,\ m-2,\ \ldots,\ k$. Show that we have the identity

$$\begin{aligned} &T_{A(m-1)} \circ T_{A(m-2)} \circ \cdots \circ T_{A(k+1)} \circ T_{A(k)}(\infty) = \\ &\quad T_{A(m-1)} \circ T_{A(m-2)} \circ \cdots \circ T_{A(k+1)}(-T_{\mathcal{M}(k)}(\infty)) = \cdots \\ &\quad = -T_{\mathcal{M}(m-1)} \circ T_{\mathcal{M}(m-2)} \circ \cdots \circ T_{\mathcal{M}(k+1)} \circ T_{\mathcal{M}(k)}(\infty). \end{aligned} \tag{2.102}$$

Thus the functional composite approximants for the A matrices are the negatives of the functional composites approximants for the $\mathcal{M}$ matrices.

We apply this result to the case of M of the form of (2.93) where

$$A = \begin{bmatrix} GF + I & G \\ F & I \end{bmatrix}.$$

The first approximant for the reverse **MCF** corresponding to A is

$$T_{A(m-1)}(\infty) = T_{A(m-1)}(-\infty) = G(m-1) + \frac{I}{F(m-1)}$$

and the reverse continued fraction determined by A converges to the negative of the reverse continued fraction determined by the $\mathcal{M}$. Thus if the sequence of functional composite approximants converges, then the "maximal solution" W_+ has

$$W_+(m) = Z_+(m)Y_+^{-1}(m) = \tag{2.103}$$
$$\lim_{k\to-\infty}\left\{G(m-1) + \frac{I}{F(m-1)} \; + \; \frac{I}{G(m-2)} + \cdots + \frac{I}{G(k)} \; + \; \frac{I}{F(k)}\right\}.$$

Exercise 2.16 (Reverse CCFs) *Apply Exercise* 2.15 *to the case of* $\mathcal{M} = C^{*-1}$ *where* C *has the form of a matrix companion matrix*

$$C = \begin{bmatrix} 0 & I \\ A & B \end{bmatrix} \qquad \textit{with nonsingular} \quad A \tag{2.104}$$

and use the fact that the Möbius transformation associated with a matrix is the same as for the negative of the matrix in order to simplify the reverse companion matrix continued fraction.

In the periodic case these results extend a result of Galois [66] which says that if one root of a quadratic equation is represented by a purely periodic continued fraction, then the negative reciprocal of the other root is obtained by reversing the order of the entries in each periodic block. These results are of particular interest in the constant coefficient case with arbitrary nonsingular E since the corresponding discrete matrix Riccati equations arise in discrete control and Kalman filtering.

Exercise 2.17 (Fibonacci Example) *Find the forward and reverse* **CCF***s and* **SCF***s associated with the Fibonacci recurrence* $y_{n+2} = y_n + y_{n+1}$. *Compare the limits of each of these four continued fractions with the roots of the chararteristic equation* $\lambda^2 = 1 + \lambda$. *Graph the four sequences of approximants as functions of* n. *Interpret the limits of these continued fractions in terms of solutions of the corresponding Riccati equations.*

One could define continued fractions with variable step size by using the system $x(t) = M(s,t)x(s)$ on a monotone sequence of real numbers t_k by $x(t_k) =$

$M(t_{k-1}, t_k)x(t_{k-1})$ or equivalently, by $x(t_{k+1}) = M(t_k, t_{k+1})x(t_k)$. The corresponding $A(t_k, t_{k+1})$ determine the associated continued fraction. These might be of value in giving continued fraction representations of functions defined in a neighborhood of a finite singularity.

2.7 CONTINUED FRACTIONS IN A NORMED RING

This section is not essential to our subsequent discussion, but in some ways it would have been cleaner to first extend our scalar continued fraction results to normed rings before we did matrix continued fractions. Then those results could be applied to the ring of real $n \times n$ matrices, either with real or complex entries and the associated matrix norm could be the matrix 1-norm, the matrix 2-norm, or the matrix Frobenius norm. (See Stewart [143] for details about matrix norms.) In each of those cases, convergence of a sequence of matrices to a matrix limit is equivalent to convergence in each entry to the corresponding entry of the matrix limit.

Recall, that for matrix continued fractions where A becomes a block matrix $\mathcal{A}$, we said that ∞ was in the domain of $T_{\mathcal{A}}$ if C was nonsingular. Also, we said that $\mathcal{Z}$ was in the domain of $T_{\mathcal{A}}$ if $C\mathcal{Z} + D$ was nonsingular.

Suppose throughout this section that $\mathcal{R} = \{S, +, \cdot\}$ is a ring with identity I and zero element Θ and $I \neq \Theta$. We say that a ring element x is a *unit* if it has a two sided multiplicative inverse. For ring elements a, b, c, and d, employ the above matrix notation of (2.1) on page 45 for A and define $T_A(z)$ for $z \in S$ by (2.2) when the ring element $cz + b$ is a unit. Make the definition $T_A(\infty) \equiv ac^{-1}$ when c is a unit. Replace 1 and 0 in our previous scalar discussion by I and Θ, respectively. Also, the scalar case terminology "nonzero" now becomes "is a unit".

Exercise 2.18 *Reinterpret the discussion of scalar continued fractions prior to the definition of convergence of continued fractions in the context of an algebraic ring with identity.*

The terminology of a "unit" in a ring seems unnatural when the ring is a ring of matrices. Perhaps better terminology for a ring element which has a two sided inverse would be *regular.* This agrees with the terminology of a "regular" matrix which is used in Germany for what is called a "nonsingular" matrix in U.S. English.

A *norm* $||\cdot||$ on a ring $\mathcal{R} = \{S, +, \cdot\}$ is a real valued function on S with the following properties:

1. (Positive Definiteness)
   $||x|| \geq 0$ for all $x \in S$ with $||x|| = 0$ if and only if $x = \Theta$.
2. (Anti-Symmetry)
   $||-x|| = ||x||$ for all $x \in S$.
3. (Consistency)
   $||xy|| \leq ||x|| \cdot ||y||$ for all $x,\ y \in S$.
4. (Triangle Inequality)
   $||x+y|| \leq ||x|| + ||y||$ for all $x,\ y \in S$.

Notice that we have not used any vector space or module structure in this definition of a norm on an arbitrary ring.

Exercise 2.19 *Show from consistency, positive definiteness, and $I \neq \Theta$, that* $||I|| \geq 1$.

Exercise 2.20 *Define $\rho : S \times S \to R$ as the real valued function $\rho(x, y) = ||x - y||$. Show that ρ is a metric on S.*

A *Cauchy sequence* in a normed ring is a sequence $\{x_k\}$ which has the property that for each positive real number ϵ, there exists an N such that $\rho(x_m, x_n) < \epsilon$ for every $m,\ n \geq N$. A sequence $\{x_k\}$ is said to *converge* if there exists a $y \in S$ such that $\rho(x_k, y) \to 0$, as $k \to \infty$. A ring $\mathcal{R}$ with norm $||\cdot||$ is said to be *complete* if every Cauchy sequence converges. We henceforth assume that our normed ring is complete.

Exercise 2.21 *Define convergence of a continued fraction in a complete normed ring.*

Exercise 2.22 *Let m be a fixed real number. Solutions*

$$[u_1(k)\ v_1(k)] \quad \textit{and} \quad [u_2(k)\ v_2(k)]$$

of

$$[u(k)\ v(k)] = [u(k-1)\ v(k-1)]A(k), \qquad \textit{for} \quad k = m,\ m+1, \ldots, \tag{2.105}$$

are called linearly independent if the only ring elements c_1 *and* c_2 *such that* $c_1[u_1(k)\ v_1(k)] + c_2[u_2(k)\ v_2(k)] = [\Theta\quad \Theta]$ *for* $k \geq m-1$ *are* $c_1 = c_2 = \Theta$. *Formulate a definition of what is meant by a nontrivial solution and what is meant by a solution which is recessive at* ∞. *Define a solution* $[u\ v]$ *as dominant if it is linearly independent from a recessive solution. Formulate and prove a Pincherle theorem in this context.*

2.8 NOTES

Our motivation for matrix continued fractions arose from the paper of Vaughan [147] which discussed discrete matrix Riccati equations of linear fractional type. Those discrete matrix Riccati equations arose in discrete Kalman Filtering and the dual problem of the discrete regulator problem of optimal control theory. Symplectic continued fractions were introduced in increasing generality in the papers [5], [7], [13]. More general matrix continued fractions were studied by Znojil [161] in problems relating to Lennard–Jones potentials. The Pincherle theorem for that context was published in [9]. That result was established independently by Levrie [97] and by Runckel [136]. We express our appreciation to L.J. Lange for informing us of the extremely interesting history and literature on continued fractions. Our study here has been directed towards continued fractions associated with discrete Hamiltonian systems. Thus we have not pursued the study of matrix continued fractions associated with higher dimension companion matrices, although that subject has been carefully developed by Levrie *et al* [96, 97]. (Paul Levrie is at the Katholieke Universiteit Leuven in Belgium.) We have also freely used the recent dissertation of Andreas Schelling [138] but have not given his extension of the parabola theorem to the setting of matrix companion continued fractions. Matrix continued fractions over an arbitrary field were studied in the book of Jones and Thron [87]. Hans J. Runckel [136] of Universität Ulm, Germany, kindly shared his unpublished notes on continued fractions over an algebraic ring with identity. Much of the literature deals with the extremely difficult problem of determining convergence regions for continued fractions. Wyman Fair [62, 63] studied continued fractions in the setting of a Banach Algebra. Excellent sources for classical results are the books of Wall [150] and Perron [115]. In particular, our result equating scalar companion matrix continued fractions with symplectic continued fractions was suggested by using Perron's odd approximants [115, page 201] on the symplectic continued fraction associated with the ratio of Bessel functions.

3

SYMPLECTIC SYSTEMS

3.1 LINEAR SYSTEMS AND THE LAGRANGE IDENTITY

In this chapter we will study the first order matrix difference equation

$$X(t+1) = M(t)X(t), \tag{3.1}$$

$t \in [a, b]$. Here $b - a$ is an integer and $M(t)$ is a given $2n \times 2n$ matrix function defined on the discrete interval $[a, b]$ and $X(t)$ is an unknown $2n \times m$ matrix function. In the special case where we have $m = 1$, we will write (3.1) as the vector equation

$$x(t+1) = M(t)x(t). \tag{3.2}$$

Theorem 3.1 *If $M(t)$ is invertible on $[a, b]$, then the* **IVP**

$$\begin{aligned} X(t+1) &= M(t)X(t) \\ X(t_0) &= X_0, \end{aligned}$$

where $t_0 \in [a, b]$ and X_0 is a given $2n \times m$ constant matrix, has a unique solution $X(t)$ defined on all of $[a, b+1]$.

Proof: This result follows from the fact that (3.1) can be uniquely solved for $X(t+1)$ and $X(t)$. Hence the matrix value X_0 of X at t_0 uniquely determines the values of $X(t)$ on the whole interval $[a, b+1]$. □

Note that in Theorem 3.1, $M(t)$ can be any square invertible matrix with real or complex entries defined on $[a, b]$. We want to assume $M(t)$ is an even order matrix and usually we will assume $M(t)$ is symplectic (defined in the next section) on $[a, b]$ so that (3.1) has all the nice properties of a formally self–adjoint system.

Define the $2n \times 2n$ constant matrix J by

$$J = \begin{bmatrix} 0 & I \\ -I & 0 \end{bmatrix},$$

where I denotes the $n \times n$ identity matrix and 0 is the $n \times n$ zero matrix. Note that the transpose of J, which we denote by the notation for the conjugate transpose J^*, satisfies $J^* = -J$. Hence the matrix J has a property similar to that of the complex number i, namely $\bar{i} = -i$. This is important in the Lagrange identity which we will prove shortly. First we define an operator L by

$$LU(t) = J\Delta U(t), \quad t \in [a, b]$$

for any $2n \times m$ matrix function $U(t)$ defined on $[a, b+1]$ with $U(t)$ having real or complex entries. We now state and prove the Lagrange identity for the operator L.

Theorem 3.2 (Lagrange Identity) *Assume matrix functions $U(t)$ and $V(t)$ are dimensions $2n \times m$ and $2n \times p$, respectively, are defined on $[a, b+1]$. Then*

$$V^*(t)(LU(t)) - (LV(t))^*U(t+1) = \Delta\{V; U\}$$

for $t \in [a, b]$, where $\{V; U\}$ is called the Lagrange bracket of $V(t)$ and $U(t)$ and is defined by

$$\{V; U\} = V^*(t)JU(t)$$

for $t \in [a, b]$. Furthermore, this bracket function satisfies

$$\{V; U\}^* = -\{U; V\}.$$

Proof: For $t \in [a, b]$, consider

$$\begin{aligned} \Delta\{V; U\} &= \Delta(V^*(t)JU(t)) \\ &= V^*(t)J\Delta U(t) + \Delta(V^*(t)J)U(t+1) \\ &= V^*(t)LU(t) - \Delta(V^*(t)J^*)U(t+1) \\ &= V^*(t)LU(t) - (LV(t))^*U(t+1). \end{aligned}$$

□

Corollary 3.3 *If $U(t)$, $V(t)$ are $2n \times m$ and $2n \times p$ matrix solutions of* (3.1) *respectively, then*

$$\Delta\{V; U\} = V^*(t)\{M^*(t)JM(t) - J\}U(t) \tag{3.3}$$

for $t \in [a, b]$.

Proof: Since $U(t)$ is defined on $[a, b+1]$ and satisfies (3.1) on $[a, b]$, we have

$$\begin{aligned} LU(t) &= J\Delta U(t) \\ &= J[U(t+1) - U(t)] \\ &= J[M(t) - I]U(t) \end{aligned}$$

for $t \in [a, b]$. By the Lagrange identity

$$\begin{aligned} \Delta\{V; U\} &= V^*(t)LU(t) - (LV(t))^*U(t+1) \\ &= V^*(t)J[M(t) - I]U(t) \\ &\quad -V^*(t)[M^*(t) - I]J^*M(t)U(t) \\ &= V^*(t)JM(t)U(t) - V^*(t)JU(t) \\ &\quad +V^*(t)M^*(t)JM(t)U(t) - V^*(t)JM(t)U(t) \\ &= V^*(t)\{M^*(t)JM(t) - J\}U(t) \end{aligned}$$

for $t \in [a, b]$. □

3.2 SYMPLECTIC MATRICES

Looking at equation (3.3) we see that it would be desirable to assume that the matrix $M(t)$ satisfies

$$M^*(t)JM(t) = J$$

for each $t \in [a, b]$. Here M^* denotes the conjugate transpose of M if M has complex entries or just the transpose if M has real entries. This leads to the following definition.

A $2n \times 2n$ constant matrix M is said to be a *symplectic matrix* provided

$$M^*JM = J. \tag{3.4}$$

We assume throughout the remainder of this chapter that the given $2n \times 2n$ matrix $M(t)$ in (3.1) is a symplectic matrix function on $[a,b]$. In this case we say (3.1) is a *symplectic system* . One might say in this case that (3.1) is *formally self-adjoint* .

Some important properties of symplectic matrices are given in the next theorem.

Theorem 3.4 *The set of all real symplectic $2n \times 2n$ constant matrices M is a multiplicative group. Also, the set of all complex symplectic $2n \times 2n$ constant matrices M is a multiplicative group. In either case, if M is partitioned into $n \times n$ blocks as*

$$M = \begin{bmatrix} E & F \\ G & H \end{bmatrix},$$

then M is symplectic if and only if M is invertible and

$$M^{-1} = J^*M^*J = \begin{bmatrix} H^* & -F^* \\ -G^* & E^* \end{bmatrix}. \tag{3.5}$$

Furthermore, $M = \begin{bmatrix} E & F \\ G & H \end{bmatrix}$ is symplectic iff (≡ if and only if)

$$E^*H - G^*F = I, \quad E^*G = G^*E, \quad F^*H = H^*F. \tag{3.6}$$

Also, M is symplectic iff

$$EH^* - FG^* = I, \quad EF^* = FE^*, \quad GH^* = HG^*. \tag{3.7}$$

Note that the form of the inverse of a symplectic matrix given in (3.5) reminds one of the rule for inverting a 2×2 real matrix with determinant 1. Here the rule for *inverting symplectic matrices* is to interchange the diagonal blocks, change the signs on the off diagonal blocks, and put * on each block entry.

Proof: Assume M is symplectic. Then

$$M^*JM = J$$

and

$$J^{-1}M^*JM = I.$$

Since $J^{-1} = J^*$, we have

$$(J^*M^*J)M = I.$$

It follows that M is invertible and

$$\begin{aligned} M^{-1} &= J^*M^*J \\ &= \begin{bmatrix} 0 & -I \\ I & 0 \end{bmatrix} \begin{bmatrix} E^* & G^* \\ F^* & H^* \end{bmatrix} \begin{bmatrix} 0 & I \\ -I & 0 \end{bmatrix} \\ &= \begin{bmatrix} 0 & -I \\ I & 0 \end{bmatrix} \begin{bmatrix} -G^* & E^* \\ -H^* & F^* \end{bmatrix} \\ &= \begin{bmatrix} H^* & -F^* \\ -G^* & E^* \end{bmatrix}. \end{aligned}$$

Thus M symplectic implies M is invertible and $M^{-1} = J^*M^*J$. Conversely, if M is invertible and $M^{-1} = J^*M^*J$ then $M^*JM = J^{*-1} = J$ and M is symplectic. Thus M is invertible and (3.5) holds if and only if M is symplectic.

To establish the group properties, note that if M is symplectic, then M^{-1} exists and (3.5) implies

$$\begin{aligned} (M^{-1})^*J(M^{-1}) &= (J^*M^*J)^*JM^{-1} \\ &= J^*MJ^2M^{-1} = -J^*MM^{-1} = J. \end{aligned}$$

Hence M^{-1} is a symplectic matrix.

Next assume M and N are symplectic. Then

$$(MN)^*J(MN) = N^*(M^*JM)N = N^*JN = J.$$

Therefore MN is symplectic.

It is easy to verify that the $2n \times 2n$ identity matrix is a symplectic matrix. The associative law for multiplication holds for symplectic matrices since it holds for matrix multiplication. Hence the set of all $2n \times 2n$ symplectic matrices forms a group under multiplication.

We next show that if $M = \begin{bmatrix} E & F \\ G & H \end{bmatrix}$ is symplectic, then (3.6) holds. Since M is symplectic, M^{-1} exists and using (3.5) in $M^{-1}M = I_{2n}$, we get that

$$\begin{bmatrix} H^* & -F^* \\ -G^* & E^* \end{bmatrix} \begin{bmatrix} E & F \\ G & H \end{bmatrix} = \begin{bmatrix} I & 0 \\ 0 & I \end{bmatrix}.$$

Therefore

$$\begin{bmatrix} H^*E - F^*G & H^*F - F^*H \\ E^*G - G^*E & E^*H - G^*F \end{bmatrix} = \begin{bmatrix} I & 0 \\ 0 & I \end{bmatrix}.$$

This implies that (3.6) holds. Note that the condition $H^*E - F^*G = I$ is the $*$ of the condition $E^*H - G^*F = I$.

Similarly, $MM^{-1} = I$ gives

$$\begin{bmatrix} E & F \\ G & H \end{bmatrix} \begin{bmatrix} H^* & -F^* \\ -G^* & E^* \end{bmatrix} = \begin{bmatrix} I & 0 \\ 0 & I \end{bmatrix}$$

and hence

$$\begin{bmatrix} EH^* - FG^* & FE^* - EF^* \\ GH^* - HG^* & HE^* - GF^* \end{bmatrix} = \begin{bmatrix} I & 0 \\ 0 & I \end{bmatrix}.$$

It follows that

$$EH^* - FG^* = I, \quad FE^* = EF^*, \quad GH^* = HG^*,$$

and consequently (3.7) holds. Also, the condition $HE^* - FG^* = I$ is the $*$ of the condition $EH^* - GF^* = I$. One could have proven (3.7) by applying (3.6) to M^{-1} given by (3.5).

The converses follow from the fact that both (3.6) and (3.7) imply that M has inverse $M^{-1} = J^*M^*J$ which implies that M is symplectic. □

3.3 SOLUTIONS OF SYMPLECTIC SYSTEMS

Note that for a symplectic system (3.1), we have from Theorem 3.4 that $M(t)$ is invertible on $[a, b]$. It follows from Theorem 3.1 that solutions of (3.1) are defined on the whole discrete interval $[a, b + 1]$.

Assume $U(t)$ and $V(t)$ are $2n \times m$ and $2n \times p$ solutions, respectively, of the symplectic system (3.1). Then from (3.3) we get that

$$\Delta\{V(t); U(t)\} = 0$$

for $t \in [a, b]$. It follows that if $U(t)$, $V(t)$ are $2n \times m$ and $2n \times p$ solutions respectively of (3.1), then

$$\{V(t); U(t)\} \equiv C$$

for $t \in [a, b+1]$ where C is a constant $p \times m$ matrix. This fact is part of the next theorem, but for future reference we interchange the roles of U and V to place them in alphabetical order.

Theorem 3.5 *Assume $U(t)$ is a $2n \times m$ solution of a symplectic system* (3.1). *Then $U(t)$ has constant rank on $[a, b+1]$. If $V(t)$ is a $2n \times p$ matrix solution of* (3.1), *then there exists an $m \times p$ constant matrix C such that*

$$\{U(t); V(t)\} \equiv U^*(t)JV(t) \equiv C \tag{3.8}$$

for $t \in [a, b+1]$.

Proof: Assume $U(t)$ is a $2n \times m$ matrix solution of (3.1). Then

$$U(t+1) = M(t)U(t)$$

for $t \in [a, b]$. Since $M(t)$ symplectic for all $t \in [a, b]$ implies $M(t)$ is invertible for all $t \in [a, b]$ (see Theorem 3.4) it follows that $U(t)$ and $U(t+1)$ have the same rank for all $t \in [a, b]$. This implies that $U(t)$ has constant rank on $[a, b+1]$. For a proof of constancy of the bracket function different from the one above the theorem, next assume $U(t)$, $V(t)$ are respectively $2n \times m$ and $2n \times p$ matrix solutions of (3.1). Then for $t \in [a, b]$,

$$\begin{aligned} \{U(t+1); V(t+1)\} &= U^*(t+1)JV(t+1) \\ &= U^*(t)M^*(t)JM(t)V(t) \\ &= U^*(t)JV(t) \\ &= \{U(t); V(t)\}. \end{aligned}$$

It follows that

$$\{U(t); V(t)\} \equiv C,$$

for $t \in [a, b+1]$, where C is an $m \times p$ constant matrix which can be expressed as $C = \{U(t_0); V(t_0)\}$, where $t_0 \in [a, b+1]$ is arbitrary. □

We next prove an important geometric property concerning symplectic systems.

Theorem 3.6 (Symplectic Flow Property) *Assume $x(t)$ is a solution of a symplectic system* (3.2) *on $[a, b+1]$. Then for any t, s in the integer interval $[a, b+1]$, there is a symplectic matrix $\Phi(t, s)$, called the transition matrix for* (3.2), *such that*

$$x(t) = \Phi(t, s)x(s). \tag{3.9}$$

Proof: First assume $a \leq s < t \leq b+1$. Considering the equations obtained from (3.2) by replacing t by s, $s+1, \ldots, t-1$ we know that (3.9) holds, where

$$\Phi(t, s) = M(t-1)M(t-2)\cdots M(s),$$

which is symplectic by Theorem 3.4.

Note that for $t = s$, equation (3.9) holds with $\Phi(t, s) = I$ which is symplectic.

Finally assume $a \leq t < s \leq b+1$, then using the equations obtained from (3.2) by replacing t by $t, t+1, \ldots, s-1$ and using the fact that symplectic matrices are invertible we have

$$\Phi(t, s) = M^{-1}(t)M^{-1}(t+1)\cdots M^{-1}(s-1)$$

which is symplectic by Theorem 3.4. □

Note that in the above proof we showed that the transition matrix $\Phi(t, s)$ for (3.2) is given by

$$\Phi(t, s) = \begin{cases} M(t-1)M(t-2)\cdots M(s), & a \leq s < t \leq b+1, \\ I, & t = s, \\ M^{-1}(t)M^{-1}(t+1)\cdots M^{-1}(s-1), & a \leq t < s \leq b+1. \end{cases}$$

Exercise 3.7 *Introduce the left matrix product notation by the recursive definition*

$$\prod_{k=s}^{t} M(k) \equiv \begin{cases} \prod_{k=s}^{s-1} M(k) \equiv I, & \textit{for } t = s-1 \\ M(t)\prod_{k=s}^{t-1} M(k), & \textit{for } t \geq s \end{cases} \tag{3.10}$$

and for nonsingular $M(k)$ make the definition

$$\prod_{k=s}^{t-1} M(k) \equiv \left\{ \prod_{k=t}^{s-1} M(k) \right\}^{-1} \quad \textit{for} \quad t < s. \tag{3.11}$$

Use this product notation and mathematical induction for each of the cases $t \geq s$ and $t \leq s$ to establish the closed form formula for the transition matrix

$$\Phi(t,s) = \prod_{k=s}^{t-1} M(k)$$

for the symplectic system (3.2). Then use the definitions of matrix powers

$$M^{-k} \equiv \{M^k\}^{-1} \quad \text{for} \quad k > 0 \quad \text{and} \quad M^0 = I \tag{3.12}$$

for M nonsingular, to find a closed form for the transition matrix when $M(t)$ is a constant (symplectic) matrix M for $t \in [a,b]$. This problem suggests a definition of a left product operator P as follows: For $M(k)$ defined for s, $s \pm 1$, $s \pm 2$, etc., inductively define the operator P by

$$\begin{aligned} P_s^s M(k) &= I, \\ P_s^t M(k) &= M(t-1)(P_s^{t-1}M(k)), \quad t = s+1, s+2, \ldots, \\ P_s^t M(k) &= [P_t^s M(k)]^{-1}, \quad t = s-1, s-2, \ldots . \end{aligned}$$

Note that for $t > s$, the product contains $t - s$ factors of functional values of M. Show that for compatible s, t, u, in any order,

$$P_s^t M(k) = \left(P_u^t M(k)\right)\left(P_s^u M(k)\right)$$

and $\Phi(t,s) = P_s^t M(k)$. Thus establish that the transition matrix function Φ has the semi-group property

$$\Phi(t,s) = \Phi(t,u)\Phi(u,s).$$

Now partition the $2n \times m$ matrix solutions $X(t)$ of (3.1) and the $2n \times 2n$ matrix coefficient $M(t)$ as block matrices

$$X(t) = \begin{bmatrix} Y(t) \\ Z(t) \end{bmatrix}, \tag{3.13}$$

$$M(t) = \begin{bmatrix} E(t) & F(t) \\ G(t) & H(t) \end{bmatrix}, \tag{3.14}$$

where $Y(t)$, $Z(t)$ are $n \times m$ matrix functions and $E(t)$, $F(t)$, $G(t)$, $H(t)$ are $n \times n$ matrix functions.

Then we can write (3.1) as the equivalent system

$$\begin{aligned} Y(t+1) &= E(t)Y(t) + F(t)Z(t) \\ Z(t+1) &= G(t)Y(t) + H(t)Z(t). \end{aligned} \tag{3.15}$$

Under our assumption that $M(t)$ is symplectic on $[a, b]$ the coefficient matrices $E(t)$, $F(t)$, $G(t)$, $H(t)$ satisfy (3.6) and (3.7) for $t \in [a, b]$. We also call (3.15), where (3.6), or equivalently (3.7), holds for $t \in [a, b]$, a symplectic system .

Of course

$$X(t) = \begin{bmatrix} Y(t) \\ Z(t) \end{bmatrix}$$

is a solution of (3.1) iff $Y(t)$, $Z(t)$ is a solution of (3.15).

We now give some examples that lead to symplectic systems. In Chapter 1 we saw that the self-adjoint scalar equation

$$\Delta[p(t)\Delta u(t-1)] + q(t)u(t) = 0$$

is equivalent to a symplectic system. In the next example we generalize this to the matrix analogue of this self-adjoint scalar equation.

Example 3.8 *Consider the second order self-adjoint matrix equation*

$$\Delta[P(t)\Delta U(t-1)] + Q(t)U(t) = 0, \tag{3.16}$$

$t \in [a+1, b+1]$, *where* $P(t)$ *and* $Q(t)$ *are* $n \times n$ *Hermitian matrix functions defined on* $[a+1, b+2]$ *and* $[a+1, b+1]$ *respectively and* $U(t)$ *is an unknown* $n \times m$ *matrix function. We also assume* $P(t)$ *is nonsingular so that all solutions* $U(t)$ *of* (3.16) *are defined on* $[a, b+2]$. *Then system* (3.16) *is equivalent to a symplectic system.*

To see this equivalence, assume $U(t)$ is a solution of (3.16) on $[a, b+2]$. Set

$$\begin{array}{ll} Y(t) = U(t), & t \in [a, b+2], \\ Z(t) = P(t)\Delta U(t-1), & t \in [a+1, b+2]. \end{array}$$

Then

$$\begin{aligned} \Delta Z(t) &= \Delta[P(t)\Delta U(t-1)] \\ &= -Q(t)U(t) \\ &= -Q(t)Y(t). \end{aligned}$$

Hence one of our desired equations is

$$Z(t+1) = -Q(t)Y(t) + I \cdot Z(t).$$

Since $Z(t) = P(t)\Delta U(t-1)$ and $P(t)$ is nonsingular,

$$\Delta U(t) = P^{-1}(t+1)Z(t+1).$$

Therefore

$$Y(t+1) = Y(t) + P^{-1}(t+1)[-Q(t)Y(t) + Z(t)]$$

and our other desired equation is

$$Y(t+1) = [I - P^{-1}(t+1)Q(t)]Y(t) + P^{-1}(t+1)Z(t).$$

Thus $Y(t)$, $Z(t)$ solves a system of the form (3.15) where

$$\begin{array}{ll} E(t) = I - P^{-1}(t+1)Q(t), & F(t) = P^{-1}(t+1) \\ G(t) = -Q(t), & H(t) = I. \end{array}$$

Since, for all $t \in [a+1, b+1]$,

$$E^*(t)H(t) - G^*(t)F(t) = I - Q(t)P^{-1}(t+1) + Q(t)P^{-1}(t+1) = I,$$

$$E^*(t)G(t) = -Q(t) + Q(t)P^{-1}(t+1)Q(t) = G^*(t)E(t),$$

and

$$F^*(t)H(t) = P^{-1}(t+1) = H^*(t)F(t),$$

we have by Theorem 3.4 that this system is a symplectic system. (Here our interval is $[a+1, b+1]$ instead of $[a, b]$.)

Conversely, it can be shown that if $Y(t)$, $Z(t)$ is a solution of the symplectic system

$$\begin{aligned} Y(t+1) &= [I - P^{-1}(t+1)Q(t)]Y(t) + P^{-1}(t+1)Z(t) \\ Z(t+1) &= -Q(t)Y(t) + I \cdot Z(t) \end{aligned} \tag{3.17}$$

then $U(t) = Y(t)$ is a solution of (3.16).

Example 3.9 *Show that* (3.6) *is equivalent to the system*

$$\begin{aligned} Y(t+1) &= I \cdot Y(t) + P^{-1}(t+1)Z(t) \\ Z(t+1) &= -Q(t+1)Y(t) + [I - Q(t+1)P^{-1}(t+1)]Z(t) \end{aligned}$$

under the change of variables

$$\begin{array}{ll} Y(t) = U(t), & t \in [a, b+2] \\ Z(t) = P(t+1)\Delta U(t), & t \in [a, b+1] \end{array}$$

and that this system is also a symplectic system.

3.4 DISCRETE LINEAR HAMILTONIAN SYSTEMS

A more general system than (3.17) which is a symplectic system is given in the next example. These linear systems arise as discrete Hamiltonian systems for quadratic functionals.

Example 3.10 *Consider the linear Hamiltonian system*

$$\Delta Y(t) = A(t)Y(t+1) + B(t)Z(t), \tag{3.18}$$

$$\Delta Z(t) = C(t)Y(t+1) - A^*(t)Z(t), \tag{3.19}$$

where we assume $B(t)$ *and* $C(t)$ *are Hermitian* $n \times n$ *matrix functions on* $[a, b]$ *and* $I - A(t)$ *is nonsingular on* $[a, b]$. *Solving equation* (3.18) *for* $Y(t+1)$ *yields*

$$Y(t+1) = [I - A(t)]^{-1}Y(t) + [I - A(t)]^{-1}B(t)Z(t).$$

Solving equation (3.19) *for* $Z(t+1)$ *and substituting for* $Y(t+1)$ *gives*

$$\begin{aligned} Z(t+1) &= C(t)Y(t+1) + (I - A^*(t))Z(t) \\ &= C(t)[I - A(t)]^{-1}Y(t) + [I - A^*(t) + C(t)[I - A(t)]^{-1}B(t)]Z(t). \end{aligned}$$

Hence we have a special case of system (3.15), *where*

$$\begin{aligned} &E(t) = [I - A(t)]^{-1}, \quad F(t) = [I - A(t)]^{-1}B(t), \\ &G(t) = C(t)[I - A(t)]^{-1}, \quad H(t) = I - A^*(t) + C(t)[I - A(t)]^{-1}B(t). \end{aligned} \tag{3.20}$$

We have just shown that the Hamiltonian system (3.18)–(3.19) can be written as a system with $E(t)$, $F(t)$, $G(t)$, $H(t)$ given by (3.20) where $E(t)$ is nonsingular.

Exercise 3.11 *Show that the system obtained in Example* 3.10 *is a symplectic system.*

We now show that certain symplectic systems can be written as discrete linear Hamiltonian systems.

Theorem 3.12 *Every symplectic system* (3.15) *with $E(t)$ nonsingular can be written as a linear Hamiltonian system* (3.18)–(3.19) *where $B(t)$ and $C(t)$ are Hermitian and $I - A(t)$ is nonsingular.*

Proof: Define $A(t)$, $B(t)$ and $C(t)$ by the equations

$$[I - A(t)]^{-1} = E(t), \quad E(t)B(t) = F(t), \quad C(t)E(t) = G(t).$$

Then

$$B(t) = E^{-1}(t)F(t) \quad \text{and} \quad C(t) = G(t)E^{-1}(t).$$

Since $M(t)$ is symplectic, the identities (3.7) give

$$E(t)F^*(t) = F(t)E^*(t).$$

Because $E(t)$ is invertible we may premultiply by E^{-1} and postmultiply by E^{*-1} to write this as

$$E^{-1}(t)F(t) = F^*(t)[E^*(t)]^{-1} = [E^{-1}(t)F(t)]^*.$$

Thus

$$B(t) = B^*(t)$$

and $B(t)$ is Hermitian. Similarly, the identities (3.6) give

$$E^*(t)G(t) = G^*(t)E(t).$$

Since $E(t)$ is invertible we may pre and post multiply by the inverses of E^* and E for

$$G(t)E^{-1}(t) = [E^*(t)]^{-1}G^*(t) = [G(t)E^{-1}(t)]^*.$$

Hence

$$C(t) = C^*(t),$$

so $C(t)$ is also Hermitian.

By the first equation in (3.15)

$$Y(t+1) = E(t)Y(t) + F(t)Z(t).$$

Since $E(t) = [I - A(t)]^{-1}$,

$$Y(t+1) = [I - A(t)]^{-1}Y(t) + F(t)Z(t).$$

Therefore

$$[I - A(t)]Y(t+1) = Y(t) + E^{-1}(t)F(t)Z(t).$$

It follows that

$$\Delta Y(t) = A(t)Y(t+1) + B(t)Z(t),$$

which is the first equation in the Hamiltonian system, namely (3.18). It remains to derive the second equation in the Hamiltonian system, namely (3.19).

By the second equation in (3.15)

$$\begin{aligned} Z(t+1) &= G(t)Y(t) + H(t)Z(t) \\ &= C(t)E(t)Y(t) + H(t)Z(t). \end{aligned}$$

Using the first equation in (3.15) we obtain

$$\begin{aligned} Z(t+1) &= C(t)[Y(t+1) - F(t)Z(t)] + H(t)Z(t) \\ &= C(t)Y(t+1) - [C(t)F(t) - H(t)]Z(t). \end{aligned}$$

It follows that

$$\Delta Z(t) = C(t)Y(t+1) - [I - H(t) + C(t)F(t)]Z(t).$$

It remains to show that

$$I - H(t) + C(t)F(t) = A^*(t).$$

To prove this we have by (3.6) that

$$E^*(t)H(t) - G^*(t)F(t) = I.$$

Solving for $H(t)$ we have that

$$H(t) = E^{*-1}(t) + E^{*-1}(t)G^*(t)F(t).$$

Hence

$$I - H(t) + C(t)F(t) = I - E^{*-1}(t) - E^{*-1}(t)G^*(t)F(t) + C^*(t)F(t)$$

$$= I - [I - A^*(t)] - E^{*-1}(t)G^*(t)F(t) + E^{*-1}(t)G^*(t)F(t) = A^*(t)$$

which is what we wanted to prove. □

Exercise 3.13 *Show that for suitable choices of* $A(t)$, $B(t)$ *and* $C(t)$ *the discrete linear Hamiltonian system* (3.18)–(3.19) *contains a system equivalent to the self-adjoint equation* (3.16).

3.5 EVEN ORDER DIFFERENCE EQUATIONS AS SYSTEMS

We now give an important example where a $2n$–th order linear self-adjoint scalar difference equation is equivalent to a symplectic system.

Example 3.14 *Consider the $2n$–th order scalar linear self-adjoint difference equation*

$$\ell_{2n}u(t) = \sum_{i=0}^{n} \Delta^i[r_i(t)\Delta^i u(t-i)] = 0, \tag{3.21}$$

$t \in [a+n, b+n]$, where the coefficient functions $r_i(t)$, $0 \le i \le n$, are real valued on $[a+n, b+n+i]$, respectively, and $r_n(t) \neq 0$ on $[a+n, b+2n]$. Because of the assumption on $r_n(t)$ equation (3.21) can be solved uniquely for $u(t-n)$ and $u(t+n)$. It follows that all solutions of (3.21) are defined on $[a, b+2n]$. Equation (3.21) is equivalent to a symplectic system of the form (3.15) for $t \in [a+n, b+n]$ (instead of $[a,b]$) where

$$y(t) = \begin{bmatrix} u(t-1) \\ \Delta u(t-2) \\ \cdots \\ \Delta^{n-1}u(t-n) \end{bmatrix}, \quad z(t) = \begin{bmatrix} (-1)^{n-1}\sum_{i=1}^{n} \Delta^{i-1}[r_i(t)\Delta^i u(t-i)] \\ \cdots \\ -\sum_{i=n-1}^{n} \Delta^{i-n+1}[r_i(t)\Delta^i u(t-i)] \\ r_n(t)\Delta^n u(t-n) \end{bmatrix}$$

for $t \in [a+n, b+n+1]$, and

$$E(t) = \begin{bmatrix} 1 & \cdots & \cdots & 1 \\ 0 & \ddots & & \vdots \\ \vdots & \ddots & \ddots & \vdots \\ 0 & \cdots & 0 & 1 \end{bmatrix}, \quad F(t) = \frac{1}{r_n(t)} \begin{bmatrix} 0 & \cdots & 0 & 1 \\ \vdots & & & \vdots \\ & \cdots & \cdots & \\ \vdots & & & \vdots \\ 0 & \cdots & 0 & 1 \end{bmatrix},$$

$$G(t) = \begin{bmatrix} (-1)^n r_0(t) & (-1)^n r_0(t) & \cdots & (-1)^n r_0(t) \\ 0 & (-1)^{n-1} r_1(t) & \cdots & (-1)^{n-1} r_1(t) \\ \vdots & & \ddots & \vdots \\ 0 & \cdots & 0 & -r_{n-1}(t) \end{bmatrix},$$

$$H(t) = \begin{bmatrix} 1 & 0 & \cdots & 0 & (-1)^n \frac{r_0(t)}{r_n(t)} \\ -1 & 1 & \ddots & & (-1)^{n-1} \frac{r_1(t)}{r_n(t)} \\ 0 & -1 & \ddots & \ddots & \vdots \\ \vdots & & \ddots & \ddots & \vdots \\ 0 & \cdots & 0 & -1 & 1 - \frac{r_{n-1}(t)}{r_n(t)} \end{bmatrix}$$

for $t \in [a+n, b+n]$.

Proof: Instead of proving this equivalence in general we now show this result is true for $n = 3$. A presentation of this example in terms of matrices A, B, C is given in Exercise 9.3 on page 333. References [6, Example 11, pg. 515] and [20] discuss the general case.

Assume $u(t)$ is a solution of the self-adjoint 6–th order linear difference equation

$$\begin{aligned} \ell_6 u(t) = \Delta^3[r_3(t)\Delta^3 u(t-3)] + \Delta^2[r_2(t)\Delta^2 u(t-2)] \\ +\Delta[r_1(t)\Delta u(t-1)] + r_0(t)u(t) = 0 \end{aligned}$$

on $[a, b+6]$. Set

$$y(t) = \begin{bmatrix} u(t-1) \\ \Delta u(t-2) \\ \Delta^2 u(t-3) \end{bmatrix},$$

$$z(t) = \begin{bmatrix} \Delta^2[r_3(t)\Delta^3 u(t-3)] + \Delta[r_2(t)\Delta^2 u(t-2)] + r_1(t)\Delta u(t-1) \\ -\Delta[r_3(t)\Delta^3 u(t-3)] - r_2(t)\Delta^2 u(t-2) \\ r_3(t)\Delta^3 u(t-3) \end{bmatrix}$$

for $t \in [a+3, b+4]$. Then, for $t \in [a+3, b+3]$,

$$\Delta y(t) = \begin{bmatrix} \Delta u(t-1) \\ \Delta^2 u(t-2) \\ \Delta^3 u(t-3) \end{bmatrix} = \begin{bmatrix} y_2(t+1) \\ y_3(t+1) \\ r_3^{-1}(t) z_3(t) \end{bmatrix}.$$

Hence

$$\Delta y(t) = \begin{bmatrix} 0 & 1 & 0 \\ 0 & 0 & 1 \\ 0 & 0 & 0 \end{bmatrix} y(t+1) + \begin{bmatrix} 0 & 0 & 0 \\ 0 & 0 & 0 \\ 0 & 0 & r_3^{-1}(t) \end{bmatrix} z(t).$$

We now have the first equation of a Hamiltonian system, namely (3.18), where

$$A = \begin{bmatrix} 0 & 1 & 0 \\ 0 & 0 & 1 \\ 0 & 0 & 0 \end{bmatrix} \quad \text{and} \quad B(t) = \begin{bmatrix} 0 & 0 & 0 \\ 0 & 0 & 0 \\ 0 & 0 & r_3^{-1}(t) \end{bmatrix}. \tag{3.22}$$

Solving for $y(t+1)$ we have

$$y(t+1) = \begin{bmatrix} 1 & 1 & 1 \\ 0 & 1 & 1 \\ 0 & 0 & 1 \end{bmatrix} y(t) + r_3^{-1}(t) \begin{bmatrix} 0 & 0 & 1 \\ 0 & 0 & 1 \\ 0 & 0 & 1 \end{bmatrix} z(t) \tag{3.23}$$

for $t \in [a+3, b+3]$. Note that (3.23) is of the form of the first equation in (3.15) with

$$E(t) = \begin{bmatrix} 1 & 1 & 1 \\ 0 & 1 & 1 \\ 0 & 0 & 1 \end{bmatrix}, \quad F(t) = r_3^{-1}(t) \begin{bmatrix} 0 & 0 & 1 \\ 0 & 0 & 1 \\ 0 & 0 & 1 \end{bmatrix}$$

for $t \in [a+3, b+3]$.

Next consider, for $t \in [a+3, b+3]$,

$$\Delta z(t) = \begin{bmatrix} \Delta^3[r_3(t)\Delta^3 u(t-3)] + \Delta^2[r_2(t)\Delta^2 u(t-2)] + \Delta[r_1(t)\Delta u(t-1)] \\ -\Delta^2[r_3(t)\Delta^3 u(t-3)] - \Delta[r_2(t)\Delta^2 u(t-2)] \\ \Delta[r_3(t)\Delta^3 u(t-3)] \end{bmatrix}.$$

Using the fact that $u(t)$ is a solution of $\ell_6 u(t) = 0$ we get that

$$\begin{aligned} \Delta z(t) &= \begin{bmatrix} -r_0(t)y_1(t+1) \\ -z_1(t) + r_1(t)y_2(t+1) \\ -z_2(t) - r_2(t)y_3(t+1) \end{bmatrix} \\ &= \begin{bmatrix} -r_0(t) & 0 & 0 \\ 0 & r_1(t) & 0 \\ 0 & 0 & -r_2(t) \end{bmatrix} y(t+1) - \begin{bmatrix} 0 & 0 & 0 \\ 1 & 0 & 0 \\ 0 & 1 & 0 \end{bmatrix} z(t) \end{aligned}$$

for $t \in [a+3, b+3]$. Note that we have the second equation of a discrete linear Hamiltonian system, namely, equation (3.19) with

$$C(t) = \begin{bmatrix} -r_0(t) & 0 & 0 \\ 0 & r_1(t) & 0 \\ 0 & 0 & -r_2(t) \end{bmatrix} \quad \text{and} \quad -A^* = -\begin{bmatrix} 0 & 0 & 0 \\ 1 & 0 & 0 \\ 0 & 1 & 0 \end{bmatrix}. \tag{3.24}$$

Thus we have shown that the discrete 6–th order self–adjoint equation is a special case of a discrete linear Hamiltonian system (3.18)–(3.19) with $B(t)$ a 3×3 matrix of rank 1. Thus it is **not** equivalent to a first order system (3.16) of Example 3.8.

Solving for $z(t+1)$ yields

$$z(t+1) = \begin{bmatrix} -r_0(t) & 0 & 0 \\ 0 & r_1(t) & 0 \\ 0 & 0 & -r_2(t) \end{bmatrix} y(t+1) + \begin{bmatrix} 1 & 0 & 0 \\ -1 & 1 & 0 \\ 0 & -1 & 1 \end{bmatrix} z(t)$$

for $t \in [a+3, b+3]$. Use (3.23) to replace $y(t+1)$ for

$$z(t+1) = \begin{bmatrix} -r_0(t) & -r_0(t) & -r_0(t) \\ 0 & r_1(t) & r_1(t) \\ 0 & 0 & -r_2(t) \end{bmatrix} y(t) + \begin{bmatrix} 1 & 0 & -\frac{r_0(t)}{r_3(t)} \\ -1 & 1 & \frac{r_1(t)}{r_3(t)} \\ 0 & -1 & 1-\frac{r_2(t)}{r_3(t)} \end{bmatrix} z(t) \quad (3.25)$$

and this equation (3.25) is of the form of the second equation in (3.15) with

$$G(t) = \begin{bmatrix} -r_0(t) & -r_0(t) & -r_0(t) \\ 0 & r_1(t) & r_1(t) \\ 0 & 0 & -r_2(t) \end{bmatrix}, \quad H(t) = \begin{bmatrix} 1 & 0 & -\frac{r_0(t)}{r_3(t)} \\ -1 & 1 & \frac{r_1(t)}{r_3(t)} \\ 0 & -1 & 1-\frac{r_2(t)}{r_3(t)} \end{bmatrix}$$

for $t \in [a+3, b+3]$. Hence we have shown that $y(t)$, $z(t)$ satisfy a system of the form (3.15).

Conversely now assume $y(t)$, $z(t)$ is a solution of (3.23), (3.25) on $[a+3, b+4]$. Then

$$y_1(t+1) = y_1(t) + y_2(t) + y_3(t) + \frac{1}{r_3(t)} z_3(t) \quad (3.26)$$

$$y_2(t+1) = y_2(t) + y_3(t) + \frac{1}{r_3(t)} z_3(t) \quad (3.27)$$

$$y_3(t+1) = y_3(t) + \frac{1}{r_3(t)} z_3(t) \quad (3.28)$$

$$z_1(t+1) = -r_0(t)[y_1(t) + y_2(t) + y_3(t)] + z_1(t) - \frac{r_0(t)}{r_3(t)} z_3(t) \quad (3.29)$$

$$z_2(t+1) = r_1(t)[y_2(t) + y_3(t)] - z_1(t) + z_2(t) + \frac{r_1(t)}{r_3(t)} z_3(t) \quad (3.30)$$

$$z_3(t+1) = -r_2(t) y_3(t) - z_2(t) + \left[1 - \frac{r_2(t)}{r_3(t)}\right] z_3(t) \quad (3.31)$$

for $t \in [a+3, b+3]$.

At this stage in the proof we know that $y_i(t)$, $z_i(t)$, $1 \leq i \leq 3$, are defined on $[a+3, b+4]$. We now show how to extend the definitions of five of these functions so that $y_i(t)$, $1 \leq i \leq 3$, are defined on $[a+3, b+7]$, $z_3(t)$ is defined on $[a+3, b+6]$ and $z_2(t)$ is defined on $[a+3, b+5]$. We will do this in such a way that (3.26)–(3.28) hold in $[a+3, b+6]$, (3.31) holds on $[a+3, b+5]$, and (3.30) holds on $[a+3, b+4]$.

To see this, note that since the right hand sides of equations (3.26)–(3.28) and (3.30), (3.31) are well defined at $t = b+4$ we can use these five equations to define $y_1(t)$, $y_2(t)$, $y_3(t)$, $z_2(t)$, $z_3(t)$ at $t = b+5$. But then the right hand sides of (3.26)–(3.28) and (3.31) are defined for $t = b+5$ and we can use these equations to define $y_1(t)$, $y_2(t)$, $y_3(t)$, and $z_3(t)$ at $t = b+6$. Finally we can use (3.26)–(3.28) to define $y_i(t)$, $1 \leq i \leq 3$ at $t = b+7$.
Using (3.26) and (3.27) we get that

$$\Delta y_1(t) = y_2(t+1) \tag{3.32}$$

for $t \in [a+3, b+6]$. From (3.27), (3.28) we obtain

$$\Delta y_2(t) = y_3(t+1) \tag{3.33}$$

on $[a+3, b+6]$. By (3.28) we have that

$$z_3(t) = r_3(t)\Delta y_3(t) \tag{3.34}$$

for $t \in [a+3, b+6]$.

Using (3.29) and (3.26) we get that

$$\Delta z_1(t) = -r_0(t)y_1(t+1) \tag{3.35}$$

for $t \in [a+3, b+3]$. From (3.30), (3.27) we obtain

$$\Delta z_2(t) = r_1(t)y_2(t+1) - z_1(t) \tag{3.36}$$

for $t \in [a+3, b+4]$. By (3.31), (3.28) we have that

$$\Delta z_3(t) = -r_2(t)y_3(t+1) - z_2(t) \tag{3.37}$$

for $t \in [a+3, b+5]$.

From (3.37) and (3.34) we get that

$$\Delta[r_3(t)\Delta y_3(t)] = -r_2(t)y_3(t+1) - z_2(t)$$

for $t \in [a+3, b+5]$. Taking the difference of both sides and using (3.36) and (3.33) we obtain

$$\Delta^2[r_3(t)\Delta y_3(t)] = -\Delta[r_2(t)\Delta y_2(t)] - r_1(t)y_2(t+1) + z_1(t)$$

for $t \in [a+3, b+4]$. Taking the difference of both sides of this equation and using (3.35) and (3.32) we get that

$$\Delta^3[r_3(t)\Delta y_3(t)] = -\Delta^2[r_2(t)\Delta y_2(t)] - \Delta[r_1(t)\Delta y_1(t)] - r_0(t)y_1(t+1) \quad (3.38)$$

for $t \in [a+3, b+3]$.

We now want to extend the domain of definition for $y_1(t)$ and $y_2(t)$ so that $y_1(t)$ is defined on $[a+1, b+7]$ and $y_2(t)$ is defined in $[a+2, b+7]$. Since the right hand sides of equations (3.32) and (3.33) make sense for $t = a+2$ we can use these equations to define $y_2(t)$ and $y_3(t)$ at $t = a+2$. But then the right hand side of equation (3.32) makes sense at $t = a+1$ so we can use equation (3.32) to define $y_1(t)$ at $t = a+1$. Hence we now have that $y_1(t)$ is defined on $[a+1, b+7]$, $y_2(t)$ is defined on $[a+2, b+7]$, (3.32) holds on $[a+1, b+6]$ and (3.33) holds on $[a+2, b+6]$.

For $t \in [a, b+6]$ define $u(t) = y_1(t+1)$, then $y_1(t) = u(t-1)$ for $t \in [a+1, b+7]$. From (3.32) we have that $y_2(t) = \Delta u(t-2)$ for $t \in [a+2, b+7]$. Then by (3.33)

$$y_3(t) = \Delta^2 u(t-3)$$

for $t \in [a+3, b+7]$. Substituting these expressions for $y_1(t)$, $y_2(t)$, and $y_3(t)$ into (3.38) we get that $u(t)$ is a solution of $\ell_6 u(t) = 0$ on $[a, b+6]$.

From (3.34) we get that

$$z_3(t) = r_3(t)\Delta^3 u(t-3)$$

for $t \in [a+3, b+6]$. Then from (3.37)

$$z_2(t) = -\Delta[r_3(t)\Delta^3 u(t-3)] - r_2(t)\Delta^2 u(t-2)$$

for $t \in [a+3, b+5]$. Finally, from (3.36) we get that

$$z_1(t) = \Delta^2[r_3(t)\Delta^3 u(t-3)] + \Delta[r_2(t)\Delta^2 u(t-2)] + r_1(t)\Delta u(t-1)$$

for $t \in [a+3, b+4]$. This completes the proof of the equivalence of (3.21) and the system (3.23), (3.25). The fact that (3.23), (3.25) is a symplectic system is left as an exercise. $\square$

Exercise 3.15 *Show that* (3.23), (3.25) *is a symplectic system. Use the observation that $I - A$ is nonsingular because it is a triangular matrix with nonzero diagonal entries.*

Exercise 3.16 *Show that the system given in Example* 3.14 *that is equivalent to* (3.21) *is a symplectic system. You may do this by showing that it is equivalent to a linear Hamiltonian system* (3.18)–(3.19).

3.6 DISCRETE JACOBI EQUATIONS

We will now show that a more general second order linear self–adjoint difference equation is a special case of a discrete Hamiltonian system. In Chapter 4 we will show that equations of the form considered in the following example arise as discrete Jacobi equations in the discrete calculus of variations.

Example 3.17 *Consider the second order matrix difference equation*

$$\Delta[P(t)\Delta U(t-1) + R^*(t)U(t)] - R(t)\Delta U(t-1) + Q(t)U(t) = 0 \qquad (3.39)$$

for $t \in [a+1, b+1]$. We assume $P(t)$ and $Q(t)$ are $n \times n$ Hermitian matrices on $[a+1, b+2]$ and $[a+1, b+1]$ respectively and $R(t)$ is an $n \times n$ matrix function defined on $[a+1, b+2]$. We further assume $P(t)$ and $P(t)+R^*(t)$ are invertible on $[a+1, b+2]$ so that solutions $U(t)$ of (3.39) *are defined on $[a, b+2]$. Of course if $R(t) = 0$,* (3.39) *reduces to* (3.16). *Then system* (3.39) *can be written as a discrete linear Hamiltonian system under the change of variables*

$$Y(t) = U(t) \qquad (3.40)$$

for $t \in [a, b+2]$,

$$Z(t) = P(t+1)\Delta U(t) + R^*(t+1)U(t+1) \qquad (3.41)$$

for $t \in [a, b+1]$. Furthermore, if $I + P^{-1}(t)R^*(t)$ is invertible on $[a+1, b+2]$, then (3.39) *can be written as a symplectic system.*

We first show that system (3.39) can be written as a Hamiltonian system. From (3.40), (3.41) we get that

$$\Delta Y(t) = -P^{-1}(t+1)R^*(t+1)Y(t+1) + P^{-1}(t+1)Z(t). \qquad (3.42)$$

Using (3.41) and (3.39) we get that

$$\Delta Z(t) = R(t+1)\Delta U(t) - Q(t+1)U(t+1)$$

for $t \in [a, b+1]$. Using (3.42) and (3.40) in this last equation we obtain

$$\begin{aligned}\Delta Z(t) &= -[Q(t+1) + R(t+1)P^{-1}(t+1)R^*(t+1)]Y(t+1) \\ &\quad +R(t+1)P^{-1}(t+1)Z(t).\end{aligned} \tag{3.43}$$

Hence $Y(t)$, $Z(t)$ solves a system of the form (3.18)–(3.19) where

$$\begin{aligned}A(t) &= -P^{-1}(t+1)R^*(t+1) \\ B(t) &= P^{-1}(t+1) \\ C(t) &= -[Q(t+1) + R(t+1)P^{-1}(t+1)R^*(t+1)].\end{aligned}$$

Since $B(t)$ and $C(t)$ are Hermitian, the system (3.42), (3.43) is a Hamiltonian system.

But then from Example 3.10 and Exercise 3.11 if $I - A(t) = I + P^{-1}(t+1)R^*(t+1)$ is invertible on $[a, b]$ then we can write (3.42), (3.43) as a symplectic system of the form (3.15) with

$$\begin{aligned}E(t) &= \left[I + P^{-1}(t+1)R^*(t+1)\right]^{-1} \\ F(t) &= E(t)P^{-1}(t+1) \\ G(t) &= [-Q(t+1) - R(t+1)P^{-1}(t+1)R^*(t+1)]E(t) \\ H(t) &= E^{*-1}(t) + G(t)P^{-1}(t+1)\end{aligned}$$

for $t \in [a, b]$.

Example 3.18 *Consider the three term recurrence relation*

$$-K(t)U(t+1) + N(t)U(t) - K^*(t-1)U(t-1) = 0 \tag{3.44}$$

for $t \in [a+1, b+1]$. Here we assume $K(t)$ is an $n \times n$ invertible matrix function defined on $[a, b+2]$ and $N(t)$ is a Hermitian matrix function on $[a+1, b+1]$. Solutions of (3.44) *are defined on $[a, b+2]$. We will show that* (3.44) *can be written in the form* (3.39) *and hence by Example* 3.17 *can be written as a symplectic system.*

To see this we first write (3.39) in the form (3.44) and note that the steps are reversible. Expanding (3.39) we get that

$$\begin{aligned}&P(t+1)\Delta U(t) + R^*(t+1)U(t+1) - P(t)\Delta U(t-1) \\ &\quad - R^*(t)U(t) - R(t)[U(t) - U(t-1)] + Q(t)U(t) = 0\end{aligned}$$

for $t \in [a+1, b+1]$. This leads to the equation

$$\begin{aligned} -[P(t+1) + R^*(t+1)]U(t+1) \\ + [P(t+1) + P(t) + R^*(t) + R(t) - Q(t)]U(t) \\ - [P(t) + R(t)]U(t-1) = 0 \end{aligned}$$

which is of the form (3.44) with

$$K(t) = P(t+1) + R^*(t+1), \quad t \in [a, b+1] \tag{3.45}$$

and

$$N(t) = P(t) + P(t+1) + R(t) + R^*(t) - Q(t), \quad t \in [a+1, b+1]. \tag{3.46}$$

Note that $N(t)$ is Hermitian on $[a+1, b+1]$.

Conversely, assume we are given equation (3.44). That is, we are given $K(t)$ and $N(t)$. Let $P(t)$ be any $n \times n$ invertible Hermitian matrix (we could take $P(t) = I$). Motivated by (3.45) we define

$$R(t) = K^*(t-1) - P(t)$$

for $t \in [a+1, b+2]$, then (3.45) holds. Similarly, motivated by (3.46) we define

$$Q(t) = P(t) + P(t+1) + R(t) + R^*(t) - N(t)$$

for $t \in [a+1, b+1]$. Note that $Q(t)$ is Hermitian on $[a+1, b+1]$ and (3.46) holds. Since the steps at the beginning of this example are reversible we get that (3.44) can be written in the form (3.39). Since $P(t)$ and $P(t) + R^*(t) = K^*(t-1)$ is invertible on $[a+1, b+2]$, we have by Example 3.17 that (3.44) under the change of variables (3.40), (3.41) can be written as the Hamiltonian system (3.42), (3.43). But $I + P^{-1}(t)R(t) = P^{-1}(t)K^*(t-1)$, (which is $K^*(t-1)$ in the case $P(t) = I$) is invertible, so again by Example 3.17 we get that $Y(t)$, $Z(t)$ satisfies a symplectic system.

3.7 DISCRETE WRONSKIANS – LIOUVILLE'S THEOREM

Let $Y_1(t)$, $Z_1(t)$ and $Y_2(t)$, $Z_2(t)$ be pairs of $n \times m$ and $n \times p$ matrix functions respectively defined on $[a, b+1]$. Then we introduce the "Wronskian" of these two pairs of matrix functions by the mnemonic notation (i.e., a "memory aid" similar to Wronskian notation)

$$\begin{Bmatrix} Y_1(t) & Y_2(t) \\ Z_1(t) & Z_2(t) \end{Bmatrix} = Y_1^*(t)Z_2(t) - Z_1^*(t)Y_2(t)$$

for $t \in [a, b]$.

We now show that if $X_i(t) = \begin{bmatrix} Y_i(t) \\ Z_i(t) \end{bmatrix}$ for $t \in [a, b+1]$, $i = 1, 2$, then

$$\{X_1(t); X_2(t)\} = \begin{Bmatrix} Y_1(t) & Y_2(t) \\ Z_1(t) & Z_2(t) \end{Bmatrix}, \tag{3.47}$$

where $\{X_1(t); X_2(t)\}$ is the Lagrange bracket function.

To see this consider

$$\begin{aligned} \{X_1(t); X_2(t)\} &= X_1^*(t)JX_2(t) \\ &= [Y_1^*(t)\ Z_1^*(t)] \begin{bmatrix} 0 & I \\ -I & 0 \end{bmatrix} \begin{bmatrix} Y_2(t) \\ Z_2(t) \end{bmatrix} \\ &= [Y_1^*(t)\ Z_1^*(t)] \begin{bmatrix} Z_2(t) \\ -Y_2(t) \end{bmatrix} \\ &= Y_1^*(t)Z_2(t) - Z_1^*(t)Y_2(t) \\ &= \begin{Bmatrix} Y_1(t) & Y_2(t) \\ Z_1(t) & Z_2(t) \end{Bmatrix}. \end{aligned}$$

Hence (3.47) holds.

The following theorem is simply a restatement of Theorem 3.5 in our "Wronskian" notation.

Theorem 3.19 (Liouville's Theorem) *Let $Y_1(t)$, $Z_1(t)$, and $Y_2(t)$, $Z_2(t)$ be pairs of $2n \times m$ and $2n \times p$ matrix functions, respectively, which satisfy the symplectic system* (3.15) (*on page* 79) *on the interval $[a, b]$, then there exists a constant $m \times p$ matrix C such that*

$$\begin{Bmatrix} Y_1(t) & Y_2(t) \\ Z_1(t) & Z_2(t) \end{Bmatrix} \equiv C$$

for $t \in [a, b+1]$.

We say that a pair of solutions $Y_1(t)$, $Z_1(t)$ and $Y_2(t)$, $Z_2(t)$ of (3.15) is a *prepared pair* of solutions of (3.5) provided

$$\begin{Bmatrix} Y_1(t) & Y_2(t) \\ Z_1(t) & Z_2(t) \end{Bmatrix} = 0$$

for all $t \in [a, b+1]$. In this case we would say that the partitioned solutions $X_1(t) = \begin{bmatrix} Y_1(t) \\ Z_1(t) \end{bmatrix}$, $X_2(t) = \begin{bmatrix} Y_2(t) \\ Z_2(t) \end{bmatrix}$ is a prepared pair of solutions of (3.1).

If $Y(t)$, $Z(t)$ is a solution of (3.15) such that

$$\begin{Bmatrix} Y(t) & Y(t) \\ Z(t) & Z(t) \end{Bmatrix} = 0$$

for $t \in [a, b+1]$, then $Y(t)$, $Z(t)$ is said to be a *prepared solution* of (3.15). (In this case we would say

$$X(t) = \begin{bmatrix} Y(t) \\ Z(t) \end{bmatrix}$$

is a *prepared* (or *self-conjoined*) matrix solution of (3.1).)

Exercise 3.20 *Show that a solution* $Y(t)$, $Z(t)$ *of* (3.15) *is a prepared solution iff* $Z^*(t)Y(t)$ *is Hermitian on* $[a, b+1]$ *iff* $Z^*(t)Y(t)$ *is Hermitian for some* $t_0 \in [a, b]$. *Show that if* $Y(t)$, $Z(t)$ *is a solution of* (3.15) *with a condition* $Y(t_0) = 0$ *or a condition* $Z(t_0) = 0$, *then* $Y(t)$, $Z(t)$ *is a prepared solution.*

Exercise 3.21 *Show that if* $U(t)$ *is an* $n \times m$ *solution of* (3.16), *where* $R \equiv 0$, *then by Exercise* 3.20 *and Theorem* 3.19

$$U^*(t)P(t+1)\Delta U(t) - \Delta U^*(t)P(t+1)U(t) \equiv C,$$

for $t \in [a, b+1]$, *where* C *is a constant* $m \times m$ *matrix. In particular,* $U(t)$ *is a prepared solution of* (3.16) *provided* $U^*(t)P(t+1)\Delta U(t)$ *is Hermitian on* $[a, b+1]$, *or equivalently,* $U^*(t)P(t+1)U(t+1)$ *is Hermitian on* $[a, b+1]$.

Exercise 3.22 *Show that if $u(t)$ is a complex valued solution of* (3.21), *then*

$$(-1)^{n-1}\overline{u(t-1)}\sum_{i=1}^{n}\Delta^{i-1}[r_i(t)\Delta^i u(t-i)]$$

$$+(-1)^{n-2}\overline{\Delta u(t-2)}\sum_{i=2}^{n}\Delta^{i-2}[r_i(t)\Delta^i u(t-i)]$$

$$+\cdots+\overline{\Delta^{n-1}u(t-n)}r_n(t)\Delta^n u(t-1)$$

$$-(-1)^{n-1}u(t-1)\sum_{i=1}^{n}\Delta^{i-1}[r_i(t)\Delta^i\overline{u(t-1)}]$$

$$-(-1)^{n-2}\Delta u(t-2)\sum_{i=2}^{n}\Delta^{i-2}[r_i(t)\Delta^i\overline{u(t-1)}]$$

$$-\cdots-\Delta^{n-1}u(t-n)r_n(t)\Delta^n\overline{u(t-1)}=c$$

where c is a complex number. State a condition which would imply that $u(t)$ is a prepared solution of (3.16).

We would like to have a theorem relating nonsingularity of the Wronskian matrix of two solutions to some kind of linear independence of those solutions. In order to do so, we need the concept of a Lagrangian Subspace introduced in the next section.

3.8 PREPARED FAMILIES – LAGRANGIAN SUBSPACES

We continue to assume that $M(t)$ is symplectic on $[a,b]$. A family of solutions $x_1(t),\ldots,x_p(t)$ of (3.2) is said to form a *prepared family* (or a conjoined family) of solutions of (3.2) provided

$$\{x_i;x_j\}=0$$

for all $1\le i,j\le p$.

Exercise 3.23 *Assume $x_1(t),\ldots,x_p(t)$ are solutions of* (3.2) *on $[a,b+1]$ and define the $2n\times p$ matrix function $X(t)$ by $X(t)=[x_1(t)\cdots x_p(t)]$, for t in $[a,b+1]$. Show that $x_1(t),\ldots,x_p(t)$ is a prepared family of solutions on $[a,b+1]$ iff $X^*(t)JX(t)=0$ for $t\in[a,b+1]$ iff $X^*(t_0)JX(t_0)=0$ for some $t_0\in[a,b+1]$.*

Exercise 3.24 *Assume $x_1(t), \ldots, x_p(t)$ is a prepared family, (i.e., a mutually conjoined family), of solutions of (3.2) defined on $[a, b+1]$ Let $X(t)$ be the $2n \times p$ matrix function $X(t) = [x_1(t) \quad \cdots \quad x_p(t)]$, for t in $[a, b+1]$, and assume that c and d are $p \times 1$ constant vectors. Then the vector solutions $X(t)c$ and $X(t)d$ are each prepared, they are a prepared pair (i.e., they are a mutually conjoined pair), since $\{X(t)c; X(t)d\} = c^*\{X(t); X(t)\}d = 0$. Thus the span of the set $\{x_1, \ldots, x_p\}$ is a linear space of mutually prepared solutions of dimension at most p.*

Theorem 3.25 *The maximum dimension of a prepared family of solutions of (3.2) is n. Furthermore, if $x_1(t), \ldots, x_p(t)$ is a prepared family of solutions of dimension p, $1 \leq p < n$, then there exist solutions $x_{p+1}(t), \ldots, x_n(t)$ of (3.2) such that $x_1(t), \ldots, x_n(t)$ is a prepared family of solutions of dimension n.*

The linear space spanned by a prepared family of n solutions of dimension n is called a *Lagrangian subspace* . Note that a Lagrangian subspace is maximal in the sense that no prepared vector solution outside the subspace may be added and be prepared (i.e., conjoined) with every vector in the Lagrangian subspace.

The proof of this theorem is quite long and involved. It might give a better overview on first reading of this chapter to proceed directly to the section on Reduction of Order. As the details of this proof are not readily accessible in the literature [13], we include the proof here for the sake of completeness.

Proof: Assume $x_1(t), \ldots, x_p(t)$ is a prepared family of solutions of (3.2) of dimension p where $p \geq 1$. Let $X(t)$ be the $2n \times p$ matrix function defined by

$$X(t) = [x_1(t) \quad \cdots \quad x_p(t)]$$

for $t \in [a, b+1]$. Fix $t_0 \in [a, b+1]$. Because solutions of symplectic systems have constant rank, the vector functions $x_1(t), \ldots, x_p(t)$ are linearly independent on $[a, b+1]$ iff $x_1(t_0), \ldots, x_p(t_0)$ are linearly independent vectors in C^{2n}. By Exercise 3.23

$$\{X;\ X\} = X^*(t)JX(t) = 0$$

for $t \in [a, b+1]$. In particular

$$X^*(t_0)JX(t_0) = 0.$$

Set $A = X^*(t_0)J$. Thus

$$AX(t_0) = 0.$$

Since the nullity of a matrix is the rank of the null space, i.e., the rank of the kernel, this last equation implies nullity $A \geq p$. Because $X(t_0)$ has rank p and J is nonsingular, A has rank p. Since

$$2n = \text{rank } A + \text{nullity } A$$

we conclued that $2n \geq p + p = 2p$ and hence $p \leq n$. Thus the maximum dimension of a prepared family of solutions is n.

Next assume $x_1(t), \ldots, x_p(t)$ is a prepared family of solutions of (3.2) of dimension p, $1 \leq p < n$. It suffices to show that we can define a solution $x_{p+1}(t)$ of (3.2) such that $x_1(t), \ldots, x_{p+1}(t)$ is a prepared family of solutions of dimension $p+1$. Let $X(t)$, t_0, and A be as in the beginning of this proof. From above we have that

$$\text{nullity} A = 2n - p.$$

Let $s = 2n - 2p$, then $s + p = 2n - p = \text{nullity} A$. Pick s vectors $q_1, \ldots, q_s$ in C^{2n} so that if we define the $2n \times s$ constant matrix Q by

$$Q = [q_1 \quad \cdots \quad q_s]$$

such that the rank of the $2n \times (p+s)$ matrix $[X(t_0) \quad Q]$ is $p+s$ and

$$A\,[X(t_0) \quad Q] = 0. \tag{3.48}$$

If u and v are vectors in C^{2n}, then we define the bracket of these two vectors by

$$\{u;\ v\} = u^* J v.$$

Equation (3.48) and the form of A gives

$$0 = A\,[X(t_0) \quad Q] = X^*(t_0) J\,[X(t_0) \quad Q] = [X^*(t_0) J X(t_0) \quad X^*(t_0) J Q]$$

and

$$\{x_i(t_0);\ q_j\} = 0$$

for $1 \leq i \leq p$, $1 \leq j \leq s$. It follows that any linear combination q of $q_1, \ldots, q_s$ satisfies

$$\{x_i(t_0);\ q\} = 0.$$

We would like to find a nontrivial linear combination q of $q_1, \ldots, q_s$ such that

$$\{q;\ q\} = q^* J q = 0.$$

Set $H = iJ$, then $H^* = -iJ^* = iJ = H$, so H is Hermitian. Define f on C^{2n} by

$$f(w) = w^* H w.$$

Note that $\bar{f} = f^* = f$ implies that f is real valued and $f(q) = 0$ iff $\{q; q\} = 0$. Hence we want to find a nontrivial linear combination q of $q_1, \ldots, q_s$ such that $f(q) = 0$. To help us do this we first find certain vectors w and w^0 so that $f(w) < 0$ and $f(w^0) > 0$.

First we find a certain vector w so that $f(w) < 0$. Assume that we can find column vectors $u \in C^p$, $v \in C^s$ such that

$$w = X(t_0)u + Qv \neq 0 \tag{3.49}$$

and

$$Jw = iw. \tag{3.50}$$

Then

$$f(w) = w^* H w = w^* iJw = -w^* w < 0$$

which is what we want. Hence we want to show that (3.49) and (3.50) hold. First note that (3.50) holds iff the components of w satisfy

$$w_j + iw_{n+j} = 0, \quad 1 \leq j \leq n.$$

This is equivalent to the matrix–vector equation

$$[I \quad iI]\, w = 0.$$

Hence for (3.49), (3.50) to hold we want to find $u \in C^p$, $v \in C^s$, not both zero vectors, so that

$$[I \quad iI]\,[X(t_0) \quad Q] \begin{bmatrix} u \\ v \end{bmatrix} = 0.$$

Since this matrix–vector equation is equivalent to n linear homogeneous equations in the $p + s$ unknowns $u_1, \ldots, u_p, v_p, \ldots, v_s$ and since $p < n$ implies $p + s = 2n - p > n$, there exist vectors u, v, not both zero vectors, such that this is true. Hence if w is given by (3.49), then $f(w) < 0$.

Similarly, there exist vectors $u^0 \in C^p, v^0 \in C^s$ such that $w^0 = X(t_0)u^0 + Qv^0$ makes $Jw^0 = -iw^0$ and $f(w^0) > 0$.

Since $AQ = X^*(t_0)JQ = 0$ and $X^*(t_0)JX(t_0) = 0$ we get that

$$f(w) = f(X(t_0)u + Qv)$$

$$\begin{aligned}
&= (u^*X^*(t_0) + v^*Q^*)iJ(X(t_0)u + Qv) \\
&= iu^*X^*(t_0)JX(t_0)u + iu^*X^*(t_0)JQv \\
&\qquad + iv^*Q^*JX(t_0)u + iv^*Q^*JQv \\
&= iv^*Q^*JQv \\
&= f(Qv).
\end{aligned}$$

Hence we have that $f(Qv) = f(w) < 0$. Similarly, it can be shown that $f(Qv^0) = f(w^0) > 0$.

Note that $Qv \neq 0$ and $Qv^0 \neq 0$. Since $f(\lambda w) = |\lambda|^2 f(w)$, Qv and Qv^0 are actually linearly independent. For $\theta \in [0,1]$ define

$$v(\theta) = \theta v + (1-\theta)v^0$$

and then define

$$g(\theta) = f(Qv(\theta)).$$

Then g is a real valued continuous function of θ on $[0,1]$. But g has a sign change since $g(0) = f(Qv^0) > 0$ and $g(1) = f(Qv) < 0$. By the intermediate value theorem there is a $\theta_1 \in (0,1)$ such that

$$g(\theta_1) = f(Qv(\theta_1)) = 0.$$

Take $q = Qv(\theta_1)$. Then q is a nontrivial linear combination of $q_1, \ldots, q_s$ satisfying

$$\{q; q\} = 0.$$

Let $x_{p+1}(t)$ be the solution of the **IVP** of system (3.2) with the initial conditon

$$x_{p+1}(t_0) = q.$$

Since $\{x_j(t_0); q\} = 0$, $1 \leq j \leq p$, $\{q; q\} = 0$ and

$$\text{rank}\,[x_1(t_0) \;\cdots\; x_p(t_0)\; q] = p+1$$

it follows that $x_1(t), \ldots, x_p(t)$, $x_{p+1}(t)$ is a prepared family of solutions of (3.2) of dimension $p+1$. □

Exercise 3.26 *In the proof of Theorem* 3.25 *prove that there is some vector* $w^0 = X(t_0)n^0 + Qv^0$ *such that* $f(w^0) > 0$.

3.9 LINEAR INDEPENDENCE AND THE WRONSKIAN TEST

We now formulate a structure in which we can attach meaning to linear independence of $2n \times n$ solutions of a symplectic system

$$X(t+1) = M(t)X(t), \qquad t \in [a,b]. \tag{3.51}$$

Suppose that $2n \times n$ matrix valued functions $X_1(t)$ and $X_2(t)$ defined on a discrete interval $[a, b+1]$ are called *linearly independent* on $[a, b+1]$ if the only $n \times n$ constant matrices C_1 and C_2 such that

$$X_1(t)C_1 + X_2(t)C_2 \equiv 0 \quad \text{on } [a, b+1]$$

are $C_1 = 0$ and $C_2 = 0$. Thus the "scalars" C_1 and C_2 come from a ring and the natural structure for the solution space of linear systems is a module instead of a vector space. We need this structure in order to establish a version of the differential equations Wronskian test for linear independence of solutions and a characterization of general solutions.

We now list the axioms for a *right unitary module* $\mathcal{M} = \{\mathcal{G}, \mathcal{R}, \cdot\}$. Suppose that $\{\mathcal{G}, +\}$ is a commutative group with addition as the group operation and zero element denoted by Θ. Suppose that $\mathcal{R}$ is a ring with identity I and zero element denoted by 0, ring multiplication is denoted by juxtaposition and addition in the ring is also denoted by $+$. Assume that right multiplication of group elements $X \in \mathcal{G}$ by ring elements $\Lambda \in \mathcal{R}$ is denoted by $X \cdot \Lambda$ and this product is defined for all X and Λ and the product is in $\mathcal{G}$.

Assume the following axioms hold for all possible group elements X, Y and ring elements Λ, Λ_1, and Λ_2:

1. $X \cdot I = X$

2. $(X \cdot \Lambda_1) \cdot \Lambda_2 = X \cdot (\Lambda_1 \Lambda_2)$

3. $X \cdot (\Lambda_1 + \Lambda_2) = X \cdot \Lambda_1 + X \cdot \Lambda_2$

4. $(X + Y) \cdot \Lambda = X \cdot \Lambda + Y \cdot \Lambda.$

If all of these conditions are satisfied, then we say that we have a *right unitary module* $\mathcal{M} = \{\mathcal{G}, \mathcal{R}, \cdot\}$.

Exercise 3.27 *Show that in a right unitary module we have $X \cdot 0 = \Theta$ for all $X \in \mathcal{G}$.*

Group elements X_k, $k = 1, \ldots, n$, are said to be *linearly independent* if the only ring elements Λ_k such that

$$\sum_{k=1}^{n} X_k \cdot \Lambda_k = \Theta$$

are $\Lambda_k = 0$, $k = 1, \ldots, n$.

The notation $R_{m \times p}$ represents the set of real $m \times p$ matrices and $C_{m \times p}$ represents the set of complex $m \times p$ matrices. We will use this module structure only when the ring $\mathcal{R}$ is either the matrix ring $R_{p \times p}$ or $C_{p \times p}$ with the usual addition and matrix multiplication. In each module the operation of multiplication of a group element X by a ring element C is the usual matrix postmultiplication $X \cdot C \equiv XC$.

Exercise 3.28 *Show that each of the following examples has the algebraic structure of a right unitary module:*

(a) $\mathcal{G} = \{R_{m \times p}, +\}$ *and* $\mathcal{R} = \{R_{p \times p}, +, \cdot\}$.

(b) $\mathcal{G} = \{C_{m \times p}, +\}$ *and* $\mathcal{R} = \{C_{p \times p}, +, \cdot\}$.

(c) *For $t \in [a, b+1]$ assume $X(t) \in R_{m \times p}$, let $\mathcal{G}$ be the group of functions*

$$\mathcal{G} = \{X(t) | t \in [a, b+1]\},$$

and let $\mathcal{R} = \{R_{p \times p}, +, \cdot\}$.

(d) *Assume that for $t \in [a, b]$, the matrix function $M(t)$ takes values in $R_{p \times p}$, let $\mathcal{S}$ be the subgroup of $\mathcal{G}$ of (c) consisting of $m \times p$ "solutions" of*

$$X(t+1) = M(t)X(t), \qquad t \in [a, b],$$

$\mathcal{R} = \{R_{p \times p}, +, \cdot\}$, *and* $\mathcal{M} = \{\mathcal{S}, \mathcal{R}, +, \cdot\}$.

(e) *Replace $R_{m \times p}$ and $R_{p \times p}$ of (c) by $C_{m \times p}$ and $C_{p \times p}$, respectively. Thus allow $X(t)$ of (c) to be in $C_{m \times p}$.*

(f) *Allow $\mathcal{S}$ in (d) to be the subgroup of $\mathcal{G}$ of (e) consisting of complex solutions.*

(g) *Allow $M(t)$ of (f) to be in $C_{p\times p}$.*

Exercise 3.29 *Suppose that $M(t)$ is a $p \times p$ nonsingular matrix function for $t \in [a, b]$. Show that the solution $X_1(t)$ with $X(a) = I_p$ is a basis for the module of examples* **(d)**, **(f)**, *and* **(g)**.

Proposition 3.30 *Suppose that $X_1(t)$ and $X_2(t)$ are in one of the modules of* **(d)**, **(f)**, *or* **(g)**. *Then $X_1(t)$ and $X_2(t)$ are linearly dependent if and only if there exist constant $p \times 1$ vectors c_1 and c_2, not both 0, which satisfy*

$$X_1(t)c_1 + X_2(t)c_2 \equiv 0. \tag{3.52}$$

Proof: Suppose $X_1(t)$ and $X_2(t)$ are linearly dependent as defined for the relevant module. Then there exist $p \times p$ constant matrices C_1 and C_2, not both zero, such that

$$X_1(t)C_1 + X_2(t)C_2 \equiv 0.$$

Then there exists a unit vector u such that $c_1 = C_1u$ and $c_2 = C_2u$ are not both zero and (3.52) holds. Conversely, if c_1 and c_2 are not both zero and satisfy (3.52), then matrices C_1 and C_2 built up by

$$C_1 = [c_1 \quad \ldots \quad c_1] \text{ and } C_2 = [c_2 \quad \ldots \quad c_2]$$

are not both zero and satisfy

$$X_1(t)C_1 + X_2(t)C_2 \equiv 0.$$

Thus $X_1(t)$ and $X_2(t)$ satisfy the definition of linear dependence for a module.□

Theorem 3.31 (Wronskian Test for Linear Independence) *Assume $M(t)$ is a $2n \times 2n$ symplectic matrix function defined on $[a, b]$. Suppose that $X_1(t)$ and $X_2(t)$ are $2n \times n$ matrix functions defined on $[a, b+1]$, each of which is a prepared solution of*

$$X(t+1) = M(t)X(t)$$

on $[a, b]$. Then $X_1(t)$ and $X_2(t)$ are linearly independent in the module of $2n \times n$ functions defined on $[a, b+1]$ if and only if the bracket function

$$\{X_1; X_2\} = X_1^* J X_2$$

is nonsingular. That is, the Wronskian matrix

$$\begin{Bmatrix} Y_1(t) & Y_2(t) \\ Z_1(t) & Z_2(t) \end{Bmatrix} = Y_1^*(t)Z_2(t) - Z_1^*(t)Y_2(t)$$

is nonsingular if and only if X_1 and X_2 are linearly independent.

Proof: Suppose that $\{X_1; X_2\} = X_1^* J X_2$ is nonsingular. If c_1 and c_2 are $n \times 1$ constant vectors such that

$$X_1 c_1 + X_2 c_2 \equiv 0,$$

then

$$0 = \{X_1; X_1 c_1 + X_2 c_2\} = \{X_1; X_1\}c_1 + \{X_1; X_2\}c_2 = \{X_1; X_2\}c_2$$

and since $\{X_1; X_2\}$ is nonsingular, we must have $c_2 = 0$. Similarly

$$0 = \{X_2; X_1 c_1 + X_2 c_2\} = \{X_2; X_1\}c_1 + \{X_2; X_2\}c_2 = -\{X_1; X_2\}^* c_1$$

implies $c_1 = 0$. Proposition 3.30 implies that X_1 and X_2 are linearly independent.

The converse requires more machinery. Suppose $\{X_1; X_2\}$ is singular. Either $X_1(t_0)$ is singular for some t_0 or it is never singular. Suppose that t_0 is such that $X_1(t_0)$ is singular. Then there exists a nonzero constant vector c_1 such that $0 = X_1(t_0)c_1$. Since solutions are uniquely determined by their initial conditions, we have $X_1(t)c_1 \equiv 0$ and the vectors c_1 and $c_2 = 0$ make $X_1 c_1 + X_2 c_2 \equiv 0$ with c_1 nonzero. Then Proposition 3.30 implies that the solutions X_1 and X_2 are linearly dependent. (Similarly, if $X_2(t)$ is singular at some point, then X_1 and X_2 are linearly dependent, but we won't need this in the proof.) The proof reduces to the case where $X_1(t)$ is nonsingular for all t, i.e., X_1 is a prepared basis. Assume that is the case and c is a nonzero $n \times 1$ constant vector such that

$$0 = \{X_1; X_2\}c \equiv \{X_1(t); X_2(t)\}c \equiv \{X_1(t); X_2(t)c\}.$$

Let x be the $2n \times 1$ solution $x(t) \equiv X_2(t)c$. Then x satisfies $\{x; x\} = 0$ since X_2 is a prepared solution. Partition X_1 by columns as $[x_1 \;\; \ldots \;\; x_n]$. Then $\{x_k; x\} = 0$ for $k = 1, \ldots, n$. But $x_1, \ldots, x_n$ is a basis for a Lagrangian subspace of maximal dimension n. Therefore, x is in the span of the columns of X_1 and there exists a constant vector c_1 such that $x(t) \equiv X(t)c_1$ and we have $X_1(t)c_1 \equiv X_2(t)c$. Set $c_2 = -c$ for

$$X_1(t)c_1 + X_2(t)c_2 \equiv 0$$

with $c_2 \neq 0$. Therefore, Proposition 3.30 implies that X_1 and X_2 are linearly dependent. □

We are now interested in constructing special linearly independent sets of solutions X_1, X_2 which have somewhat the role of exponential dichotomies.

3.10 REDUCTION OF ORDER

Assume $Y_0(t)$, $Z_0(t)$ is a solution of the symplectic system (3.15) such that $Y_0(t)$ is invertible on $[a, b+1]$. Then we define

$$\mathcal{S}_0(t) = \sum_{s=a}^{t-1} Y_0^{-1}(s+1)F(s)Y_0^{*-1}(s) \tag{3.53}$$

for $a \leq t \leq b+1$. We use the summation convention that when the lower index of a summation exceeds the upper index by 1, then the sum is defined to be 0. That is, we make the convention that

$$\sum_{s=a}^{a-1} D(s) = 0.$$

Therefore in the above definition we understand that $\mathcal{S}_0(a)$ is the zero matrix.

We now state and prove the important reduction of order theorem.

Theorem 3.32 (Reduction of Order Theorem) *Assume $Y_0(t)$, $Z_0(t)$ is a prepared $n \times n$ matrix solution of* (3.15) *such that $Y_0(t)$ is invertible on $[a, b+1]$. If $Y(t)$, $Z(t)$ is an $n \times m$ matrix solution of* (3.15), *then*

$$Y(t) = Y_0(t)[P + \mathcal{S}_0(t)Q] \tag{3.54}$$

$$Z(t) = Z_0(t)[P + \mathcal{S}_0(t)Q] + Y_0^{*-1}(t)Q \tag{3.55}$$

for $t \in [a, b+1]$, where $\mathcal{S}_0(t)$ is given by (3.53),

$$P = Y_0^{-1}(a)Y(a) \tag{3.56}$$

and

$$Q = \begin{Bmatrix} Y_0(t) & Y(t) \\ Z_0(t) & Z(t) \end{Bmatrix}. \tag{3.57}$$

Conversely, if P and Q are constant $n \times m$ matrices and $Y(t)$, $Z(t)$ are defined by (3.54), (3.55) *respectively, then $Y(t)$, $Z(t)$ is a solution of* (3.15) *and therefore equations* (3.56), (3.57) *hold. Furthermore $Y(t)$, $Z(t)$ is a prepared solution iff P^*Q is Hermitian.*

Proof: Assume $Y_0(t)$, $Z_0(t)$ is a prepared $n \times n$ matrix solution of (3.15) such that $Y_0(t)$ is invertible on $[a, b+1]$. Assume $Y(t)$, $Z(t)$ is an $n \times m$ matrix solution of (3.15) and let Q be the constant matrix

$$Q = \begin{Bmatrix} Y_0(t) & Y(t) \\ Z_0(t) & Z(t) \end{Bmatrix}.$$

Then

$$Y_0^*(t)Z(t) - Z_0^*(t)Y(t) \equiv Q.$$

Solving for $Z(t)$ we get that

$$Z(t) = Y_0^{*-1}(t)Q + Y_0^{*-1}(t)Z_0^*(t)Y(t). \tag{3.58}$$

Since $Y_0(t)$, $Z_0(t)$ is a prepared solution of (3.15),

$$Y_0^*(t)Z_0(t) = Z_0^*(t)Y_0(t) \qquad \text{i.e.,} \quad Z_0^*(t)Y_0(t) = Y_0^*(t)Z_0(t),$$

for $t \in [a, b+1]$. Premultiply by $Y_0^{*-1}(t)$ and postmultiply by $Y_0^{-1}(t)$ to obtain

$$Y_0^{*-1}(t)Z_0^*(t) = Z_0(t)Y_0^{-1}(t)$$

for $t \in [a, b+1]$. Using this last equation and (3.58) we obtain

$$Z(t) = Z_0(t)Y_0^{-1}(t)Y(t) + Y_0^{*-1}(t)Q. \tag{3.59}$$

Since $Y(t)$, $Z(t)$ is a solution of (3.15),

$$\begin{aligned} Y(t+1) &= E(t)Y(t) + F(t)Z(t) \\ &= E(t)Y(t) + F(t)Z_0(t)Y_0^{-1}(t)Y(t) + F(t)Y_0^{*-1}(t)Q \\ &= [E(t)Y_0(t) + F(t)Z_0(t)]Y_0^{-1}(t)Y(t) + F(t)Y_0^{*-1}(t)Q \\ &= Y_0(t+1)Y_0^{-1}(t)Y(t) + F(t)Y_0^{*-1}(t)Q. \end{aligned}$$

Multiplying both sides by $Y_0^{-1}(t+1)$ we get that

$$Y_0^{-1}(t+1)Y(t+1) = Y_0^{-1}(t)Y(t) + Y_0^{-1}(t+1)F(t)Y_0^{*-1}(t)Q.$$

Hence

$$\Delta(Y_0^{-1}(t)Y(t)) = Y_0^{-1}(t+1)F(t)Y_0^{*-1}(t)Q.$$

Summing both sides from a to $t-1$ we get that

$$Y_0^{-1}(t)Y(t) = P + \sum_{s=a}^{t-1} Y_0^{-1}(s+1)F(s)Y_0^{*-1}(s)Q$$

where P is given by (3.56). Solving for $Y(t)$ we get that

$$Y(t) = Y_0(t)[P + \mathcal{S}_0(t)Q]$$

which is equation (3.54). Using this last equation and (3.59) we get that (3.55) holds.

Next we prove the converse statement. Again assume $Y_0(t)$, $Z_0(t)$ is a prepared solution of (3.15) such that $Y_0(t)$ is invertible in $[a, b+1]$. We first show that if we define $Y_1(t)$, $Z_1(t)$ by

$$\begin{aligned} Y_1(t) &= Y_0(t)\mathcal{S}_0(t) \\ Z_1(t) &= Z_0(t)\mathcal{S}_0(t) + Y_0^{*-1}(t) \end{aligned}$$

for $t \in [a, b+1]$, then $Y_1(t)$, $Z_1(t)$ satisfies system (3.15) on page 79 for t in $[a, b]$.

Consider, for $t \in [a, b]$,

$$\begin{aligned} Y_1(t+1) &= Y_0(t+1)\mathcal{S}_0(t+1) \\ &= Y_0(t+1)[\mathcal{S}_0(t) + Y_0^{-1}(t+1)F(t)Y_0^{*-1}(t)]. \end{aligned}$$

From (3.15) we get that

$$\begin{aligned} Y_1(t+1) &= E(t)Y_0(t)\mathcal{S}_0(t) + F(t)Z_0(t)\mathcal{S}_0(t) + F(t)Y_0^{*-1}(t) \\ &= E(t)Y_1(t) + F(t)Z_1(t). \end{aligned}$$

Hence $Y_1(t)$, $Z_1(t)$ satisfies the first equation in (3.15) for $t \in [a, b]$.

Next consider, for $t \in [a, b]$,

$$\begin{aligned} Z_1(t+1) &= Z_0(t+1)\mathcal{S}_0(t+1) + Y_0^{*-1}(t+1) \\ &= Z_0(t+1)[\mathcal{S}_0(t) + Y_0^{-1}(t+1)F(t)Y_0^{*-1}(t)] + Y_0^{*-1}(t+1). \end{aligned}$$

From (3.15) we obtain

$$\begin{aligned} Z_1(t+1) &= G(t)Y_0(t)\mathcal{S}_0(t) + H(t)Z_0(t)\mathcal{S}_0(t) \\ &\quad + Z_0(t+1)Y_0^{-1}(t+1)F(t)Y_0^{*-1}(t) + Y_0^{*-1}(t+1) \\ &= G(t)Y_1(t) + H(t)Z_0(t)\mathcal{S}_0(t) \\ &\quad + Z_0(t+1)Y_0^{-1}(t+1)F(t)Y_0^{*-1}(t) + Y_0^{*-1}(t+1). \end{aligned}$$

Hence to show that the second equation in (3.15) holds it suffices to show that

$$Z_0(t+1)Y_0^{-1}(t+1)F(t)Y_0^{*-1}(t) + Y_0^{*-1}(t+1) = H(t)Y_0^{*-1}(t) \qquad (3.60)$$

holds for $t \in [a, b]$.

To verify (3.60) consider, for $t \in [a, b]$,

$$\begin{bmatrix} Y_0(t+1) \\ Z_0(t+1) \end{bmatrix} = M(t) \begin{bmatrix} Y_0(t) \\ Z_0(t) \end{bmatrix}.$$

It follows that

$$\begin{bmatrix} Y_0(t) \\ Z_0(t) \end{bmatrix} = M^{-1}(t) \begin{bmatrix} Y_0(t+1) \\ Z_0(t+1) \end{bmatrix}.$$

Using (3.5) we get that

$$\begin{bmatrix} Y_0(t) \\ Z_0(t) \end{bmatrix} = \begin{bmatrix} H^*(t) & -F^*(t) \\ -G^*(t) & E^*(t) \end{bmatrix} \begin{bmatrix} Y_0(t+1) \\ Z_0(t+1) \end{bmatrix}$$

for $t \in [a, b]$. Hence

$$Y_0(t) = H^*(t)Y_0(t+1) - F^*(t)Z_0(t+1).$$

Taking the conjugate transpose of both sides we get that

$$\begin{aligned} Y_0^*(t) &= Y_0^*(t+1)H(t) - Z_0^*(t+1)F(t) \\ &= Y_0^*(t+1)H(t) - Z_0^*(t+1)Y_0(t+1)Y_0^{-1}(t+1)F(t) \\ &= Y_0^*(t+1)H(t) - Y_0^*(t+1)Z_0(t+1)Y_0^{-1}(t+1)F(t) \end{aligned}$$

since $Y_0(t)$, $Z_0(t)$ is a prepared solution. Thus

$$Y_0^*(t+1)H(t) = Y_0^*(t) + Y_0^*(t+1)Z_0(t+1)Y_0^{-1}(t+1)F(t)$$

which implies that (3.60) holds.

Hence

$$\begin{aligned} Y_1(t) &= Y_0(t)\mathcal{S}_0(t) \\ Z_1(t) &= Z_0(t)\mathcal{S}_0(t) + Y_0^{*-1}(t) \end{aligned}$$

is a solution of (3.15) on $[a, b+1]$. Let P, Q be $n \times m$ constant matrices, then $Y_0(t)P$, $Z_0(t)P$ and $Y_1(t)Q$, $Z_1(t)Q$ are solutions of (3.15). Since the sum of two solutions is a solution we get that

$$\begin{aligned} Y(t) &= Y_0(t)[P + \mathcal{S}_0(t)Q] \\ Z(t) &= Z_0(t)[P + \mathcal{S}_0(t)Q] + Y_0^{*-1}(t)Q \end{aligned}$$

is a solution of (3.15), which is one of the results that we wanted to prove.

Next consider

$$\begin{aligned}
\begin{Bmatrix} Y_0(t) & Y(t) \\ Z_0(t) & Z(t) \end{Bmatrix} &= \begin{Bmatrix} Y_0(a) & Y(a) \\ Z_0(a) & Z(a) \end{Bmatrix} \\
&= Y_0^*(a)Z(a) - Z_0^*(a)Y(a) \\
&= Y_0^*(a)[Z_0(a)P + Y_0^{*-1}(a)Q] - Z_0^*(a)Y_0(a)P \\
&= [Y_0^*(a)Z_0(a) - Z_0^*(a)Y_0(a)]P + Q = Q
\end{aligned}$$

since $Y_0(t)$, $Z_0(t)$ is a prepared solution. Hence (3.57) holds. Letting $t = a$ in equation (3.54) we get that (3.56) holds.

Finally consider

$$\begin{aligned}
\begin{Bmatrix} Y(t) & Y(t) \\ Z(t) & Z(t) \end{Bmatrix} &= \begin{Bmatrix} Y(a) & Y(a) \\ Z(a) & Z(a) \end{Bmatrix} = Y^*(a)Z(a) - Z^*(a)Y(a) \\
&= [Y_0(a)P]^*[Z_0(a)P + Y_0^{*-1}(a)Q] \\
&\quad - [Z_0(a)P + Y_0^{*-1}(a)Q]^*Y_0(a)P \\
&= P^*Y_0^*(a)Z_0(a)P + P^*Q \\
&\quad - P^*Z_0^*(a)Y_0(a)P - Q^*P \\
&= P^*[Y_0^*(a)Z_0(a) - Z_0^*(a)Y_0(a)]P \\
&\quad + P^*Q - Q^*P \\
&= P^*Q - Q^*P
\end{aligned}$$

since $Y_0(t)$, $Z_0(t)$ is a prepared solution of (3.15). Hence $Y(t)$, $Z(t)$ is a prepared solution of (3.15) iff $P^*Q = Q^*P$. □

Theorem 3.33 *Assume $Y(t)$, $Z(t)$ is a prepared solution of* (3.15) *such that $Y(t)$ is nonsingular on $[a, b+1]$. Then*

$$D(t) \equiv Y^{-1}(t+1)F(t)Y^{*-1}(t) \tag{3.61}$$

is Hermitian on $[a,b]$. *Hence* $S_0(t)$ *in the reduction of order theorem is Hermitian on* $[a,b+1]$.

Proof: First we prove this result for the case when $F(t)$ is nonsingular in $[a,b]$. From (3.15)

$$Y(t+1)=E(t)Y(t)+F(t)Z(t).$$

Solving for $Z(t)$ we get that

$$Z(t)=F^{-1}(t)Y(t+1)-F^{-1}(t)E(t)Y(t).$$

Multiplying by $Y^*(t)$ we get that

$$\begin{aligned} Y^*(t)Z(t) &= Y^*(t)F^{-1}(t)Y(t+1)-Y^*(t)F^{-1}(t)E(t)Y(t) \\ &= D^{-1}(t)-Y^*(t)F^{-1}(t)E(t)Y(t). \end{aligned}$$

Hence

$$D^{-1}(t)=Y^*(t)Z(t)+Y^*(t)F^{-1}(t)E(t)Y(t). \tag{3.62}$$

Since $M(t)$ is symplectic we have by (3.7)

$$E(t)F^*(t)=F(t)E^*(t)$$

for $t\in[a,b]$. It follows that

$$F^{-1}(t)E(t)=E^*(t)F^{*-1}(t)$$

for $t\in[a,b]$. Hence from (3.62), using this last equation and the fact that $Y(t)$, $Z(t)$ is a prepared solution, we get that $D^{-1}(t)$ and therefore $D(t)$ is Hermitian on $[a,b]$.

Now we give a proof where we do not assume that $F(t)$ is nonsingular in $[a,b]$. Fix $t_0\in[a,b]$. We will show that $D(t_0)$ is Hermitian. Since $t_0\in[a,b]$ is arbitrary it would then follow that $D(t)$ is Hermitian on $[a,b]$. To see that $D(t_0)$ is Hermitian, set

$$\begin{aligned} Y_1(t) &= Y(t)S(t) \\ Z_1(t) &= Z(t)S(t)+Y^{*-1}(t) \end{aligned}$$

for $t\in[t_0,b+1]$, where

$$S(t)=\sum_{s=t_0}^{t-1}Y^{-1}(s+1)F(s)Y^{*-1}(s) \tag{3.63}$$

for $t \in [t_0, b+1]$. By the reduction of order theorem with a replaced by t_0, $Q = I$ and $P = 0$ we get that $Y_1(t)$, $Z_1(t)$ is a solution of (3.15) and by

$$\begin{Bmatrix} Y(t) & Y_1(t) \\ Z(t) & Z_1(t) \end{Bmatrix} = Q = I$$

it follows that

$$Y^*(t)Z_1(t) - Z^*(t)Y_1(t) = I \tag{3.64}$$

for $t \in [a, b+1]$.

Since $P^*Q = 0$ is Hermitian, the reduction of order theorem implies that $Y_1(t)$, $Z_1(t)$ is a prepared solution of (3.15). Using (3.64) and the fact that the identities $Y^*(t)Z(t) = Z^*(t)Y(t)$ and $Y_1^*(t)Z_1(t) = Z_1^*(t)Y_1(t)$ hold, we conclude from Theorem 3.4 that the $2n \times 2n$ matrix function

$$\begin{bmatrix} Y(t) & Y_1(t) \\ Z(t) & Z_1(t) \end{bmatrix}$$

is symplectic for $t \in [t_0, b+1]$. It follows from (3.7) that

$$Y(t)Y_1^*(t) = Y_1(t)Y^*(t)$$

for $t \in [t_0, b+1]$. Letting $t = t_0 + 1$ we obtain

$$Y(t_0+1)Y_1^*(t_0+1) = Y_1(t_0+1)Y^*(t_0+1).$$

It follows that

$$Y^{-1}(t_0+1)Y_1(t_0+1) = Y_1^*(t_0+1)Y^{*-1}(t_0+1).$$

But, from the definition of $Y_1(t)$,

$$\begin{aligned} Y^{-1}(t_0+1)Y_1(t_0+1) &= \mathcal{S}(t_0+1) \\ &= Y^{-1}(t_0+1)F(t_0)Y_0^{*-1}(t_0) \\ &= D(t_0) \end{aligned}$$

and hence $D(t_0)$ is Hermitian. □

Exercise 3.34 *Introduce the summation convention by the recursive definition*

$$\sum_{k=s}^{t} D(k) \equiv \begin{cases} \sum_{k=s}^{s-1} D(k) \equiv 0, & \textit{for } t = s-1 \\ D(t) + \sum_{k=s}^{t-1} D(k), & \textit{for } t \geq s \end{cases} \tag{3.65}$$

and

$$\sum_{k=s}^{t-1} D(k) \equiv -\left\{\sum_{k=t}^{s-1} D(k)\right\} \quad for \quad t < s. \tag{3.66}$$

Then establish the following generalization of the reduction of order theorem, Theorem 3.32. For Y_0, Z_0 as in that theorem, for t_0 a point of $[a, b]$, and for $D(t)$ as above, show that relations (3.54) and (3.55) generalize, respectively, to

$$Y(t) = Y_0(t)\left[P + \sum_{s=t_0}^{t-1} D(s)\right] \tag{3.67}$$

and

$$Z(t) = Z_0(t)\left[P + \sum_{s=t_0}^{t-1} D(s)\right] + Y_0^{*-1}(t)Q, \tag{3.68}$$

where

$$P = Y_0^{-1}(t_0)Y(t_0) \quad and \quad Q = \{X_0,\ X\}. \tag{3.69}$$

This suggests the following summation operator S_s^t recursively defined by

$$\begin{aligned} S_s^s D(k) &= 0, \\ S_s^t D(k) &= (S_s^{t-1} D(k)) + D(t-1), \quad t = s+1, s+2, \dots, \\ S_s^t D(k) &= -\left[S_t^s D(k)\right], \quad t = s-1, s-2, \dots . \end{aligned}$$

Note that for $t > s$, the sum contains $t - s$ terms of functional values of D. Show that for compatible s, t, u, in any order,

$$S_s^t D(k) = (S_s^u D(k)) + \left(S_u^t D(k)\right).$$

Also, show that we have the Fundamental Theorem:

If $\Delta F(k) \equiv f(k)$, then $S_s^t f(k) = F(t) - F(s)$.

Observe that in this notation, the reduction of order result reads much like the continuous case:

$$Y(t) = Y_0(t)\left[P + S_{t_0}^t D_0(s)\right] \tag{3.70}$$

and

$$Z(t) = Z_0(t)\left[P + S_{t_0}^t D_0(s)\right] + Y_0^{*-1}(t)Q \tag{3.71}$$

where $D_0(s) = Y^{-1}(s+1)F(s)Y_0^{*-1}(s)$.

It is interesting to note that in the scalar case our summation convention is suggested by taking the logarithm of our product notation of Exercise 3.7 on page 78.

3.11 DOMINANT AND RECESSIVE SOLUTIONS

We now want to consider (3.15) on an infinite discrete interval of the form $[a,\infty) \equiv \{a, a+1, a+2, \cdots\}$. Thus we will now consider the system

$$\begin{aligned} Y(t+1) &= E(t)Y(t) + F(t)Z(t) \\ Z(t+1) &= G(t)Y(t) + H(t)Z(t) \end{aligned} \tag{3.72}$$

for $t \in [a,\infty)$. Here we assume that the $2n \times 2n$ matrix function

$$M(t) \equiv \begin{bmatrix} E(t) & F(t) \\ G(t) & H(t) \end{bmatrix} \tag{3.73}$$

is symplectic on $[a,\infty)$. Every $n \times m$ solution $Y(t)$, $Z(t)$ of (3.72) is defined on $[a,\infty)$.

An $n \times n$ prepared solution $Y(t)$, $Z(t)$ of (3.72) is said to be a *dominant solution* at ∞ provided there is an integer $t_0 \in [a,\infty)$ such that $Y(t)$ is invertible on $[t_0,\infty)$ and

$$\sum_{s=t_0}^{\infty} D(s) = \sum_{s=t_0}^{\infty} Y^{-1}(s+1)F(s)Y^{*-1}(s) \tag{3.74}$$

converges (to a Hermitian matrix with finite entries).

Theorem 3.35 *Assume* (3.72) *has a dominant solution* $Y(t)$, $Z(t)$ *at* ∞. *Then for each solution* $Y_1(t)$, $Z_1(t)$ *of* (3.72), *there exists an* $n \times n$ *matrix* C *such that*

$$\lim_{t\to\infty} Y^{-1}(t)Y_1(t) = C.$$

Proof: From the definition that $Y(t)$, $Z(t)$ is a dominant solution at ∞, we know that it is a prepared solution and there is an integer $t_0 \in [a,\infty)$ such that $Y(t)$ is invertible on $[t_0,\infty)$. Assume $Y_1(t)$, $Z_1(t)$ is a solution of (3.72). By the reduction of order theorem with a replaced by t_0 we get that there are $n \times n$ matrices P and Q such that

$$Y_1(t) = Y(t)[P + \mathcal{S}(t)Q]$$

for $t \in [t_0,\infty)$, where

$$\mathcal{S}(t) = \sum_{s=t_0}^{t-1} D(s)$$

for $t \in [t_0, \infty)$. It follows that

$$\begin{aligned} \lim_{t\to\infty} Y^{-1}(t)Y_1(t) &= \lim_{t\to\infty}[P + S(t)Q] \\ &= P + \sum_{s=t_0}^{\infty} D(s)Q \equiv C. \end{aligned}$$

□

If $Y(t)$, $Z(t)$ is an $n \times n$ prepared matrix solution of (3.72) such that

$$\text{rank } X(t) = \text{rank} \begin{bmatrix} Y(t) \\ Z(t) \end{bmatrix} = n$$

for all $t \in [a, \infty)$, then we say that $X(t)$ is a *prepared basis* for (3.72). (More correctly, but less succinctly, the columns of $X(t)$ are a set of mutually prepared (or conjoined) solutions which provides a basis for a Lagrangian subspace.)

Theorem 3.36 *Assume $Y_1(t)$, $Z_1(t)$, and $Y_2(t)$, $Z_2(t)$ are $n \times n$ prepared solutions of* (3.72) *such that the $n \times n$ constant matrix*

$$C \equiv \begin{Bmatrix} Y_1(t) & Y_2(t) \\ Z_1(t) & Z_2(t) \end{Bmatrix}$$

is nonsingular, then each of the solutions

$$X_1(t) = \begin{bmatrix} Y_1(t) \\ Z_1(t) \end{bmatrix}, \qquad X_2(t) = \begin{bmatrix} Y_2(t) \\ Z_2(t) \end{bmatrix}$$

is a prepared basis for (3.72).

Proof: To show that $X_2(t)$ is a prepared basis we have to show that the rank of $X_2(t)$ is n for all $t \in [a, \infty)$. To see this suppose t_0 is in $[a, \infty)$ and assume α is an $n \times 1$ constant vector such that

$$X_2(t_0)\alpha = 0. \tag{3.75}$$

Then

$$\begin{aligned} C\alpha &= \begin{Bmatrix} Y_1(t_0) & Y_2(t_0) \\ Z_1(t_0) & Z_2(t_0) \end{Bmatrix} \alpha \\ &= \{X_1; X_2\}\alpha \\ &\equiv X_1^*(t_0) J X_2(t_0)\alpha = 0 \end{aligned}$$

by (3.75). Since C is nonsingular it follows that $\alpha = 0$.

Hence rank $X_2(t) = n$ for $t \in [a, \infty)$ and $X_2(t)$ is a prepared basis for (3.72). Since

$$\begin{Bmatrix} Y_2(t) & Y_1(t) \\ Z_2(t) & Z_1(t) \end{Bmatrix} = -C^*$$

is a nonsingular matrix we have from the above argument that $X_1(t)$ is also a prepared basis for (3.72). □

We say that a solution $Y_0(t)$, $Z_0(t)$ of (3.72) is *recessive at* ∞ provided $X_0(t) = \begin{bmatrix} Y_0(t) \\ Z_0(t) \end{bmatrix}$ is a prepared basis of (3.1) and provided that whenever $Y(t)$, $Z(t)$ is a solution of (3.72) with

$$C \equiv \begin{Bmatrix} Y_0(t) & Y(t) \\ Z_0(t) & Z(t) \end{Bmatrix}$$

a nonsingular $n \times n$ (constant) matrix, it follows that there is a $t_0 \geq a$ such that $Y(t)$ is nonsingular for $t \geq t_0$ and

$$\lim_{t \to \infty} Y^{-1}(t) Y_0(t) = 0.$$

The definition of a recessive solution X_0 of (3.1) involves other solutions X such that $\{X_0; X\}$ is nonsingular. The next remark establishes the existence of such solutions by linear algebra methods. However the solution X produced by those methods is not necessarily prepared even if X_0 is prepared. After this remark we then give another remark, which shows how to produce a prepared basis X from a prepared basis X_0 such that $\{X_0; X\}$ is nonsingular.

Remark 3.37 *Suppose that X_0 is a $2n \times n$ solution of* (3.1) *such that the rank of $X_0(t)$ is n for one point t_0 (and hence all t). Since J is nonsingular the $n \times 2n$ matrix $A = X_0^*(t_0)J$ has rank n and nullity of $2n - n = n$. Let $f_1, \cdots, f_n$ in C^n be a basis for the range of A and let $g_1, \cdots, g_n$ be vectors in C^{2n} such that $f_i = Ag_i$. Set $F = [f_1 \ \cdots \ f_n]$. Then the matrix $G = [g_1 \ \cdots \ g_n]$ has linearly independent columns and the solution $X(t)$ with $X(t_0) = G$ has*

$$\{X_0; X\}(t_0) = X_0^*(t_0) J X(t_0) = AG = F,$$

which is nonsingular.

Remark 3.38 *Suppose that X_0 is a prepared basis for* (3.1). *Select a point t_0. Since $X_0(t_0)$ is a $2n \times n$ matrix of full column rank of n, the $n \times n$ matrix $X_0^*(t_0)X_0(t_0)$ is positive definite and hence nonsingular. Let $X(t)$ be the solution determined by the initial conditions*

$$Y(t_0) = -Z_0(t_0), \qquad Z(t_0) = Y_0(t_0). \tag{3.76}$$

Then

$$\begin{aligned}\{X_0(t_0);\; X(t_0)\} &= \begin{Bmatrix} Y_0(t_0) & -Z_0(t_0) \\ Z_0(t_0) & Y_0(t_0) \end{Bmatrix} \\ &= Y_0^*(t_0)Y_0(t_0) + Z_0^*(t_0)Z_0(t_0) \\ &= [Y_0^*(t_0) \quad Z_0^*(t_0)] \begin{bmatrix} Y_0(t_0) \\ Z_0(t_0) \end{bmatrix} = X_0^*(t_0)X_0(t_0) > 0,\end{aligned}$$

i.e., is positive definite, and hence $\{X_0; X\}(t_0)$ is nonsingular. The matrix $X(t_0)$ has linearly independent columns, since $X(t_0)c = 0$ implies $-Z(t_0)c = 0 = Y_0(t_0)c$ and thus c must be 0 because $X_0(t_0)$ has linearly independent columns. Also,

$$\{X(t_0); X(t_0)\} = \begin{Bmatrix} -Z_0(t_0) & -Z_0(t_0) \\ Y_0(t_0) & Y_0(t_0) \end{Bmatrix} = -Z_0^*(t_0)Y_0(t_0) + Y_0^*(t_0)Z_0(t_0) = 0,$$

because X_0 is prepared, and the solution $X(t)$ is a prepared basis.

We start our discussion of recessive solutions with some unusual examples where we can explicitly determine recessive solutions or show that none exist.

Example 3.39 *Suppose that $M = I_{2n}$, i.e.,*

$$M = \begin{bmatrix} I & 0 \\ 0 & I \end{bmatrix}.$$

Then M is symplectic and all solutions of $x(t+1) = Mx(t)$ are constant. If X_0 is recessive at ∞ and X is a solution such that $\{X_0; X\}$ is nonsingular, then Y must be nonsingular and

$$\lim_{t\to\infty} Y^{-1}(t)Y_0(t) = 0 = Y^{-1}Y_0$$

implies that $Y_0 = 0$. Thus each recessive solution X_0 must have the form

$$X_0(t) \equiv \begin{bmatrix} 0 \\ K \end{bmatrix} \tag{3.77}$$

for some nonsingular constant matrix K. *Conversely, if* X_0 *is defined this way and* X *is a solution such that* $\{X_0; X\}$ *is nonsingular, then*

$$\left\{\begin{matrix} 0 & Y(t) \\ K & Z(t) \end{matrix}\right\} = -K^*Y(t)$$

implies that $Y(t) \equiv Y$ *is nonsingular. Furthermore,*

$$Y^{-1}(t)Y_0(t) \equiv Y^{-1}Y_0 \equiv 0 \to 0, \quad \text{as} \quad t \to \infty$$

and X_0 *of the form* (3.77) *is recessive.*

Exercise 3.40 *Suppose that* $M = -I_{2n}$, *i.e.,*

$$M = \begin{bmatrix} -I & 0 \\ 0 & -I \end{bmatrix}.$$

Then M *is symplectic and all solutions of* $X(t+1) = MX(t)$ *are of period* 2. *Show that if* X_0 *is recessive at* ∞, *then there must exist a nonsingular constant matrix* K *such that* $X_0(t)$ *has the form*

$$Y_0(t) \equiv 0, \qquad Z_0(t) = \begin{cases} K, & \text{for even } t \\ -K, & \text{for odd } t. \end{cases} \tag{3.78}$$

Conversely, for each choice of nonsingular K, *the solution* $X_0(t)$ *defined in this way is recessive at* ∞.

Exercise 3.41 *Show that if a symplectic system has the property that every solution is periodic, then each recessive solution* X_0 *must have* $Y_0(t) \equiv 0$. *Note that a periodic function which has limit of* 0 *must be identically* 0.

The next example shows that we do not always have existence of solutions which are recessive at ∞.

Exercise 3.42 *Suppose that*

$$M = \begin{bmatrix} 0 & I \\ -I & 0 \end{bmatrix} = J.$$

Then M *is symplectic and all solutions of* $X(t+1) = MX(t)$ *are of period* 4. *Show that if* X_0 *were a solution which is recessive at* ∞, *then we would have* $X_0(t) \equiv 0$ *contrary to a recessive solution being a prepared basis. Thus this system does not have a solution which is recessive at* ∞.

Theorem 3.43 *Let $Y_0(t)$, $Z_0(t)$ be a solution of (3.72) which is recessive at ∞.*

(i) *If K is a nonsingular $n \times n$ constant matrix, then the solution $Y_0(t)K$, $Z_0(t)K$ is also recessive at ∞.*

(ii) *If $Y(t)$, $Z(t)$ is a prepared solution of (3.72) such that*

$$Q \equiv \begin{Bmatrix} Y(t) & Y_0(t) \\ Z(t) & Z_0(t) \end{Bmatrix}$$

is nonsingular, then $Y(t)$, $Z(t)$ is dominant at ∞.

Proof: Assume $Y_0(t)$, $Z_0(t)$ is a recessive solution of (3.72) at ∞ and K is a nonsingular constant matrix. We now show that $Y_0(t)K$, $Z_0(t)K$ is a recessive solution of (3.72) at ∞. Clearly $Y_0(t)K$, $Z_0(t)K$ is a solution of (3.72) on $[a, \infty)$. Since

$$\begin{Bmatrix} Y_0(t)K & Y_0(t)K \\ Z_0(t)K & Z_0(t)K \end{Bmatrix} = K^* Y_0^*(t) Z_0(t) K - K^* Z_0^*(t) Y_0(t) K$$

$$= K^*[Y_0^*(t) Z_0(t) - Z_0^*(t) Y_0(t)]K$$

$$= K^* \begin{Bmatrix} Y_0(t) & Y_0(t) \\ Z_0(t) & Z_0(t) \end{Bmatrix} K = 0,$$

the solution $Y_0(t)K$, $Z_0(t)K$ is a prepared solution of (3.72). Furthermore,

$$\text{rank } (X_0(t)K) = \text{rank } X_0(t) = n$$

for $t \in [a, \infty)$ and $X_0(t)K$ is a prepared basis for (3.72).

Next assume $Y(t)$, $Z(t)$ is a solution of (3.72) such that

$$C = \begin{Bmatrix} Y_0(t)K & Y(t) \\ Z_0(t)K & Z(t) \end{Bmatrix}$$

is a nonsingular matrix. Because the bracket function is a skew form

$$\begin{Bmatrix} Y_0(t) & Y(t) \\ Z_0(t) & Z(t) \end{Bmatrix} = K^{*-1} C$$

is nonsingular. Since $Y_0(t)$, $Z_0(t)$ is a recessive solution at ∞ it follows that there is a $t_0 \geq a$ such that $Y(t)$ is nonsingular on $[t_0, \infty)$ and

$$\lim_{t \to \infty} Y^{-1}(t) Y_0(t) = 0.$$

But this implies

$$\lim_{t\to\infty} Y^{-1}(t)[Y_0(t)K] = 0$$

and so (i) holds.

In order to prove that (ii) holds assume $Y(t)$, $Z(t)$ is a prepared solution of (3.72) such that

$$Q \equiv \begin{Bmatrix} Y(t) & Y_0(t) \\ Z(t) & Z_0(t) \end{Bmatrix}$$

is nonsingular. Again, because the bracket function is a skew form

$$\begin{Bmatrix} Y_0(t) & Y(t) \\ Z_0(t) & Z(t) \end{Bmatrix} = -Q^*$$

is nonsingular. Since $Y_0(t)$, $Z_0(t)$ is a recessive solution at ∞ it follows that there is an integer $t_0 \geq a$ such that $Y(t)$ is invertible on $[t_0, \infty)$. By the reduction of order theorem with the roles of Y and Y_0 reversed,

$$Y_0(t) = Y(t)[P + \mathcal{S}(t)Q]$$

where $P = Y^{-1}(t_0)Y_0(t_0)$ and

$$\mathcal{S}(t) = \sum_{s=t_0}^{t-1} D(s) = \sum_{s=t_0}^{t-1} Y^{-1}(s+1)F(s)Y^{*-1}(s)$$

for $t \geq t_0$. But

$$\lim_{t\to\infty} [P + \mathcal{S}(t)Q] = \lim_{t\to\infty} Y^{-1}(t)Y_0(t) = 0$$

gives

$$\sum_{s=t_0}^{\infty} D(s) = -PQ^{-1}$$

and consequently $Y(t)$, $Z(t)$ is dominant at ∞. □

3.12 THE NORMAL BASIS THEOREM

If each of $Y_1(t)$, $Z_1(t)$ and $Y_2(t)$, $Z_2(t)$ is a prepared solution of (3.72) on page 113 such that

$$\begin{Bmatrix} Y_1(t) & Y_2(t) \\ Z_1(t) & Z_2(t) \end{Bmatrix} \tag{3.79}$$

is nonsingular, then we know from the Wronskian Test, i.e., Theorem 3.31 on page 103 that they are linearly independent solutions. We will soon see that it is appropriate to call such a pair of solutions a *fundamental set of solutions* of (3.72). That is, every solution can be represented uniquely in terms of these two solutions. If, in addition,

$$\begin{Bmatrix} Y_1(t) & Y_2(t) \\ Z_1(t) & Z_2(t) \end{Bmatrix} = I$$

then we say $Y_1(t)$, $Z_1(t)$ and $Y_2(t)$, $Z_2(t)$ is a *normal pair of solutions* of (3.72).

Exercise 3.44 *Show that if* $Y_1(t)$, $Z_1(t)$ *and* $Y_2(t)$, $Z_2(t)$ *is a fundamental set of solutions of* (3.72) *and* Q *is defined by*

$$Q = \begin{Bmatrix} Y_1(t) & Y_2(t) \\ Z_1(t) & Z_2(t) \end{Bmatrix}$$

then $Y_1(t)$, $Z_1(t)$ *and* $Y_2(t)Q^{-1}$, $Z_2(t)Q^{-1}$ *is a normal pair of solutions of equation* (3.72).

Exercise 3.45 *Show that if* $Y_1(t)$, $Z_1(t)$ *is a* $2n \times n$ *solution of* (3.72) *which is a prepared basis, then there exists a solution* Y_2, Z_2 *such that* X_1 *and* X_2 *is a normal pair of solutions. Use Remark* 3.38 *on page* 116 *and the above exercise.*

Exercise 3.46 *Show that if* $Y_1(t)$, $Z_1(t)$ *and* $Y_2(t)$, $Z_2(t)$ *is a normal pair of solutions of* (3.72), *then the* $2n \times 2n$ *matrix function*

$$\begin{bmatrix} Y_1(t) & Y_2(t) \\ Z_1(t) & Z_2(t) \end{bmatrix}$$

is symplectic for all $t \geq a$.

The following result shows that normal pairs of solutions play a role analogous to orthonormal bases.

Theorem 3.47 (Normal Basis Theorem) *Assume* $Y_1(t)$, $Z_1(t)$ *and* $Y_2(t)$, $Z_2(t)$ *is a normal pair of solutions of* (3.72). *If* Γ_1 *and* Γ_2 *are* $n \times m$ *constant matrices, then*

$$Y(t) = Y_1(t)\Gamma_1 + Y_2(t)\Gamma_2 \tag{3.80}$$

$$Z(t) = Z_1(t)\Gamma_1 + Z_2(t)\Gamma_2 \tag{3.81}$$

is a solution of (3.72), *where*

$$\Gamma_1 = -\begin{Bmatrix} Y_2(t) & Y(t) \\ Z_2(t) & Z(t) \end{Bmatrix}, \quad \Gamma_2 = \begin{Bmatrix} Y_1(t) & Y(t) \\ Z_1(t) & Z(t) \end{Bmatrix}, \tag{3.82}$$

and

$$\begin{Bmatrix} Y(t) & Y(t) \\ Z(t) & Z(t) \end{Bmatrix} = \Gamma_1^*\Gamma_2 - \Gamma_2^*\Gamma_1. \tag{3.83}$$

Conversely, if $Y(t)$, $Z(t)$ *(each being* $n \times m$*) is a solution of* (3.72), *then* $Y(t)$, $Z(t)$ *are given by* (3.80) *and* (3.81) *when* Γ_1, Γ_2 *are defined by* (3.82).

Proof: Assume $Y_1(t)$, $Z_1(t)$ and $Y_2(t)$, $Z_2(t)$ is a normal pair of solutions of (3.72) and $Y(t)$, $Z(t)$ are defined by (3.80) and (3.81) where Γ_1 and Γ_2 are $n \times m$ constant matrices. Clearly $Y(t)$, $Z(t)$ is a solution of (3.72). Consider

$$\begin{aligned} \begin{Bmatrix} Y_1(t) & Y(t) \\ Z_1(t) & Z(t) \end{Bmatrix} &= \begin{Bmatrix} Y_1(t) & Y_1(t)\Gamma_1 + Y_2(t)\Gamma_2 \\ Z_1(t) & Z_1(t)\Gamma_1 + Z_2(t)\Gamma_2 \end{Bmatrix} \\ &= \begin{Bmatrix} Y_1(t) & Y_1(t) \\ Z_1(t) & Z_1(t) \end{Bmatrix}\Gamma_1 + \begin{Bmatrix} Y_1(t) & Y_2(t) \\ Z_1(t) & Z_2(t) \end{Bmatrix}\Gamma_2 \\ &= \Gamma_2. \end{aligned}$$

Similarly,

$$\begin{Bmatrix} Y_2(t) & Y(t) \\ Z_2(t) & Z(t) \end{Bmatrix} = -\Gamma_1.$$

Next consider

$$\begin{aligned} \begin{Bmatrix} Y(t) & Y(t) \\ Z(t) & Z(t) \end{Bmatrix} &= \{X(t); X(t)\} \\ &= \{X_1(t)\Gamma_1 + X_2(t)\Gamma_2; X_1(t)\Gamma_1 + X_2(t)\Gamma_2\} \\ &= \left\{[X_1(t) \quad X_2(t)]\begin{bmatrix}\Gamma_1 \\ \Gamma_2\end{bmatrix}; [X_1(t) \quad X_2(t)]\begin{bmatrix}\Gamma_1 \\ \Gamma_2\end{bmatrix}\right\} \\ &= [\Gamma_1^* \quad \Gamma_2^*]\,[X_1(t) \quad X_2(t)]^*\, J\,[X_1(t) \quad X_2(t)]\begin{bmatrix}\Gamma_1 \\ \Gamma_2\end{bmatrix}. \end{aligned}$$

By Exercise 3.46 $[X_1(t)\ X_2(t)]$ is symplectic for all $t \geq a$. Hence

$$\begin{Bmatrix} Y(t) & Y(t) \\ Z(t) & Z(t) \end{Bmatrix} = [\Gamma_1^* \quad \Gamma_2^*]\, J \begin{bmatrix}\Gamma_1 \\ \Gamma_2\end{bmatrix}$$

$$\begin{aligned} &= [\Gamma_1^* \quad \Gamma_2^*] \begin{bmatrix} \Gamma_2 \\ -\Gamma_1 \end{bmatrix} \\ &= \Gamma_1^*\Gamma_2 - \Gamma_2^*\Gamma_1 \end{aligned}$$

and (3.83) holds.

Conversely, assume $Y(t)$, $Z(t)$ is an $n \times m$ solution of (3.72). Since $Y_1(t)$, $Z_1(t)$ and $Y_2(t)$, $Z_2(t)$ is a normal pair of solutions of (3.72) we have by Exercise 3.46 that

$$\begin{bmatrix} Y_1(t) & Y_2(t) \\ Z_1(t) & Z_2(t) \end{bmatrix}$$

is symplectic for all $t \geq a$. Fix $t_0 \geq a$. Since

$$\begin{bmatrix} Y_1(t_0) & Y_2(t_0) \\ Z_1(t_0) & Z_2(t_0) \end{bmatrix}$$

is symplectic, then it is invertible. Hence we can uniquely define $n \times m$ constant matrices Γ_1 and Γ_2 by

$$\begin{bmatrix} Y(t_0) \\ Z(t_0) \end{bmatrix} = \begin{bmatrix} Y_1(t_0) & Y_2(t_0) \\ Z_1(t_0) & Z_2(t_0) \end{bmatrix} \begin{bmatrix} \Gamma_1 \\ \Gamma_2 \end{bmatrix}.$$

It follows that

$$\begin{aligned} Y(t_0) &= Y_1(t_0)\Gamma_1 + Y_2(t_0)\Gamma_2 \\ Z(t_0) &= Z_1(t_0)\Gamma_1 + Z_2(t_0)\Gamma_2. \end{aligned}$$

Define

$$\begin{aligned} Y_3(t) &= Y_1(t)\Gamma_1 + Y_2(t)\Gamma_2 \\ Z_3(t) &= Z_1(t)\Gamma_1 + Z_2(t)\Gamma_2, \end{aligned}$$

then $Y_3(t)$, $Z_3(t)$ is a solution of (3.72). Since initial value problems for (3.72) have unique solutions

$$\begin{aligned} Y(t) &= Y_3(t) = Y_1(t)\Gamma_1 + Y_2(t)\Gamma_2 \\ Z(t) &= Z_3(t) = Z_1(t)\Gamma_1 + Z_2(t)\Gamma_2. \end{aligned}$$

By the first part of this theorem Γ_1 and Γ_2 are given by (3.82). □

Now the terminology of a fundamental set of solutions is justified because such a pair of solutions may be adjusted to a normal basis.

Theorem 3.48 *If X_1 and X_2 are $2n \times n$ matrix functions defined on $[a, b+1]$ each of which is a prepared solution on $[a, b]$, then the following conditions are equivalent:*

(a) *X_1 and X_2 are linearly independent in the solution module.*

(b) *The Wronskian of X_1 and X_2 is nonsingular.*

(c) *If $X(t)$ is a $2n \times n$ solution, then there exist unique $n \times n$ matrices C_1 and C_2 such that*

$$X(t) \equiv X_1(t)C_1 + X_2(t)C_2.$$

Exercise 3.49 *Prove Theorem* 3.48.

Theorem 3.50 *Assume $Y_2(t)$, $Z_2(t)$ is a solution of* (3.72) *which is dominant at ∞ with $Y_2(t)$ invertible for $t \geq t_0 \geq a$. Define a solution by*

$$\begin{aligned} Y_0(t) &= Y_2(t)\mathcal{S}_2(t) \\ Z_0(t) &= Z_2(t)\mathcal{S}_2(t) - Y_2^{*-1}(t) \end{aligned}$$

for $t \geq t_0$, where

$$\mathcal{S}_2(t) = \sum_{s=t}^{\infty} D_2(s)$$

where $D_2(t) = Y_2^{-1}(t+1)F(t)\, Y_2^{*-1}(t)$ for $t \geq t_0$. Then $Y_0(t)$, $Z_0(t)$ is recessive at ∞. Furthermore $Y_0(t)$, $Z_0(t)$ and $Y_2(t)$, $Z_2(t)$ is a normal pair of solutions of (3.72). *Finally if $Y(t)$, $Z(t)$ is any solution of* (3.72), *then*

$$\begin{aligned} Y(t) &= Y_2(t)\Gamma_1 - Y_0(t)\Gamma_2 \\ Z(t) &= Z_2(t)\Gamma_1 - Z_0(t)\Gamma_2 \end{aligned}$$

for $t \geq t_0$, where

$$\Gamma_1 = \begin{Bmatrix} Y_0(t) & Y(t) \\ Z_0(t) & Z(t) \end{Bmatrix}, \quad \Gamma_2 = \begin{Bmatrix} Y_2(t) & Y(t) \\ Z_2(t) & Z(t) \end{Bmatrix}.$$

Proof: Assume $Y_2(t)$, $Z_2(t)$ is a dominant solution of (3.72) at ∞ with $Y_2(t)$ invertible for $t \geq t_0$. It follows that the series

$$\sum_{s=t_0}^{\infty} D_2(s)$$

converges to a Hermitian matrix. Hence we can define

$$\mathcal{S}_2(t) = \sum_{s=t}^{\infty} D_2(s)$$

for $t \geq t_0$. We wish to use the reduction of order theorem, Theorem 3.32 on page 105 with a replaced by t_0, use arbitrary $b > t_0$, and Y_0, Z_0 of that theorem is replaced by Y_2, Z_2. Then use the summation operator S of Exercise 3.34 on page 111 to write

$$\begin{aligned} Y_0(t) &= Y_2(t)[P + S_{t_0}^t D_2(s)Q] \\ Z_0(t) &= Z_2(t)[P + S_{t_0}^t D_2(s)Q] - Y_2^{*-1}(t). \end{aligned}$$

Choose $P = S_{t_0}^{\infty} D_2(s)$ and $Q = -I$ and use the additivity property of S to get

$$\begin{aligned} Y_0(t) &= Y_2(t)[[S_{t_0}^{\infty} D_2(s) - S_{t_0}^t D_2(s)] = [S_t^{\infty} D_2(s)] = \\ Z_0(t) &= Z_2(t)[S_t^{\infty} D_2(s)] - Y_2^{*-1}(t) \end{aligned}$$

for $t \geq t_0$. which in our present notation is

$$Y_0(t) = Y_2(t)\mathcal{S}_2(t),$$

and

$$Z_0(t) = Z_2(t)\mathcal{S}_2(t) - Y_2^{*-1}(t)$$

for $t \geq t_0$. Since $P^*Q = -\mathcal{S}_2(t_0)$ is Hermitian from Theorem 3.33, we get from the reduction of order theorem that $Y_0(t)$, $Z_0(t)$ is a prepared solution of (3.72). Since (by the reduction of order theorem)

$$\begin{Bmatrix} Y_2(t) & Y_0(t) \\ Z_2(t) & Z_0(t) \end{Bmatrix} = Q = -I$$

is invertible we have by Theorem 3.36 $Y_0(t)$, $Z_0(t)$ is a prepared basis for (3.72). Since

$$\begin{Bmatrix} Y_0(t) & Y_2(t) \\ Z_0(t) & Z_2(t) \end{Bmatrix} = I$$

the pair of solutions $Y_0(t)$, $Z_0(t)$ and $Y_2(t)$, $Z_2(t)$ is a normal pair. To finish seeing that $Y_0(t)$, $Z_0(t)$ is recessive at ∞ assume $Y(t)$, $Z(t)$ is a solution of (3.72) such that

$$\begin{Bmatrix} Y_0(t) & Y(t) \\ Z_0(t) & Z(t) \end{Bmatrix}$$

is nonsingular. By the Normal Basis Theorem

$$Y(t) = Y_2(t)\Gamma_1 - Y_0(t)\Gamma_2$$

where

$$\Gamma_1 = \begin{Bmatrix} Y_0(t) & Y(t) \\ Z_0(t) & Z(t) \end{Bmatrix}, \quad \Gamma_2 = \begin{Bmatrix} Y_2(t) & Y(t) \\ Z_2(t) & Z(t) \end{Bmatrix}.$$

Thus

$$\begin{aligned} Y_2^{-1}(t)Y(t) &= \Gamma_1 - Y_2^{-1}(t)Y_0(t)\Gamma_2 \\ &= \Gamma_1 - \mathcal{S}_2(t)\Gamma_2. \end{aligned}$$

It follows that

$$\lim_{t\to\infty} Y_2^{-1}(t)Y(t) = \Gamma_1$$

which is nonsingular. Pick an integer $t_1 \geq t_0$ such that $Y(t)$ is invertible for $t \geq t_1$. It follows that

$$\lim_{t\to\infty} Y^{-1}(t)Y_0(t) = \lim_{t\to\infty} [Y^{-1}(t)Y_2(t)\mathcal{S}_2(t)] = \Gamma_1 \cdot 0 = 0.$$

Hence $Y_0(t)$, $Z_0(t)$ is recessive at ∞. The remainder of this theorem follows from the Normal Basis Theorem. □

3.13 THE CONNECTION THEOREM

Theorem 3.51 (The Connection Theorem) *For $t_0 \geq a$ let solutions $Y_1(t, t_0)$, $Z_1(t, t_0)$ and $Y_2(t, t_0)$, $Z_2(t, t_0)$ of (3.72) be determined by the initial conditions*

$$\begin{aligned} Y_1(t_0, t_0) &= I, \quad Z_1(t_0, t_0) = 0 \\ Y_2(t_0, t_0) &= 0, \quad Z_2(t_0, t_0) = I. \end{aligned}$$

Then these two solutions are a normal pair and the following are equivalent.

(a) *$Y_2(t, t_0)$, $Z_2(t, t_0)$ is dominant at ∞.*

(b) *$Y_2(t, t_0)$ is invertible for all sufficiently large t and there exists a Hermitian matrix $\Omega(t_0)$ with finite entries such that*

$$\lim_{t\to\infty} Y_2^{-1}(t, t_0)Y_1(t, t_0) = \Omega(t_0). \tag{3.84}$$

(c) *There exists a solution $Y_0(t)$, $Z_0(t)$ which is recessive at ∞ and has $Y_0(t_0)$ nonsingular.*

Furthermore, if **(a)**, **(b)**, *or* **(c)** *holds, then for any recessive solution $Y_0(t)$, $Z_0(t)$ at ∞ it follows that*

$$Z_0(t_0)Y_0^{-1}(t_0) = -\Omega(t_0).$$

Proof: First note that

$$\begin{aligned}\left\{\begin{array}{cc} Y_1(t,t_0) & Y_2(t,t_0) \\ Z_1(t,t_0) & Z_2(t,t_0)\end{array}\right\} &\equiv \left\{\begin{array}{cc} Y_1(t_0,t_0) & Y_2(t_0,t_0) \\ Z_1(t_0,t_0) & Z_2(t_0,t_0)\end{array}\right\} \\ &= \left\{\begin{array}{cc} I & 0 \\ 0 & I\end{array}\right\} = I.\end{aligned}$$

Hence $Y_1(t,t_0)$, $Z_1(t,t_0)$ and $Y_2(t,t_0)$, $Z_2(t,t_0)$ form a normal pair of solutions of (3.72).

Now we show that **(a)** implies **(c)**. To this end assume $Y_2(t,t_0)$, $Z_2(t,t_0)$ is a solution of (3.72) which is dominant at ∞. Then there is an integer $t_1 > t_0$ such that $Y_2(t,t_0)$ is invertible for $t \geq t_1$. As in Theorem 3.50 define

$$\begin{aligned}Y_0(t) &= Y_2(t,t_0)\mathcal{S}_2(t,t_0) \\ Z_0(t) &= Z_2(t,t_0)\mathcal{S}_2(t,t_0) - Y_2^{*-1}(t,t_0)\end{aligned}$$

for $t \geq t_1$, where

$$\mathcal{S}_2(t,t_0) = \sum_{s=t}^{\infty} D_2(s,t_0)$$

and

$$D_2(t,t_0) = Y_2^{-1}(t+1,t_0)F(t)Y_2^{*-1}(t,t_0)$$

for $t \geq t_1$. By Theorem 3.50 $Y_0(t)$, $Z_0(t)$ is a solution of (3.72) which is recessive at ∞. Since local solutions of symplectic systems always extend uniquely, we may extend the domain of definition of $Y_0(t)$, $Z_0(t)$ to $[a,\infty)$. Since $Y_1(t,t_0)$, $Z_1(t,t_0)$ and $Y_2(t,t_0)$, $Z_2(t,t_0)$ is a normal pair of solutions there are constant matrices C_1, C_2 such that

$$\begin{aligned}Y_0(t) &= Y_1(t,t_0)C_1 + Y_2(t,t_0)C_2 \\ Z_0(t) &= Z_1(t,t_0)C_1 + Z_2(t,t_0)C_2.\end{aligned}$$

In particular, the Normal Basis Theorem implies that

$$C_1 = -\left\{ \begin{matrix} Y_2(t,t_0) & Y_0(t) \\ Z_2(t,t_0) & Z_0(t) \end{matrix} \right\} = I.$$

Then choose t as t_0 for

$$Y_0(t_0) = Y_1(t_0,t_0)C_1 + Y_2(t_0,t_0)C_2 = C_1 = I.$$

Thus $Y_0(t_0)$ is nonsingular and **(c)** holds.

Next we show that **(c)** implies **(b)** and **(a)**. Assume **(c)** holds, that is assume $Y_0(t)$, $Z_0(t)$ is a recessive solution of (3.72) at ∞ with $Y_0(t_0)$ nonsingular. Since $Y_1(t,t_0)$, $Z_1(t,t_0)$ and $Y_2(t,t_0)$, $Z_2(t,t_0)$ is a normal pair of solutions of (3.72) there are constant matrices C_1, C_2 such that

$$Y_0(t) = Y_1(t,t_0)C_1 + Y_2(t,t_0)C_2 \tag{3.85}$$

$$Z_0(t) = Z_1(t,t_0)C_1 + Z_2(t,t_0)C_2 \tag{3.86}$$

where

$$C_1 = -\left\{ \begin{matrix} Y_2(t,t_0) & Y_0(t) \\ Z_2(t,t_0) & Z_0(t) \end{matrix} \right\}.$$

It follows from (3.85) and (3.86) that $Y_0(t_0) = C_1$ and $Z_0(t_0) = C_2$. Since $Y_0(t_0)$ is nonsingular, C_1 is also. Since $Y_0(t)$, $Z_0(t)$ is a recessive solution at ∞ and

$$\left\{ \begin{matrix} Y_0(t) & Y_2(t,t_0) \\ Z_0(t) & Z_2(t,t_0) \end{matrix} \right\} = C_1^*$$

is nonsingular, $Y_2(t,t_0)$ must be nonsingular for large t (because of the definition of recessive) and we use (3.85) to obtain

$$\lim_{t\to\infty} Y_2^{-1}(t,t_0)Y_0(t) = \lim_{t\to\infty}[Y_2^{-1}(t,t_0)Y_1(t,t_0)C_1 + C_2] = 0.$$

Since C_1 is nonsingular it follows that

$$\begin{aligned} \lim_{t\to\infty} Y_2^{-1}(t,t_0)Y_1(t,t_0) &= -C_2C_1^{-1} \\ &= -Z_0(t_0)Y_0^{-1}(t_0) \\ &\equiv \Omega(t_0). \end{aligned}$$

By Exercise 3.46

$$\begin{bmatrix} Y_1(t,t_0) & Y_2(t,t_0) \\ Z_1(t,t_0) & Z_2(t,t_0) \end{bmatrix}$$

is symplectic for all $t \geq a$. Using Theorem 3.4 we get that

$$Y_1(t,t_0)Y_2^*(t,t_0) = Y_2(t,t_0)Y_1^*(t,t_0)$$

for all $t \geq a$. It follows from this that

$$Y_2^{-1}(t,t_0)Y_1(t,t_0)$$

is Hermitian for all sufficiently large t. Hence

$$\Omega(t_0) = \lim_{t\to\infty} Y_2^{-1}(t,t_0)Y_1(t,t_0)$$

is Hermitian (because limits of Hermitian sequences must be Hermitian) and **(b)** holds. From above

$$\begin{bmatrix} Y_2(t,t_0) & Y_0(t) \\ Z_2(t,t_0) & Z_0(t) \end{bmatrix}$$

is nonsingular. Hence by Theorem 3.43, $Y_2(t,t_0)$, $Z_2(t,t_0)$ is a dominant solution at ∞. Thus **(a)** holds.

To prove that **(a)**, **(b)** and **(c)** are equivalent it remains to show that **(b)** implies **(c)**. Assume $Y_2(t,t_0)$ is invertible for all $t \geq t_1$ and

$$\lim_{t\to\infty} Y_2^{-1}(t,t_0)Y_1(t,t_0) = \Omega(t_0)$$

where $\Omega(t_0)$ is a Hermitian matrix.

Set

$$\begin{aligned} Y_0(t) &= Y_1(t,t_0) - Y_2(t,t_0)\Omega(t_0) &(3.87)\\ Z_0(t) &= Z_1(t,t_0) - Z_2(t,t_0)\Omega(t_0). &(3.88) \end{aligned}$$

Then $Y_0(t)$, $Z_0(t)$ is a solution of (3.72) on page 113 and

$$\begin{aligned} Y_0(t_0) &= I \\ Z_0(t_0) &= -\Omega(t_0). \end{aligned}$$

Note that

$$\left\{ \begin{matrix} Y_0(t) & Y_0(t) \\ Z_0(t) & Z_0(t) \end{matrix} \right\} = \left\{ \begin{matrix} Y_0(t_0) & Y_0(t_0) \\ Z_0(t_0) & Z_0(t_0) \end{matrix} \right\}$$
$$= \left\{ \begin{matrix} I & I \\ -\Omega(t_0) & -\Omega(t_0) \end{matrix} \right\} = -\Omega(t_0) + \Omega^*(t_0) = 0$$

since $\Omega(t_0)$ is Hermitian. Hence $Y_0(t)$, $Z_0(t)$ is a prepared solution of (3.72). Also

$$\operatorname{rank}\begin{bmatrix} Y_0(t) \\ Z_0(t) \end{bmatrix} = \operatorname{rank}\begin{bmatrix} Y_0(t_0) \\ Z_0(t_0) \end{bmatrix} = \operatorname{rank}\begin{bmatrix} I \\ -\Omega(t_0) \end{bmatrix} = n.$$

Hence $Y_0(t)$, $Z_0(t)$ is a prepared basis for (3.72).

Next assume $Y(t)$, $Z(t)$ is a solution of (3.72) such that

$$\left\{ \begin{matrix} Y_0(t) & Y(t) \\ Z_0(t) & Z(t) \end{matrix} \right\}$$

is nonsingular. Let C_1 and C_2 be constant matrices such that

$$\begin{aligned} Y(t) &= Y_1(t,t_0)C_1 + Y_2(t,t_0)C_2 \\ Z(t) &= Z_1(t,t_0)C_1 + Z_2(t,t_0)C_2. \end{aligned}$$

Consider the Wronskian

$$\begin{aligned} \left\{ \begin{matrix} Y_0(t) & Y(t) \\ Z_0(t) & Z(t) \end{matrix} \right\} &= \left\{ \begin{matrix} Y_0(t) & Y_1(t,t_0)C_1 + Y_2(t,t_0)C_2 \\ Z_0(t) & Z_1(t,t_0)C_1 + Z_2(t,t_0)C_2 \end{matrix} \right\} \\ &= \left\{ \begin{matrix} Y_0(t) & Y_1(t,t_0) \\ Z_0(t) & Z_1(t,t_0) \end{matrix} \right\} C_1 + \left\{ \begin{matrix} Y_0(t) & Y_2(t,t_0) \\ Z_0(t) & Z_2(t,t_0) \end{matrix} \right\} C_2 \end{aligned}$$

Using (3.87) and (3.88) we get that

$$\begin{aligned} \left\{ \begin{matrix} Y_0(t) & Y(t) \\ Z_0(t) & Z(t) \end{matrix} \right\} &= \left\{ \begin{matrix} -Y_2(t,t_0)\Omega(t_0) & Y_1(t,t_0) \\ -Z_2(t,t_0)\Omega(t_0) & Z_1(t,t_0) \end{matrix} \right\} C_1 \\ &\quad + \left\{ \begin{matrix} Y_1(t,t_0) & Y_2(t,t_0) \\ Z_1(t,t_0) & Z_2(t,t_0) \end{matrix} \right\} C_2 \\ &= \Omega^*(t_0)C_1 + C_2 \\ &= \Omega(t_0)C_1 + C_2. \end{aligned}$$

Hence $\Omega(t_0)C_1 + C_2$ is nonsingular.

Using the above formula for $Y(t)$ to obtain

$$\lim_{t\to\infty} Y_2^{-1}(t,t_0)Y(t) = \lim_{t\to\infty}[Y_2^{-1}(t,t_0)Y_1(t,t_0)C_1 + C_2] = \Omega(t_0)C_1 + C_2$$

which implies that $Y(t)$ is nonsingular for all sufficiently large t.

Next consider

$$
\begin{aligned}
Y^{-1}(t)Y_0(t) &= Y^{-1}(t)[Y_1(t,t_0) - Y_2(t,t_0)\Omega(t_0)] \\
&= [Y_1(t,t_0)C_1 + Y_2(t,t_0)C_2]^{-1}[Y_1(t,t_0) - Y_2(t,t_0)\Omega(t_0)] \\
&= \{Y_2(t,t_0)[Y_2^{-1}(t,t_0)Y_1(t,t_0)C_1 + C_2]\}^{-1}[Y_1(t,t_0) - Y_2(t,t_0)\Omega(t_0)] \\
&= [Y_2^{-1}(t,t_0)Y_1(t,t_0)C_1 + C_2]^{-1}[Y_2^{-1}(t,t_0)Y_1(t,t_0) - \Omega(t_0)] \\
&\to [\Omega(t_0)C_1 + C_2]^{-1}[\Omega(t_0) - \Omega(t_0)] = 0.
\end{aligned}
$$

Thus $Y_0(t)$, $Z_0(t)$ is a recessive solution of (3.72) at ∞.

Finally if **(a)**, **(b)** or **(c)** holds, then they all hold. In the above proof of **(c)** implies **(b)** we proved that

$$Z_0(t_0)Y_0^{-1}(t_0) = -\Omega(t_0)$$

and the proof is complete. □

3.14 ESSENTIAL UNIQUENESS OF RECESSIVE SOLUTIONS

Theorem 3.52 (Essential Uniqueness of Recessive Solutions) *Assume equation* (3.72) *has a solution* $Y_0(t)$, $Z_0(t)$ *which is recessive at* ∞ *and has* $Y_0(t_0)$ *nonsingular at some* t_0. *If* $\hat{Y}_0(t)$, $\hat{Z}_0(t)$ *is also a recessive solution at* ∞ *which has* $\hat{Y}_0(t_0)$ *nonsingular, then there exists a nonsingular constant matrix* K *such that*

$$\hat{Y}_0(t) \equiv Y_0(t)K \tag{3.89}$$

$$\hat{Z}_0(t) \equiv Z_0(t)K. \tag{3.90}$$

Proof: From Theorem 3.43 on page 118 the solutions

$$Y_3(t) \equiv Y_0(t)Y_0^{-1}(t_0), \qquad Z_3(t) \equiv Z_0(t)Y_0^{-1}(t_0)$$

and

$$Y_4(t) \equiv \hat{Y}_0(t)\hat{Y}_0^{-1}(t_0), \qquad Z_4(t) \equiv \hat{Z}_0(t)\hat{Y}_0^{-1}(t_0)$$

are recessive at ∞. Furthermore $Y_3(t_0) = I = Y_4(t_0)$. If we could show that $Z_3(t_0) = Z_4(t_0)$, then by the uniqueness of solutions of **IVPs** it would follow that

$$\begin{aligned} Y_3(t) &\equiv Y_4(t) \\ Z_3(t) &\equiv Z_4(t). \end{aligned}$$

Hence

$$\begin{aligned} Y_0(t)Y_0^{-1}(t_0) &\equiv \hat{Y}_0(t)\hat{Y}_0^{-1}(t_0) \\ Z_0(t)Y_0^{-1}(t_0) &\equiv \hat{Z}_0(t)\hat{Y}_0^{-1}(t_0). \end{aligned}$$

It follows from this that

$$\begin{aligned} Y_0(t) &\equiv \hat{Y}_0(t)K \\ Z_0(t) &\equiv \hat{Z}_0(t)K \end{aligned}$$

where $K = \hat{Y}_0^{-1}(t_0)Y_0(t_0)$ is nonsingular as desired. Hence it remains to show that $Z_3(t_0) = Z_4(t_0)$. To this end let $Y_1(t,t_0)$, $Z_1(t,t_0)$ and $Y_2(t,t_0)$, $Z_2(t,t_0)$ be the normal pair of solutions defined in Theorem 3.51. Let C_1, C_2, $\hat{C}_1$, $\hat{C}_2$ be $n \times n$ constant matrices such that

$$\begin{aligned} Y_3(t) &= Y_1(t,t_0)C_1 + Y_2(t,t_0)C_2 \\ Z_3(t) &= Z_1(t,t_0)C_1 + Z_2(t,t_0)C_2 \end{aligned}$$

and

$$\begin{aligned} Y_4(t) &= Y_1(t,t_0)\hat{C}_1 + Y_2(t,t_0)\hat{C}_2 \\ Z_4(t) &= Z_1(t,t_0)\hat{C}_1 + Z_2(t,t_0)\hat{C}_2. \end{aligned}$$

Let $t = t_0$ in these last four equations for

$$\begin{aligned} C_1 &= I, \quad C_2 = Z_3(t_0) \\ \hat{C}_1 &= I, \quad \hat{C}_2 = Z_4(t_0). \end{aligned}$$

Hence

$$\begin{aligned} Y_3(t) &= Y_1(t,t_0) + Y_2(t,t_0)Z_3(t_0) \\ Y_4(t) &= Y_1(t,t_0) + Y_2(t,t_0)Z_4(t_0). \end{aligned}$$

Note that

$$\left\{ \begin{array}{cc} Y_3(t) & Y_2(t,t_0) \\ Z_3(t) & Z_2(t,t_0) \end{array} \right\} \equiv \left\{ \begin{array}{cc} I & 0 \\ Z_3(t_0) & I \end{array} \right\} = I$$

and

$$\left\{ \begin{array}{cc} Y_4(t) & Y_2(t,t_0) \\ Z_4(t) & Z_2(t,t_0) \end{array} \right\} = \left\{ \begin{array}{cc} I & 0 \\ Z_4(t_0) & I \end{array} \right\} = I$$

are nonsingular. Since $Y_3(t)$, $Z_3(t)$ and $Y_4(t)$, $Z_4(t)$ are recessive at ∞, we have

$$\lim_{t\to\infty} Y_2^{-1}(t,t_0)Y_3(t) = \lim_{t\to\infty}[Y_2^{-1}(t,t_0)Y_1(t,t_0) + Z_3(t_0)] = 0$$

and

$$\lim_{t\to\infty} Y_2^{-1}(t,t_0)Y_4(t) = \lim_{t\to\infty}[Y_2^{-1}(t,t_0)Y_1(t,t_0) + Z_4(t_0)] = 0.$$

It follows that

$$Z_3(t_0) = -\lim_{t\to\infty} Y_2^{-1}(t,t_0)Y_1(t,t_0) = Z_4(t_0)$$

and the proof is complete. □

3.15 ASYMPTOTIC BEHAVIOR OF SOLUTIONS

Theorem 3.53 (Asymptotic Behavior) *Assume* (3.72) *has a solution* $Y_0(t)$, $Z_0(t)$ *which is recessive at* ∞ *with* $Y_0(t_0)$ *nonsingular for some* $t_0 \geq a$. *Let* $Y_2(t,t_0)$, $Z_2(t,t_0)$ *and* $\Omega(t_0)$ *be as in Theorem* 3.51. *If* $Y(t)$, $Z(t)$ *is an* $n \times m$ *matrix solution of* (3.72), *then*

$$\lim_{t\to\infty} Y_2^{-1}(t,t_0)Y(t) = \Omega(t_0)Y(t_0) + Z(t_0). \tag{3.91}$$

Proof: Let C_1 and C_2 be constant $n \times m$ matrices such that

$$\begin{aligned} Y(t) &= Y_1(t,t_0)C_1 + Y_2(t,t_0)C_2 \\ Z(t) &= Z_1(t,t_0)C_1 + Z_2(t,t_0)C_2. \end{aligned}$$

Using the initial conditions at t_0 we get that

$$Y(t) = Y_1(t,t_0)Y(t_0) + Y_2(t,t_0)Z(t_0).$$

It follows that

$$\begin{aligned} \lim_{t\to\infty} Y_2^{-1}(t,t_0)Y(t) &= \lim_{t\to\infty}[Y_2^{-1}(t,t_0)Y_1(t,t_0)Y(t_0) + Z(t_0)] \\ &= \Omega(t_0)Y(t_0) + Z(t_0) \end{aligned}$$

by Theorem 3.51. □

3.16 THE OLVER–REID CONSTRUCTION

We say that (3.72) has the *unique two point property* on $[a, \infty)$ provided that whenever $y_1(t)$, $z_1(t)$ and $y_2(t)$, $z_2(t)$ are solutions of (3.72) with

$$\begin{aligned} y_1(t_1) &= y_2(t_1) \\ y_1(t_2) &= y_2(t_2), \end{aligned}$$

where $a \le t_1 < t_2 < \infty$, it follows that $y_1(t) \equiv y_2(t)$ and $z_1(t) \equiv z_2(t)$ on $[a, \infty)$.

Exercise 3.54 *Show that if* (3.72) *has the unique two point property on* $[a, \infty)$*, then if* $Y_1(t)$*,* $Z_1(t)$ *and* $Y_2(t)$*,* $Z_2(t)$ *are* $n \times m$ *solutions of* (3.72) *with*

$$\begin{aligned} Y_1(t_1) &= Y_2(t_1) \\ Y_1(t_2) &= Y_2(t_2) \end{aligned}$$

where $a \le t_1 < t_2 < \infty$*, then* $Y_1(t) \equiv Y_2(t)$ *and* $Z_1(t) \equiv Z_2(t)$ *on* $[a, \infty)$*.*

Theorem 3.55 *If* (3.72) *has the unique two point property on* $[a, \infty)$*, then there is a unique solution of* (3.72) *satisfying the boundary conditions*

$$Y(t_1) = I, \quad Y(t_2) = 0$$

where $a \le t_1 < t_2 < \infty$*.*

Proof: Let $Y_1(t)$, $Z_1(t)$ be the solution satisfying the initial condition

$$Y_1(t_2) = 0, \quad Z_1(t_2) = I.$$

Assume $Y_1(t_1)$ is singular. Then there is a nonzero vector u such that

$$Y_1(t_1)u = 0.$$

Let

$$y_1(t) = Y_1(t)u, \quad z_1(t) = Z_1(t)u,$$

then $y_1(t)$, $z_1(t)$ is a solution of (3.72) on $[a, \infty)$. Since

$$\begin{aligned} y_1(t_1) &= Y_1(t_1)u &= 0 \\ y_1(t_2) &= Y_1(t_2)u &= 0 \end{aligned}$$

and the trivial solution $y(t) \equiv 0$, $z(t) \equiv 0$, satisfies these boundary conditions, we get by the two point uniqueness assumption that $y_1(t) \equiv 0$, $z_1(t) \equiv 0$. But this implies

$$0 = z_1(t_2) = Z_1(t_2)u = u$$

which is a contradiction. Hence $Y_1(t_1)$ is nonsingular. It follows that

$$\begin{aligned} Y(t) &= Y_1(t)Y_1^{-1}(t_1) \\ Z(t) &= Z_1(t)Y_1^{-1}(t_1) \end{aligned}$$

is the desired solution. □

Theorem 3.56 (Olver–Reid Recessive Solution Construction) *Assume equation* (3.72) *has the unique two point property on* $[t_0, \infty)$ *and suppose there exists a solution* $Y_0(t)$, $Z_0(t)$ *of* (3.72) *which is recessive at* ∞ *and has* $Y_0(t_0)$ *nonsingular. Let* $Y(t,n)$, $Z(t,n)$ *be the solution of the system* (3.72) *which satisfies the boundary conditions*

$$Y(t_0, n) = I, \qquad Y(n,n) = 0. \tag{3.92}$$

Then

$$\begin{bmatrix} Y(t,n) \\ Z(t,n) \end{bmatrix} \to \begin{bmatrix} Y_0(t)Y_0^{-1}(t_0) \\ Z_0(t)Y_0^{-1}(t_0) \end{bmatrix} \tag{3.93}$$

as $n \to \infty$; *that is,* $Y(t,n)$, $Z(t,n)$ *converges to the recessive solution which has initial condition of* $Y(t_0) = I$.

Proof: Let $Y_2(t,t_0)$, $Z_2(t,t_0)$ be the solution of the **IVP** (3.72), $Y_2(t_0,t_0) = 0$, $Z_2(t_0,t_0) = I$. By Theorem 3.51 $Y_2(t,t_0)$, $Z_2(t,t_0)$ is dominant at ∞. It follows that $Y_2(t,t_0)$ is nonsingular for all sufficiently large t. For all n sufficiently large define

$$\begin{aligned} C_0 &= Y_0^{-1}(t_0) \\ C_2(n) &= -Y_2^{-1}(n,t_0)Y_0(n)Y_0^{-1}(t_0) \end{aligned}$$

and set

$$\begin{aligned} Y_3(t,n) &= Y_0(t)C_0 + Y_2(t,t_0)C_2(n) \\ Z_3(t,n) &= Z_0(t)C_0 + Z_2(t,t_0)C_2(n). \end{aligned}$$

Then $Y_3(t,n)$, $Z_3(t,n)$ is a solution of (3.72) satisfying the boundary conditions

$$\begin{aligned} Y_3(t_0,n) &= I \\ Y_3(n,n) &= 0. \end{aligned}$$

By the unique two point property (see Exercise 3.54)

$$Y(t,n) = Y_3(t,n) = Y_0(t)Y_0^{-1}(t_0) - Y_2(t,t_0)C_2(n)$$
$$Z(t,n) = Z_3(t,n) = Z_0(t)Y_0^{-1}(t_0) - Z_2(t,t_0)C_2(n).$$

Note that

$$\left\{ \begin{array}{cc} Y_0(t) & Y_2(t,t_0) \\ Z_0(t) & Z_2(t,t_0) \end{array} \right\} = \left\{ \begin{array}{cc} Y_0(t_0) & 0 \\ Z_0(t_0) & I \end{array} \right\} = Y_0^*(t_0)$$

which is nonsingular. Since $Y_0(t)$, $Z_0(t)$ is recessive at ∞ it follows that

$$\lim_{t\to\infty} Y_2^{-1}(t,t_0)Y_0(t) = 0.$$

This implies that

$$\lim_{n\to\infty} C_2(n) = 0.$$

It follows that

$$\begin{aligned} \lim_{n\to\infty} Y(t,n) &= Y_0(t)Y_0^{-1}(t_0) \\ \lim_{n\to\infty} Z(t,n) &= Z_0(t)Y_0^{-1}(t_0). \end{aligned}$$

By Theorem 3.43, $Y_0(t)Y_0^{-1}(t_0)$, $Z_0(t)Y_0^{-1}(t_0)$ is recessive at ∞. □

For sufficient conditions for the existence of recessive solutions at ∞, we will first develop the discrete variational theory. It turns out that, as in the continuous case, eventual disconjugacy implies existence of a recessive solution with $Y_0(t)$ nonsingular for large t.

3.17 ASSOCIATED RICCATI EQUATIONS

The Olver–Reid construction of Theorem 3.56 on page 134 tells us that under the unique two point property, if

$$X = \begin{bmatrix} Y \\ Z \end{bmatrix}$$

is a solution which is recessive at ∞ and has $Y(t_0)$ nonsingular, then the value of $Z(t_0)Y^{-1}(t_0)$ is somehow of special interest. This same expression arises in the Connection Theorem 3.51 on page 125. Those results motivate the study of the discrete matrix Riccati equations associated with a symplectic system. In the scalar continuous case, the choice of Riccati variable w for

$$y'' + p(t)y' + q(t)y = 0$$

is motivated by the logarithmic derivative, i.e., if $w = \{\ln(y)\}' = y'/y$, then w satisfies a logistics differential equation

$$w' = -[pw + q + w^2].$$

Note that $\ln y = \int w$ gives

$$y = ce^{\int w}.$$

When the coefficients are constant the constant solutions of the Riccati equation reduce to the usual roots of the characterisic equation. Thus the Riccati equation may be regarded as a generalization of the characteristic equation. For formally self adjoint scalar differential equations

$$-(ry')' + p(t)y = 0$$

the substitution $w = z/y$, for z the momentum variable $z = ry'$, gives the associated Riccati equation of

$$w' = p - \frac{1}{r}w^2.$$

Thus the Hamiltonian systems viewpoint suggests a discrete Riccati variable

$$W(t) = Z(t)Y^{-1}(t). \tag{3.94}$$

Suppose that for $t \geq a$ a pair of $2n \times n$ functions $Y(t)$, $Z(t)$ satisfies a system (recall equation (3.72) on page 113)

$$\begin{aligned} Y(t+1) &= E(t)Y(t) + F(t)Z(t) \\ Z(t+1) &= G(t)Y(t) + H(t)Z(t). \end{aligned} \tag{3.95}$$

We assume throughout that the coefficient matrix $M(t)$ of (3.73) on page 113, i.e.,

$$M(t) = \begin{bmatrix} E(t) & F(t) \\ G(t) & H(t) \end{bmatrix}, \tag{3.96}$$

is symplectic and also assume that $E(t)$ is nonsingular when it is useful to do so. These are reasonable assumptions since the discrete linear Hamiltonian systems have these properties. We will see at the end of Chapter 4 that when variable step sizes are introduced, that if the step size is sufficiently small, then the discrete Jacobi equation can be written as such a symplectic system.

In the literature of calculus of variations, Riccati variables have often been labeled with the letter w in the scalar case and W in the matrix case. In control and filtering literature, the letter P is frequently used for solutions of matrix Riccati equations. We will primarily use $W(t) = Z(t)Y^{-1}(t)$ in later chapters, but here we will list some discrete matrix Riccati equations which arise in filtering, control, and noncooperative game theory. Thus, for now, we define a Riccati variable $P(t)$ by

$$P(t) = Z(t)Y^{-1}(t). \tag{3.97}$$

Assume $Y(t)$ is nonsingular for $t \geq t_0 \geq a$. Then

$$\begin{aligned} P(t+1) &= Z(t+1)Y^{-1}(t+1) \\ &= (G(t)Y(t) + H(t)Z(t))(E(t)Y(t) + F(t)Z(t))^{-1} \\ &= \left[G(t) + H(t)Z(t)Y^{-1}(t)\right] Y(t)(E(t)Y(t) + F(t)Z(t))^{-1} \end{aligned}$$

and we have the discrete Riccati equation , labeled by **DRE** ,

$$P(t+1) = (G(t) + H(t)P(t))(E(t) + F(t)P(t))^{-1}. \tag{3.98}$$

The assumptions that $M(t)$ is symplectic and $E(t)$ is nonsingular allow us to solve for H in the identity $E^*H - G^*F = I$ to obtain

$$H = E^{*-1}[I + G^*F] = E^{*-1} + E^{*-1}G^*F = E^{*-1} + GE^{-1}F.$$

Consequently

$$\begin{aligned} &(G + HP)(E + FP)^{-1} \\ &\quad = \left[G + (E^{*-1} + GE^{-1}F)P\right](E + FP)^{-1} \\ &\quad = \left[GE^{-1}(E + FP) + E^{*-1}P\right](E + FP)^{-1} \\ &\quad = GE^{-1} + E^{*-1}P(E + FP)^{-1} \end{aligned}$$

and we have the statement **(i)** in following theorem.

Theorem 3.57 (Associated Discrete Riccati Equations) *Assume $M(t)$ is symplectic and $E(t)$ is nonsingular for $t \geq a$.*

(i) *Suppose $Y(t)$, $Z(t)$ satisifies system (3.95) for $t \geq a$ with $Y(t)$ nonsingular for $t \geq t_0 \geq a$ and a Riccati variable P is defined by $P(t) = Z(t)Y^{-1}(t)$ for $t \geq t_0$. Then $E(t) + F(t)P(t)$ is nonsingular for $t \geq t_0$ and P satisfies the discrete matrix Riccati equation*

$$P(t+1) = G(t)E^{-1}(t) + E^{*-1}(t)P(t)\left[E(t) + F(t)P(t)\right]^{-1} \tag{3.99}$$

for $t \geq t_0$, with $G(t)E^{-1}(t)$ and $E^{-1}(t)F(t)$ Hermitian. Also, $P(t)$ is Hermitian if and only if $Y(t)$, $Z(t)$ is prepared.

(ii) *Conversely, if $P(t)$ is an $n \times n$ matrix function defined for $t \geq t_0$ such that $E(t) + F(t)P(t)$ is nonsingular on $[t_0, \infty)$, the discrete matrix Riccati equation (3.99) holds for t in $[t_0, \infty)$, and $Y(t_0)$ is a nonsingular matrix, then $Y(t)$ recursively defined on (t_0, ∞) by*

$$Y(t+1) = [E(t) + F(t)P(t)]Y(t) \tag{3.100}$$

and $Z(t)$ is defined on $[t_0, \infty)$ by $Z(t) \equiv P(t)Y(t)$, imply that $Y(t), Z(t)$ satisfies the discrete system (3.95) for $t \geq t_0$ and $Y(t)$ is nonsingular on $[t_0, \infty)$.

Exercise 3.58 *Prove part* **(ii)** *of the above theorem.*

Remark 3.59 *The discrete matrix Riccati equation (3.99) may be written as*

$$P(t+1) = G(t)E^{-1}(t) + E^{*-1}(t)P(t)\left[I + E^{-1}(t)F(t)P(t)\right]^{-1} E^{-1}(t). \tag{3.101}$$

Exercise 3.60 *Assume $M(t)$ is symplectic and $E(t)$ is nonsingular for $t \geq a$. Show the following results.*

(i) *Suppose $Y(t)$, $Z(t)$ satisifies system (3.95) for $t \geq a$ with $Z(t)$ nonsingular for $t \geq t_0 \geq a$ and a dual Riccati variable Q is defined by $Q(t) = -Y(t)Z^{-1}(t)$ for $t \geq t_0$. Then $G(t)E^{-1}(t)Q(t+1) + I$ is nonsingular for $t \geq t_0$ and Q satisfies the dual discrete matrix Riccati equation*

$$Q(t) = E^{-1}(t)F(t) + E^{-1}(t)Q(t+1)\left[G(t)E^{-1}(t)Q(t+1) + I\right]^{-1} E^{*-1}(t) \tag{3.102}$$

for $t \geq t_0$, with $E^{-1}(t)F(t)$ and $G(t)E^{-1}(t)$ Hermitian. Also, $Q(t)$ is Hermitian if and only if $Y(t)$, $Z(t)$ is prepared.

(ii) *Conversely, if $Q(t)$ is an $n \times n$ matrix function defined for $t \geq t_0$ such that $G(t)E^{-1}(t)Q(t+1) + I$ is nonsingular for t on $[t_0, \infty)$, the dual discrete matrix Riccati equation (3.102) holds for t in $[t_0, \infty)$, and $Z(t_0)$ is a nonsingular matrix, then Z implicitly defined by the recurrence*

$$Z(t) = E^*(t)\left[G(t)E^{-1}(t)Q(t+1) + I\right] Z(t+1) \tag{3.103}$$

for $t \geq t_0$ and the function Y defined on $[t_0, \infty)$ by $Y(t) \equiv -Q(t)Z(t)$, imply that $Y(t), Z(t)$ satisfies the discrete system (3.95) *for $t \geq t_0$ and $Z(t)$ is nonsingular on $[t_0, \infty)$.*

A major difference between the Riccati equations for differential systems (which have right hand side which is polynomial) and these discrete Riccati equations (arising from symplectic systems) is that these discrete Riccati equations are intrinsically not polynomial since they involve linear fractional forms. These linear fractional forms give rise to continued fraction representations of discrete Riccati equations.

Discrete matrix Riccati equations have arisen in several equivalent forms, depending upon the application. For the remainder of this chapter, we will assume that the coefficient matrices have real entries. We do so from variational motivation and because the examples arising in control and filtering usually have real coefficient matrices. In order to provide some unification of these types of equations, compare the coefficients in equation (3.99) with the coefficients in equation (3.20) on page 82 where $E = [I - A]^{-1} \equiv \hat{A}$, $GE^{-1} = C$, and $E^{-1}F = B$, which arise from discrete linear Hamiltonian systems. We introduce a generic discrete Riccati equation, which we will call the *discrete variational Riccati equation,* labeled by **DVRE** ,

$$P(t+1) = \hat{C}(t) + \hat{A}^T(t)P(t)\left[I + \hat{B}(t)P(t)\right]^{-1}\hat{A}(t) \tag{3.104}$$

and a generic dual discrete Riccati equation

$$Q(t) = \hat{B}(t) + \hat{A}(t)Q(t+1)\left[I + \hat{C}(t)Q(t+1)\right]^{-1}\hat{A}^T(t) \tag{3.105}$$

where $\hat{B}$ and $\hat{C}$ are symmetric, $\hat{A}$ is nonsingular, and all matrices are square and of order n.

In order to see that the DVRE, i.e., equation (3.104), contains many forms given in the literature, introduce a generic linear fractional function defined by

$$\mathcal{F}[P] = \hat{C} + \hat{A}^T P(I + \hat{B}P)^{-1}\hat{A}. \tag{3.106}$$

The above discrete Riccati equation (3.104) now takes the form

$$P(t+1) = \mathcal{F}[P(t)] \tag{3.107}$$

Note that we still allow the coefficients to be dependent upon t. If the coefficients are constant and P is constant, then the equation, denoted by **DARE** ,

$$P = \mathcal{F}[P] \tag{3.108}$$

is called the *discrete algebraic Riccati equation.* Under further restrictions on the coefficient matrices, but still allowing variable coefficients, we can show equivalent forms of the polynomial $\mathcal{F}[P]$ and hence obtain several other forms of discrete Riccati equations. The remaining results of this section are not necessary to reading subsequent sections, but provide some unification for various discrete Riccati equations arising in applications.

Proposition 3.61 (Some Matrix Identities) *Given that $\hat{B}$ and $\hat{C}$ are symmetric and $\hat{A}$ is nonsingular.*

(a) *Suppose $\hat{B}$ is nonsingular and $R = \hat{B}^{-1}$. Then $I + \hat{B}P$ is nonsingular if and only if $R + P$ is nonsingular and*

$$\mathcal{F}[P] = \hat{C} + \hat{A}^T P[R + P]^{-1} R\hat{A}. \tag{3.109}$$

(b) *Suppose $\hat{B}$ is nonsingular and $R = \hat{B}^{-1}$. Then the function $\mathcal{F}[P]$ of* **(a)** *may be written as*

$$\mathcal{F}[P] = \hat{C} + \hat{A}^T P\hat{A} - \hat{A}^T P[R + P]^{-1} P\hat{A}. \tag{3.110}$$

(c) *Suppose R is an $m \times m$ nonsingular symmetric matrix, B is $n \times m$, and $\hat{B} = BR^{-1}B^T$. Then $I + BR^{-1}B^T P$ is nonsingular if and only if $R + B^T PB$ is nonsingular and*

$$\mathcal{F}[P] = \hat{C} + \hat{A}^T P\hat{A} - \hat{A}^T PB[R + B^T PB]^{-1} B^T P\hat{A}. \tag{3.111}$$

Note that if $m = n$ and $B = I_n$, then this latter form is equivalent to that of part **(b)**.

(d) *Suppose R is an $m \times m$ nonsingular symmetric matrix. For $A_{n\times n}$, $B_{n\times m}$, $L_{n\times m}$ and symmetric $C_{n\times n}$, set*

$$\begin{aligned} \hat{A} &= A - BR^{-1}L^T \\ \hat{B} &= BR^{-1}B^T \\ \hat{C} &= C - LR^{-1}L^T. \end{aligned} \tag{3.112}$$

Then $I_n + BR^{-1}B^TP$ is nonsingular if and only if $R + B^TPB$ is nonsingular and

$$\mathcal{F}[P] = C + A^TPA - (A^TPB + L)[R + B^TPB]^{-1}(L^T + B^TPA). \quad (3.113)$$

Proof: **(a)** The identity

$$I + \hat{B}P = I + R^{-1}P = R^{-1}(R + P) = \hat{B}(R + P)$$

implies that $I + \hat{B}P$ is nonsingular iff $(R + P)$ is nonsingular and

$$P(I + \hat{B}P)^{-1} = P[\hat{B}(R + P)]^{-1} = P(R + P)^{-1}R. \quad (3.114)$$

(b) Observe the identity

$$P(R + P)^{-1}R = P(R + P)^{-1}[(R + P) - P] = P - P(R + P)^{-1}P. \quad (3.115)$$

(c) Verify the following poorly motivated, but useful, identity

$$BR^{-1}(R + B^TPB) = B + BR^{-1}B^TPB = (I + BR^{-1}B^TP)B. \quad (3.116)$$

Suppose that $I + BR^{-1}B^TP$ is nonsingular and v is a vector such that

$$(R + B^TPB)v = 0.$$

Then

$$0 = BR^{-1}(R + B^TPB)v = (I + BR^{-1}B^TP)Bv$$

and $I + BR^{-1}B^TP$ nonsingular imply $Bv = 0$ which implies from the assumed property of v that

$$0 = (R + B^TPB)v = Rv + B^TPBv = Rv$$

and $v = 0$. Thus $R + B^TPB$ is nonsingular. Conversely, suppose $R + B^TPB$ is nonsingular and w is a column vector such that $w^T(I + BR^{-1}B^TP) = 0$. From identity (3.116) we have $0 = w^TBR^{-1}(R + B^TPB)$ with the latter factor nonsingular. Hence $w^TBR^{-1} = 0$ which gives $0 = w^T + w^TBR^{-1}B^TP = w^T$ and we conclude that the matrix $I + BR^{-1}B^TP$ is nonsingular. Thus we have shown the desired equivalence on nonsingularities. In order to see the alternative form of $\mathcal{F}[P]$, observe that

$$\begin{aligned}&P(I+BR^{-1}B^TP)^{-1}\\&\quad= P(I+BR^{-1}B^TP)^{-1}[I+BR^{-1}B^TP-BR^{-1}B^TP]\\&\quad= P-P(I+BR^{-1}B^TP)^{-1}BR^{-1}B^TP.\end{aligned}\tag{3.117}$$

But from (3.116) we have the identity

$$(I+BR^{-1}B^TP)^{-1}BR^{-1}=B(R+B^TPB)^{-1}\tag{3.118}$$

and from (3.117)

$$P(I+BR^{-1}B^TP)^{-1}=P-PB(R+B^TPB)^{-1}B^TP\tag{3.119}$$

and the alternative form holds. Since these steps are reversible, part **(c)** is true.

(d) Set $S=(R+B^TPB)^{-1}$ and $T=B^TP\hat{A}$. Then

$$(R+B^TPB)S=I=S(R+B^TPB)$$

gives $B^TPBS=I-RS$ and $SB^TPB=I-SR$. We focus on the last term in (3.111), i.e., we write

$$\mathcal{F}[P]=\hat{C}+\hat{A}^TP\hat{A}+U,$$

where

$$U=-\hat{A}^TPBST=-(A^T-LR^{-1}B^T)PBST=-[A^TPBS-LR^{-1}(B^TPBS)]T.$$

But $B^TPBS=I-RS$ gives the form

$$U=-[(A^TPB+L)S-LR^{-1}]T=-(A^TPB+L)ST+LR^{-1}T.$$

However,

$$\begin{aligned}ST&=SB^TP(A-BR^{-1}L^T)=SB^TPA-(SB^TPB)R^{-1}L^T\\&=SB^TPA-(I-SR)R^{-1}L^T=S(B^TPA+L^T)-R^{-1}L^T,\end{aligned}$$

$$U=-(A^TPB+L)\left[S(B^TPA+L^T)-R^{-1}L^T\right]+LR^{-1}T,$$

and

$$\mathcal{F}[P]=C-LR^{-1}L^T+(A-BR^{-1}L^T)^TP(A-BR^{-1}L^T)+U,$$

which expands and simplifies to the form of $\mathcal{F}[P]$ given in **(d)**. □

Exercise 3.62 (Riccati Difference Equations) *Suppose that* $\hat{A} \equiv I$, $\hat{B}$ *is nonsingular,* $\hat{B}$ *is symmetric,* $\hat{C}$ *is symmetric, and* $R = \hat{B}^{-1}$. *Use part* **(c)** *of Proposition* 3.61 *to show that the discrete Riccati equation*

$$P(t+1) = \hat{C}(t) + P(t)(I + \hat{B}(t)P(t))^{-1}$$

may be written as a difference equation

$$\Delta P(t) = \hat{C}(t) - P(t)[R(t) + P(t)]^{-1}P(t).$$

Now we show that it is always possible to transform $\hat{A}$ to the identity matrix. Let us back up to the symplectic system (3.95) (on page 136) with symplectic coefficient matrix $M(t)$ of (3.96) having nonsingular $E(t)$. We wish to determine a nonsingular matrix $D(t)$ such that the symplectic transformation

$$\begin{bmatrix} \hat{Y}(t) \\ \hat{Z}(t) \end{bmatrix} = \begin{bmatrix} D(t) & 0 \\ 0 & D^{*-1}(t) \end{bmatrix} \begin{bmatrix} Y(t) \\ Z(t) \end{bmatrix} \tag{3.120}$$

takes us to a system $\hat{X}(t+1) = \hat{M}(t, t+1)\hat{X}(t)$ with $\hat{E}(t) \equiv I$.

Exercise 3.63 (Transformation to $\hat{E} = I$) *Let* $D(t)$ *be the solution of the initial value problem*

$$D(t_0) = I \qquad \text{and} \qquad D(t+1) = D(t)E(t).$$

Then the transformation (3.120) *takes the symplectic system* (3.95) *to a system* $\hat{X}(t+1) = \hat{M}(t, t+1)\hat{X}(t)$ *with*

$$\hat{M}(t, t+1) = \begin{bmatrix} I & \hat{F}(t, t+1) \\ \hat{G}(t, t+1) & I + \hat{G}(t, t+1)\hat{F}(t, t+1) \end{bmatrix}, \tag{3.121}$$

where

$$\hat{F}(t, t+1) = D(t+1)F(t)D^*(t), \qquad \hat{G}(t, t+1) = D^{*-1}(t+1)G(t)D^{-1}(t). \tag{3.122}$$

This transformation makes it possible to study systems with $E = I$, or for the generic Riccati equations, to have $\hat{A} = I$, or the discrete Hamiltonian systems can just as well have $A = 0$. But this is not without some additional complications. First of all, the coefficients are now functions of two variables (which may be more natural from a variational viewpoint). More important, however, is that a constant coefficient system is transformed to a system with variable coefficients. This complicates autonomous systems, exponential dichotomies, and the numerical algorithm of Section 3.20.

Notice that this transformation shows the utility of a more natural class of symplectic systems with coefficient matrix M depending upon t and $t+1$. This example motivates the next section on transformations of symplectic systems.

3.18 TRANSFORMATIONS OF SYMPLECTIC SYSTEMS

Example 3.63 leads us to consider systems of the form

$$X(t+1) = M(t,t+1)X(t) \tag{3.123}$$

where $M(t,t+1)$ is symplectic for each $t \in [a,b]$.

Suppose that $N(t)$ is nonsingular on $[a,b+1]$ and variables $X(t)$ and $\hat{X}(t)$ are related by

$$X(t) = N(t)\hat{X}(t) \tag{3.124}$$

for $t \in [a,b+1]$. This leads to a system

$$\hat{X}(t+1) = \hat{M}(t,t+1)\hat{X}(t) \tag{3.125}$$

for $t \in [a,b]$, where

$$\hat{M}(t,t+1) = N^{-1}(t+1)M(t,t+1)N(t). \tag{3.126}$$

Theorem 3.64 (Transformations of Symplectic Systems) *Suppose the matrix coefficient $M(t,t+1)$ is symplectic and $N(t)$ is nonsingular. Then $X(t)$ satisfies* (3.123) *for $t \in [a,b]$ if and only if $\hat{X}(t)$ satisfies* (3.125) *for $t \in [a,b]$. Furthermore,*

(a) *if $N(t)$ is symplectic on $[a,b+1]$, then the new coefficient matrix $\hat{M}(t,t+1)$ is symplectic for $t \in [a,b]$;*

(b) *if M and N are constant, then $\hat{M}$ is similar to M with $\hat{M} = N^{-1}MN$.*

Proof: Use (3.124) with t replaced by $t+1$ and (3.123) to obtain two ways of writing $X(t+1)$ for

$$N(t+1)\hat{X}(t+1) = M(t,t+1)X(t) = M(t,t+1)N(t)\hat{X}(t)$$

which gives equation (3.125) with the form of $\hat{M}$ in (3.126). □

3.19 DISCRETE FLOQUET THEORY

Exercise 3.65 (Discrete Floquet Theory) *Suppose that $M(t)$ is square, (not necessarily symplectic), nonsingular, and periodic with period $T > 1$. Let Q be the matrix defined by*

$$Q = M(T-1)M(T-2)\cdots M(0). \tag{3.127}$$

If $X(t)$ is a solution of $X(t+1) = M(t)X(t)$, then for each integer k, we have

$$X(kT+j) = M(j-1)M(j-2)\cdots M(0)Q^k X(0), \qquad j = 1,2,\ldots,T-1,$$

and $X(kT) = Q^k X(0)$. This suggests a transformation

$$X(kT+j) = N(kT+j)\hat{X}(k), \tag{3.128}$$

where N is periodic of period T with

$$N(j) = \begin{cases} I, & \text{for } j = 0, \\ M(j-1)M(j-2)\cdots M(0), & j = 1,\ldots T-1 \end{cases} \tag{3.129}$$

and $\hat{X}(k) = Q^k X(0)$. Thus $\hat{X}$ satisfies the constant coefficient **IVP**

$$\hat{X}(k+1) = Q\hat{X}(k) \qquad \text{and} \qquad \hat{X}(0) = X(0).$$

(a) *Verify that*

$$\hat{M}(k) \equiv N^{-1}(kT+j+1)M(kT+j)N(kT+j) = Q.$$

(b) *Show that if μ is an eigenvalue of Q with eigenvector v, then the column vector solution x of $x(t+1) = M(t)x(t)$ with $x(0) = v$ is given by*

$$x(kT+j) = \mu^k N(j)v, \qquad \text{for} \quad j = 0,\ldots,T-1.$$

Thus $x(kT+j) = \mu^k x(j)$ and $x(t+T) = \mu x(t)$.

(c) *Verify that if $M(t)$ is of period T, then there exists a periodic matrix function $N(j)$ of period T and a constant matrix Q such that each solution $X(t)$ has representation*

$$X(kT+j) = N(j)Q^k X(0), \tag{3.130}$$

for integer k and for $j = 0, \ldots, T-1$. Furthermore N and Q are independent of X. Note that this reduces to the constant coefficient result of Exercise 3.7 *on page* 78 *if $T = 1$.*

After the next section, we will use this result to show that discrete Riccati equations (3.98) have periodic solutions when they arise from periodic symplectic systems.

3.20 EXPONENTIAL DICHOTOMIES

Proposition 3.66 (Eigenvalues of Symplectic Matrices) *Suppose that M is a constant symplectic matrix with real entries. Let J be the $2n \times 2n$ block matrix used at the start to this chapter. Then*

(i) *M is nonsingular and hence 0 is not an eigenvalue of M.*

(ii) *If λ is an eigenvalue of M with eigenvector v, then $\bar{\lambda}$ is an eigenvalue of M with eigenvector $\bar{v}$.*

(iii) *If λ is an eigenvalue of M with eigenvector v, then λ^{-1} is an eigenvalue of M with left eigenvector $w^T = (Jv)^T$.*

(iv) *If M has no eigenvalues on the unit circle, then $M_{2n\times 2n}$ has n eigenvalues outside the unit circle and n eigenvalues inside the unit circle.*

Proof: Since symplectic matrices are nonsingular, they must have nonzero eigenvalues. From $Mv = \lambda v$ we have $\lambda^{-1}v = M^{-1}v$. But, since M is symplectic and has real entries,

$$\lambda^{-1}v = M^{-1}v = J^T M^T J v = J^T M^T (Jv).$$

Use the fact that $J^T = J^{-1}$ to pull the first factor on the right side to the left side. Then take the transpose of both sides for

$$\lambda^{-1}(Jv)^T = (Jv)^T M$$

and $w^T = (Jv)^T$ is a left eigenvector for λ^{-1}. □

Theorem 3.67 (Exponential Dichotomies) *Suppose that $M_{2n\times 2n}$ is a constant symplectic matrix with real entries which has no eigenvalues on the unit circle and has distinct eigenvalues. Let Λ_+ be a diagonal matrix of eigenvalues of M which lie outside the unit circle and let Λ_- be a diagonal matrix of eigenvalues of M inside the unit circle. Then there exists a nonsingular matrix $N_{2n\times 2n}$ whose columns are eigenvectors such that*

$$\hat{M} = N^{-1}MN = \begin{bmatrix} \Lambda_+ & 0 \\ 0 & \Lambda_- \end{bmatrix} \equiv \Lambda, \quad say, \quad N \equiv \begin{bmatrix} U_+ & U_- \\ V_+ & V_- \end{bmatrix}. \tag{3.131}$$

Assume nonsingular U_+ and U_-. Let $P_+ = V_+(U_+)^{-1}$ and $P_- = V_-(U_-)^{-1}$, which are constant real symmetric solutions of the associated matrix Riccati equation which is equivalent to the discrete algebraic Riccati equation **(DARE)**

$$PE + PFP - G - HP = 0. \tag{3.132}$$

If $P_+ - P_-$ is nonsingular, then the solutions defined for integer t by

$$X_+(t) = \begin{bmatrix} U_+ \\ V_+ \end{bmatrix} \Lambda_+^t, \qquad X_-(t) = \begin{bmatrix} U_- \\ V_- \end{bmatrix} \Lambda_-^t, \tag{3.133}$$

constitute a fundamental set of solutions with the following properties:

(a) *X_+ is dominant at ∞ and recessive at $-\infty$,*

(b) *X_- is recessive at ∞ and dominant at $-\infty$.*

Note that we could assume that $\Lambda_- = \Lambda_+^{*-1}$, which would make $\hat{M}$ symplectic, but this may cause extra computational work because standard eigenvalue packages may not provide the eigenvalues in that order.

Exercise 3.68 (Independence of ordering of eigenvalues) *Show that P_+ and P_- are independent of the order of eigenvalues in Λ_+ and Λ_- and choice of eigenvectors. Note that the assumption of distinct eigenvalues makes the eigenspaces one dimensional. Start by replacing Λ_+ in*

$$M \begin{bmatrix} U_+ \\ V_+ \end{bmatrix} = \begin{bmatrix} U_+ \\ V_+ \end{bmatrix} \Lambda_+ \tag{3.134}$$

with $\mathcal{P}_{ij}\tilde{\Lambda}_+\mathcal{P}_{ij}$ where $\mathcal{P}_{ij}$ is obtained from I_n by interchanging the ith and jth columns. Also, this equation may be postmultiplied by a diagonal matrix $D_{n\times n}$ with nonzero diagonal entries in order to rescale the eigenvectors. Note that diagonal matrices commute.

Proof of Theorem 3.67: The bracket functions are

$$\{X_+; X_+\} = \{X_+; X_+\}(t) = (\Lambda_+^*)^t (U_+^* V_+ - V_+^* U_+) \Lambda_+^t \to 0,$$

as $t \to -\infty$, and

$$\{X_-; X_-\} = \{X_-; X_-\}(t) = (\Lambda_-^*)^t (U_-^* V_- - V_-^* U_-) \Lambda_-^t \to 0,$$

as $t \to \infty$. But these bracket functions are constant, hence they must be 0 and each of X_+ and X_- is a prepared solution. Evaluation of these bracket functions at $t = 0$ implies that P_+ and P_- are Hermitian. Also,

$$\{X_+; X_-\} = \{X_+; X_-\}(0) = U_+^* V_- - V_+^* U_- = U_+^* (P_- - P_+) U_-$$

is nonsingular and we have a fundamental set of solutions. We show that X_- is recessive at ∞. Suppose that X is a $2n \times n$ solution such that $\{X; X_-\}$ nonsingular. Use the normal pair $X_1 = X_-$, $X_2 = X_+ \{X_-; X_+\}$ for the representation

$$X = -X_- \{X_2; X\} + X_2 \{X_-; X\}.$$

Thus,

$$Y(t) = -U_- \Lambda_-^t + \{X_2; X\} + U_+ \Lambda_+^t \{X_-; X_+\}^{-1} \{X_-; X\}$$

and

$$\begin{aligned}
&\Lambda_+^{-t} U_+^{-1} Y(t) \\
&\quad = -\Lambda_+^{-t} U_+^{-1} U_- \Lambda_-^t + \{X_2; X\} + \{X_-; X_+\}^{-1} \{X_-; X\} \\
&\quad \to \{X_-; X_+\}^{-1} \{X_-; X\}
\end{aligned}$$

as $t \to \infty$ which is nonsingular. Thus $Y(t)$ is nonsingular for large t. Also,

$$Y^{-1}(t) Y_-(t) = [\Lambda_+^{-t} U_+^{-1} Y(t)]^{-1} [\Lambda_+^{-t} U_+^{-1} U_- \Lambda_-^t] \to 0$$

and $X(t)$ is recessive at ∞.

Since $X_+(t)$ is prepared and linearly independent of $X_-(t)$, then from Theorem 3.43 (on page 118) the solution $X_+(t)$ is dominant at ∞. The fact that P_- has real entries comes from the Connection Theorem (Theorem 3.51 on page 125) since X_- is a recessive solution with $Y_-(0) = U_-$ nonsingular. But part **(b)** of that theorem is in real arithmetic. Thus $\Omega(0) = -P_-$ has real entries.

The dual results at $-\infty$ can be seen by time reversal. Since the coefficient matrix M is constant we have

$$X(t) = MX(t-1) \qquad \text{and} \qquad X(-t) = MX(-t-1).$$

Introduce $\tilde{X}$ by time reversal $\tilde{X}(t) = X(-t)$. Then $\tilde{X}(t+1) = \tilde{M}\tilde{X}(t)$ for $\tilde{M} = M^{-1}$. At this point of the proof, we assume the eigenvalues are ordered so that $\Lambda_- = \Lambda_+^{-1}$. Then

$$N^{-1}MN = \Lambda \qquad \text{iff} \qquad N^{-1}\tilde{M}N = \Lambda^{-1} = \begin{bmatrix} \Lambda_- & 0 \\ 0 & \Lambda_+ \end{bmatrix}$$

and we have solutions

$$\tilde{X}_+(t) = \begin{bmatrix} U_+ \\ V_+ \end{bmatrix} \Lambda_-^t, \qquad \tilde{X}_-(t) = \begin{bmatrix} U_- \\ V_- \end{bmatrix} \Lambda_+^t, \tag{3.135}$$

which are respectively recessive at ∞ and dominant at ∞ and P_+ is real symmetric. Thus $X_+(t) = \tilde{X}_+(-t)$ is recessive at $-\infty$ and $X_-(t) = \tilde{X}_-(-t)$ is dominant at $-\infty$. □

Exercise 3.69 (A Numerical Example) *Numerically verify the hypotheses of Theorem* 3.67 *for the following example associated with the scalar fourth order equation*

$$\Delta^4 y_{n-2} + y_n = 0$$

where (see Example 3.14 *on page* 85*)*

$$A = \begin{bmatrix} 0 & 1 \\ 0 & 0 \end{bmatrix}, \quad B = \begin{bmatrix} 0 & 0 \\ 0 & 1 \end{bmatrix}, \quad C = \begin{bmatrix} 1 & 0 \\ 0 & 0 \end{bmatrix},$$

and

$$E = \begin{bmatrix} 1 & 1 \\ 0 & 1 \end{bmatrix}, \quad F = \begin{bmatrix} 0 & 1 \\ 0 & 1 \end{bmatrix}, \quad G = \begin{bmatrix} 1 & 1 \\ 0 & 0 \end{bmatrix}, \quad H = \begin{bmatrix} 1 & 1 \\ -1 & 1 \end{bmatrix}.$$

Verify that all eigenvalues of M are determined by $\lambda \approx 1.62 + 1.30i$ by complex conjugates and reciprocals. Also verify numerically that N has complex entries, but P_+ is real symmetric and has positive eigenvalues and P_- is real symmetric with negative eigenvalues. Thus $P_+ > 0 > P_-$. (For example, use Matlab's $[N, D] =$ eig (M) to produce N and $\Lambda = D$.)

3.21 PERIODIC SYMPLECTIC SYSTEMS

Exercise 3.70 (Periodic Symplectic Systems) *Suppose that $M(t)$ is real symplectic, periodic with period $T > 1$, and Q the matrix of Exercise* 3.65 *has* *no eigenvalues on the unit circle and has distinct eigenvalues. Then for* $N(j)$ *the matrix of Exercise* 3.65 *partitioned by*

$$N(j) = \begin{bmatrix} N_{11}(j) & N_{12}(j) \\ N_{21}(j) & N_{22}(j) \end{bmatrix},$$

let Q *play the role of* M, *let* R *play the role of* N, *and let* k *play the role of* t *in Theorem* 3.67. *Assume that* U_- *and* U_+ *are nonsingular. Show that we have solutions*

$$X_+(kT+j) = N(j) \begin{bmatrix} U_+ \\ V_+ \end{bmatrix} \Lambda_+^k, \qquad X_-(kT+j) = N(j) \begin{bmatrix} U_- \\ V_- \end{bmatrix} \Lambda_-^k \tag{3.136}$$

and periodic solutions of the discrete Riccati equation (3.98) *given by*

$$\begin{aligned} P_+(kT+j) &= Z_+(kT+j)Y_+^{-1}(kT+j) \\ &= [N_{21}(j)U_+ + N_{22}(j)V_+][N_{11}(j)U_+ + N_{12}(j)V_+]^{-1} \end{aligned}$$

and

$$\begin{aligned} P_-(kT+j) &= Z_-(kT+j)Y_-^{-1}(kT+j) \\ &= [N_{21}(j)U_- + N_{22}(j)V_-][N_{11}(j)U_- + N_{12}(j)V_-]^{-1} \end{aligned}$$

if all the inverses exist. These can also be written as

$$P_+(kT+j) = [N_{21}(j) + N_{22}(j)\hat{P}_+][N_{11}(j) + N_{12}(j)\hat{P}_+]^{-1} \tag{3.137}$$

and

$$P_-(kT+j) = [N_{21}(j) + N_{22}(j)\hat{P}_-][N_{11}(j) + N_{12}(j)\hat{P}_-]^{-1} \tag{3.138}$$

where $\hat{P}_+$ *and* $\hat{P}_-$ *are for the corresponding constant coefficient equation of the Floquet transformation.*

3.22 NOTES

Second order systems of the form of equation (3.16) on page 80, three term recurrences of the form of equation (3.44) on page 92 and symplectic systems of the form (3.15) on page 80 have been studied by ourselves with valuable assistance from coauthors Clark, Heifetz, Hooker, Patula, Peil, and Ridenhour in [2, 4, 5, 6, 7, 8, 11, 12, 13, 14, 15, 16, 17, 18, 19, 20, 116, 117, 124, 120, 122, 113]. Closely related work is that of Chen and Erbe [41], Erbe and Yan [54,

55, 56]. In the earlier papers, strong sign conditions are assumed in a manner parallel to the continuous theory, but we now realize that such sign conditions are not necessary. Bohner [27] has relaxed the assumption of nonsingularity on our B matrix, so as to include higher order scalar equations as part of a general Reid Roundabout Theorem.

For references in the equivalence of higher order systems with special Hamiltonian systems discussed in Example 3.14, see Ahlbrandt [6, Equation 99] with $P_j(t) = (-1)^{n-j} r_j(t)$, Erbe [54, 55], and Ahlbrandt and Peterson [20, Example 2]. A special case (the two term equation) of $\ell_{2n}u(t) = 0$ is studied by Peil and Peterson [113].

Discrete matrix Riccati equations of the form DVRE in (3.104) on page 139 were used by Iglesias and Glover [83, Lemma 2.1], Nikoukhah *et al* [107, equation (3.13)], Stoorvogel [144, Lemma 2.2], as the predictor equation (5.17) by Yaesh and Shaked [157], by Tsai *et al* [146, eq. (5)], in reverse form by Yaesh and Shaked [156, eq. 10], and by Ionescu and Weiss [84] for the case of $\hat{B} = BR^{-1}B^T$.

Discrete Riccati equations of the form $P = \mathcal{F}[P]$ with $\mathcal{F}[P]$ of the form given in (3.111) of Proposition 3.61, part **(c)** on page 140 were studied by Ionescu and Weiss [84], Chan *et al* [39, eq. (2.14)], Hench and Laub [78, eq. (5)], Komaroff [90, eq. (1)], Mori *et al* [103, eq. 1], Nicoloa [106, eq. (3)], Pappas *et al* [109, eq. (1)], Stoorvogel [145, eq. (3.3)], Tsai *et al* [146, eq. (6)], Walker [149, eq. (9) and Theorem 2], Wimmer *et al* [152, 153, 112], and Xie *et al* [155, eqs. (14) and (16)].

Equations of the form (3.111) were employed by Byrnes [37, eq. (1.4)], Ionescu and Weiss [84], Jonckheere and Silverman [86, eq. (5a)], Lancaster *et al* [94, eq (1.1)], and Ran and Vreugdenhill [129, eq. (0.2)]. Reverse time versions of (3.111) were used by Halany and Ionescu [69, eq. (6)] and Silverman [140, eq. (13)].

Many other forms of discrete Riccati equations can be found in the literature. For example, see [38], [158, eq. (4)], and [159, eq. (24a)]. In particular, Yaesh and Shaked show that these Riccati equations also contain the discrete game Riccati equation [156, eq. (16)].

Finally, the recent book of Lancaster and Rodman [95] is a detailed account of the subject of algebraic Riccati equations for both the continuous (CARE) and discrete cases (DARE).

4

DISCRETE VARIATIONAL THEORY

4.1 THE DISCRETE VARIATIONAL PROBLEM

We begin this chapter by describing a simple fixed endpoint discrete variational problem. We initially consider fixed step sizes of length 1. However, in Section 4.7 we will let the step sizes be of variable length. Assume $f(t, y, r)$ for each t in the discrete interval $[a+1, b+2]$ is of class C^2 with respect to the components of the n dimensional vector variables y and r. We define a set of admissible functions by

$$\mathcal{F} = \{y : [a, b+2] \to \mathrm{R}^n : y(a) = \alpha, \quad y(b+2) = \beta\}$$

where α and β are given column n vectors.

We then define $J : \mathcal{F} \to \mathrm{R}$ by

$$J[y] = \sum_{t=a+1}^{b+2} f(t, y(t), \Delta y(t-1)). \tag{4.1}$$

Our main problem of interest is to extremize J subject to $y \in \mathcal{F}$. For a study of this problem in the scalar case see Chapter 8 of [88].

We say J has a local minimum at $y_0 \in \mathcal{F}$ on $\mathcal{F}$ provided there is a $\delta > 0$ such that

$$J[y] \geq J[y_0]$$

for all $y \in \mathcal{F}$ whenever

$$\|y - y_0\| \equiv \max\{|y(t) - y_0(t)| : a \leq t \leq b+2\} < \delta$$

where $|\cdot|$ is some norm on R^n. We say that J has a *proper local minimum* at y_0 in $\mathcal{F}$ if there is a $\delta > 0$ such that

$$J[y] \geq J[y_0]$$

for all $y \in \mathcal{F}$ which satisfy $\|y - y_0\| < \delta$ and the only y in this neighborhood for which $J[y] = J[y_0]$ is $y = y_0$. There are similar definitions for local maximum, proper local maximum, global minimum, proper global minimum, global maximum and proper global maximum at y_0 on $\mathcal{F}$.

We define a set of *admissible variations* by

$$\mathcal{A} = \{\eta : [a, b+2] \to \mathrm{R}^n \text{ such that } \eta(a) = \eta(b+2) = 0\}.$$

Note that if $y \in \mathcal{F}$, $\eta \in \mathcal{A}$ and ε is a real number then $y + \varepsilon\eta \in \mathcal{F}$.

We will use the notation

$$f_y(t,y,r) = \begin{bmatrix} f_{y_1}(t,y,r) \\ f_{y_2}(t,y,r) \\ \vdots \\ f_{y_n}(t,y,r) \end{bmatrix}, \qquad f_r(t,y,r) = \begin{bmatrix} f_{r_1}(t,y,r) \\ f_{r_2}(t,y,r) \\ \vdots \\ f_{r_n}(t,y,r) \end{bmatrix}$$

to denote the gradients of f with respect to y and with respect to r respectively. For $y \in \mathcal{F}$, the first variation of J along y is defined by

$$\begin{aligned} J_1[\eta] &= J_1[\eta; y] \\ &= \sum_{a+1}^{b+2} \{ f_y^T(t, y(t), \Delta y(t-1))\eta(t) \\ &\qquad + f_r^T(t, y(t), \Delta y(t-1))\Delta\eta(t-1)\} \end{aligned}$$

for $\eta \in \mathcal{A}$, where $f_y^T(t,y,r)$ denotes the transpose of $f_y(t,y,r)$.

We now give our first necessary condition for J to have a local extremum on $\mathcal{F}$.

Theorem 4.1 *If J has a local extremum on $\mathcal{F}$ at y_0, then the first variation of J along y_0 satisfies*

$$J_1[\eta; y_0] = 0$$

for all $\eta \in \mathcal{A}$.

Proof: Assume J has a local extremum on $\mathcal{F}$ at $y_0 \in \mathcal{F}$. Fix $\eta \in \mathcal{A}$ and define $\phi : \mathrm{R} \to \mathrm{R}$ by

$$\begin{aligned}\phi(\varepsilon) &= J[y_0 + \varepsilon\eta] \\ &= \sum_{t=a+1}^{b+2} f(t, y_0(t) + \varepsilon\eta(t), \Delta y_0(t-1) + \varepsilon\Delta\eta(t-1)).\end{aligned}$$

Note that ϕ has a local extremum at the interior point $\varepsilon = 0$ and hence $\phi'(0) = 0$. Differentiating ϕ with respect to ε we obtain

$$\begin{aligned}\phi'(\varepsilon) = \sum_{t=a+1}^{b+2} \{&f_y^T(t, y_0(t) + \varepsilon\eta(t), \Delta y_0(t-1) + \varepsilon\Delta\eta(t-1))\eta(t) \\ &+ f_r^T(t, y_0(t) + \varepsilon\eta(t), \Delta y_0(t-1) + \varepsilon\Delta\eta(t-1))\Delta\eta(t-1)\}.\end{aligned} \tag{4.2}$$

Setting $\varepsilon = 0$ and using $\phi'(0) = 0$ we get

$$J_1[\eta; y_0] = 0.$$

Since this argument works for each fixed $\eta \in \mathcal{A}$ we get that $J_1[\eta; y_0] = 0$ for all $\eta \in \mathcal{A}$. □

Next we prove a lemma which will be useful for several later results.

Lemma 4.2 *Assume y and η are arbitrary real n–dimensional vector functions defined on $[a, b+2]$, then*

$$\begin{aligned}J_1[\eta] &= J_1[\eta; y] \\ &= \sum_{t=a+1}^{b+1} \{f_y^T(t, y(t), \Delta y(t-1)) - \Delta\left[f_r^T(t, y(t), \Delta y(t-1))\right]\}\eta(t) \\ &\quad - f_r^T(a+1, y(a+1), \Delta y(a))\eta(a) \\ &\quad + \{f_y^T(b+2, y(b+2), \Delta y(b+1)) \\ &\quad + f_r^T(b+2, y(b+2), \Delta y(b+1))\}\eta(b+2).\end{aligned}$$

Proof: Assume y and η are real n–dimensional vector functions defined on $[a, b+2]$. Consider

$$\begin{aligned} J_1[\eta; y] = & \textstyle\sum_{t=a+1}^{b+1}\{f_y^T(t, y(t), \Delta y(t-1))\eta(t) \\ & + [f_r^T(t, y(t), \Delta y(t-1))]\Delta\eta(t-1)\} \\ & + f_y^T(b+2, y(b+2), \Delta y(b+1))\eta(b+2) \\ & + f_r^T(b+2, y(b+2), \Delta y(b+1))\Delta\eta(b+1). \end{aligned}$$

Using the summation by parts formula

$$\sum_{t=a+1}^{b+1} u^T(t)\Delta v(t-1) = u^T(t)v(t-1)|_{t=a+1}^{b+2} - \sum_{t=a+1}^{b+1} [\Delta u^T(t)]v(t)$$

on the second term under the sum we obtain

$$\begin{aligned} J_1[\eta] = & \textstyle\sum_{t=a+1}^{b+1}\{f_y^T(t, y(t), \Delta y(t-1)) - \Delta f_r^T(t, y(t), \Delta y(t-1))\}\eta(t) \\ & +[f_r^T(t, y(t), \Delta y(t-1))\eta(t-1)]_{a+1}^{b+2} \\ & +f_y^T(b+2, y(b+2), \Delta y(b+1))\eta(b+2) \\ & +f_r^T(b+2, y(b+2), \Delta y(b+1))\Delta\eta(b+1). \end{aligned}$$

Simplifying this last equation gives the desired result. □

Next we give another necessary condition for J to have a local extremum at $y_0 \in \mathcal{F}$ on $\mathcal{F}$.

Theorem 4.3 *If J has a local extremum at $y_0 \in \mathcal{F}$ on $\mathcal{F}$, then $y_0(t)$ satisfies the (vector) Euler–Lagrange equation*

$$f_y(t, y(t), \Delta y(t-1)) - \Delta(f_r(t, y(t), \Delta y(t-1)) = 0, \tag{4.3}$$

on $[a+1, b+1]$.

Proof: Assume J has a local extremum on $\mathcal{F}$ at y_0. By Theorem 4.1

$$J_1[\eta; y_0] = 0 \qquad \text{for all } \eta \in \mathcal{A}$$

and Lemma 4.2 with y replaced by y_0 we use the end conditions on η of $\eta(a) = 0 = \eta(b+2)$ to obtain

$$\sum_{t=a+1}^{b+1} \{f_y^T(t, y_0(t), \Delta y_0(t-1)) - \Delta \left[f_r^T(t, y_0(t), \Delta y_0(t-1))\right]\}\eta(t) = 0 \quad (4.4)$$

for all $\eta \in \mathcal{A}$. Fix $s \in [a+1, b+1]$ and fix $i \in \{1, \ldots, n\}$. Then define η_s on $[a, b+2]$ by

$$\eta_s(t) = \begin{cases} e_i, & t = s \\ 0, & \text{otherwise}, \end{cases}$$

where e_i is the unit vector of R^n in the i–th direction. Since $\eta_s \in \mathcal{A}$ we get from (4.4) with η replaced by η_s that

$$\{f_y^T(s, y_0(s), \Delta y_0(s-1)) - \Delta \left[f_r^T(s, y_0(s), \Delta y_0(s-1))\right]\}e_i = 0.$$

Since this is true for each i, $1 \le i \le n$, we have

$$f_y^T(s, y_0(s), y_0(s-1)) - \Delta \left[f_r^T(s, y_0(s), \Delta y_0(s-1))\right] = 0.$$

Taking the transpose of both sides and since $s \in [a+1, b+1]$ is arbitrary we get that $y_0(t)$ satisfies the Euler–Lagrange equation (4.3) on $[a+1, b+1]$. □

Exercise 4.4 *Consider the functional*

$$J[y] = \sum_{t=1}^{300} \left\{ \left(\frac{1}{6}\right)^{t-1} [\Delta y(t-1)]^2 - 2\left(\frac{1}{6}\right)^t y^2(t) \right\}$$

for y in the class of real valued functions defined on $[0, 300]$ which satisfy the end conditions $y(0) = 0$, $y(300) = 1$. Assume that J has a local minimum at $y_0(t)$. Find $y_0(t)$.

For a given real number α let us denote by $\mathcal{F}_1$ the class of n–dimensional vector functions y defined on $[a, b+2]$ which satisfy

$$\mathcal{F}_1 \equiv \{y : \ y(a) = \alpha\}.$$

Theorem 4.5 *If J has a local extremum at $y_0 \in \mathcal{F}_1$, then $y_0(t)$ satisfies the Euler–Lagrange equation* (4.3) *for $t \in [a+1, b+1]$, $y_0(a) = \alpha$ and y_0 satisfies the transversality condition*

$$f_y(b+2, y(b+2), \Delta y(b+1)) + f_r(b+2, y(b+2), \Delta y(b+1)) = 0. \quad (4.5)$$

Proof: In a manner similar to the proof of Theorem 4.1 with $\mathcal{F}$ replaced by $\mathcal{F}_1$ and $\mathcal{A}$ replaced by

$$\mathcal{A}_1 = \{\eta : [a, b+2] \to \mathrm{R}^n \text{ such that } \eta(a) = 0\}$$

one obtains that

$$J_1[\eta; y_0] = 0$$

for all $\eta \in \mathcal{A}_1$. From Lemma 4.2 we obtain, using $\eta(a) = 0$, that

$$\begin{aligned} J_1[\eta; y_0] = \quad & \textstyle\sum_{t=a+1}^{b+1}[\{f_y^T(t, y_0(t), \Delta y_0(t-1)) \\ & -\Delta\left[f_r^T(t, y_0(t), \Delta y_0(t-1))\right]\}\eta(t)] \\ & + \{f_y^T(b+2, y_0(b+2), \Delta y_0(b+1)) \\ & + f_r^T(b+2, y_0(b+2), \Delta y_0(b+1))\}\eta(b+2) = 0 \end{aligned} \tag{4.6}$$

for all $\eta \in \mathcal{A}_1$.

Define η_b by

$$\eta_b(t) = \begin{cases} 0, & t \in [a, b+1] \\ e_i, & t = b+2, \end{cases}$$

where $i \in \{1, \ldots, n\}$ and e_i is the elementary vector $e_i = (\delta_{i,j})$ in R^n. Since $\eta_b \in \mathcal{A}_1$ we have by (4.6) that

$$\{f_y^T(b+2, y_0(b+2), \Delta y_0(b+1)) + f_r^T(b+2, y_0(b+2), \Delta y_0(b+1))\}e_i = 0.$$

Since this is true for $1 \le i \le n$, we easily get from this the transversality condition (4.5). This transversality condition and (4.6) gives us the equation (4.4). Taking $\eta_s(t)$ as in the proof of Theorem 4.3 and proceeding as in that proof we conclude that $y_0(t)$ is a solution of the Euler–Lagrange equation (4.3) on $[a+1, b+1]$. □

Exercise 4.6 *Assume*

$$J[y] = \sum_{t=1}^{500}[\Delta y(t-1)]^2$$

subject to y being a real valued function defined on $[0, 500]$ satisfying $y(0) = 1$ has a local minimum at $y_0(t)$. Find $y_0(t)$ using Theorem 4.5. Explain why your answer makes sense.

For β a fixed real number, let us denote by $\mathcal{F}_2$ the class of n–dimensional vector functions y defined on $[a, b+2]$ by

$$\mathcal{F}_2 \equiv \{y : \; y(b+2) = \beta\}.$$

In a manner similar to the proof of Theorem 4.5 we can prove the following theorem.

Theorem 4.7 *If J has a local extremum at $y_0 \in \mathcal{F}_2$, then y_0 is a solution of the Euler–Lagrange equation* (4.3) *on $[a, b+2]$, $y_0(b+2) = \beta$, and y_0 satisfies the transversality condition*

$$f_r(a+1, y(a+1), \Delta y(a)) = 0. \tag{4.7}$$

Exercise 4.8 *Prove Theorem* 4.7.

Exercise 4.9 *Assume*

$$J[y] = \sum_{t=1}^{1000} \left\{ [\Delta y(t-1)]^2 + \frac{1}{2} y^2(t) \right\}$$

subject to y being a real valued function defined on $[0, 1000]$ which satisfies $y(1000) = 2$ has a local minimum at $y_0(t)$. Find $y_0(t)$.

Theorem 4.10 *If J has a local extremum in the set $\mathcal{F}_3 \equiv \{$real n–dimensional vector functions y defined on $[a, b+2]\}$ at $y_0 \in \mathcal{F}_3$, then y_0 is a solution of the Euler–Lagrange equation on $[a, b+2]$ and $y_0(t)$ satisfies the transversality conditions* (4.5), (4.7).

Exercise 4.11 *Prove Theorem* 4.10.

Exercise 4.12 *Assume*

$$J[y] = \sum_{t=1}^{100} \left(\frac{1}{2}\right)^{t-1} [\Delta y(t-1)]^2$$

subject to $y(t)$ being a real valued function on $[0, 100]$ has a local minimum at $y_0(t)$. Find $y_0(t)$.

4.2 THE SECOND VARIATION

Assume J has a local extremum on $\mathcal{F}$ at y_0. We will now do some calculations that lead to the definition of the second variation of J along $y_0(t)$.

Fix $\eta \in \mathcal{A}$ and define ϕ as in the proof of Theorem 4.1. Differentiating both sides of (4.2) with respect to ε and setting $\varepsilon = 0$ we obtain

$$\begin{aligned}\phi''(0) = \sum_{t=a+1}^{b+2} \quad & \{\eta^T(t) f_{yy}(t, y_0(t), \Delta y_0(t-1))\eta(t) \\ & + \eta^T(t) f_{yr}(t, y_0(t), \Delta y_0(t-1))\Delta\eta(t-1) \\ & + \Delta\eta^T(t-1) f_{ry}(t, y_0(t), \Delta y_0(t-1))\eta(t) \\ & + \Delta\eta^T(t-1) f_{rr}(t, y_0(t), \Delta y_0(t-1))\Delta\eta(t-1)\}\end{aligned}$$

where $f_{yr}(t, y, r)$ is the matrix defined by

$$f_{yr}(t, y, r) = \begin{bmatrix} f_{y_1 r_1}(t, y, r) & f_{y_1 r_2}(t, y, r) & \cdots & f_{y_1 r_n}(t, y, r) \\ \cdots & & & \cdots \\ \cdots & & & \cdots \\ f_{y_n r_1}(t, y, r) & f_{y_n r_2}(t, y, r) & \cdots & f_{y_n r_n}(t, y, r) \end{bmatrix}$$

where $f_{y_i r_j}(t, y, r)$ denotes the partial $\frac{\partial^2}{\partial y_i \partial r_j} f(t, y, r)$. The matrices $f_{yy}(t, y, r)$, $f_{ry}(t, y, r)$, and $f_{rr}(t, y, r)$ are defined in a similar manner.

We then introduce the notation

$$P(t) = P(t; y_0(t)) = f_{yy}(t, y_0(t), \Delta y_0(t-1)) \tag{4.8}$$

$$Q(t) = Q(t; y_0(t)) = f_{yr}(t, y_0(t), \Delta y_0(t-1)) \tag{4.9}$$

$$R(t) = R(t; y_0(t)) = f_{rr}(t, y_0(t), \Delta y_0(t-1)). \tag{4.10}$$

Note that $P(t)$ and $R(t)$ are symmetric matrix functions but $Q(t)$ in general is not symmetric. With this notation we can write

$$\begin{aligned}\phi''(0) = \sum_{t=a+1}^{b+2} \quad & \{\eta^T(t) P(t)\eta(t) + \eta^T(t) Q(t)\Delta\eta(t-1) \\ & + \Delta\eta^T(t-1) Q^T(t)\eta(t) + \Delta\eta^T(t-1) R(t)\Delta\eta(t-1)\}.\end{aligned}$$

We then define the second variation of J along y_0 on the set of admissible variations $\mathcal{A}$ by

$$J_2[\eta] = J_2[\eta; y_0] = \sum_{t=a+1}^{b+2} 2w(t, \eta(t), \Delta\eta(t-1))$$

where $w(t, \eta, \zeta)$, is defined by

$$2w(t, \eta, \zeta) = \eta^T P(t)\eta + \eta^T Q(t)\zeta + \zeta^T Q^T(t)\eta + \zeta^T R(t)\zeta \tag{4.11}$$

for $n \times 1$ vector variables η and ζ.

From the above calculations we have that for each fixed $\eta \in \mathcal{A}$

$$\phi''(0) = J_2[\eta; y_0].$$

We now can state and easily prove the following theorem.

Theorem 4.13 *Assume J on $\mathcal{F}$ has a local minimum $\{$maximum$\}$ at y_0, then the second variation of J along y_0 satisfies*

$$J_2[\eta; y_0] \geq 0 \quad \{\leq 0\}$$

for all $\eta \in \mathcal{A}$.

Proof: Assume J on $\mathcal{F}$ has a local minimum at y_0. It follows that the function $\phi(\varepsilon)$ as defined above has a local minimum at the interior point of its domain $\varepsilon = 0$. It follows from this that

$$\phi''(0) = J_2[\eta; y_0] \geq 0$$

for each fixed $\eta \in \mathcal{A}$. The proof in the local maximum case is similar. □

The Euler–Lagrange equation for the second variation of J along y_0 is called the Jacobi equation for J. We now prove the following important result.

Theorem 4.14 *The Jacobi equation of J along y_0 is the generalized self–adjoint difference equation*

$$\Delta\left[R(t)\Delta\eta(t-1) + Q^T(t)\eta(t)\right] - \left[Q(t)\Delta\eta(t-1) + P(t)\eta(t)\right] = 0. \tag{4.12}$$

(Do not forget that the coefficient matrices in this equation depend on y_0.)

Proof: We need to find the Euler–Lagrange equation for

$$J_2[\eta] = \sum_{t=a+1}^{b+2} 2w(t, \eta(t), \Delta\eta(t+1))$$

where $w(t, \eta, \zeta)$ is given by equation (4.11). We want to calculate $w_\eta(t, \eta, \zeta)$ and $w_\zeta(t, \eta, \zeta)$. To find $w_\eta(t, \eta, \zeta)$ note that for $1 \le i \le n$

$$2w_{\eta_i} = e_i^T P(t)\eta + \eta^T P(t) e_i + e_i^T Q(t)\zeta + \zeta^T Q^T(t) e_i$$

where e_i is the $n \times 1$ unit vector in the i–th direction. It follows that $2w_{\eta_i} = e_i^T [2P(t)\eta + 2Q(t)\zeta]$. Hence

$$w_{\eta_i} = e_i^T [P(t)\eta + Q(t)\zeta]$$

for $1 \le i \le n$. When the i–th row of two matrices are the same for each i then the matrices are the same. Hence

$$w_\eta = P(t)\eta + Q(t)\zeta. \tag{4.13}$$

To find w_ζ we first note that for $1 \le i \le n$

$$\begin{aligned} 2w_{\zeta_i} &= \eta^T Q(t) e_i + e_i^T Q^T(t)\eta + e_i^T R(t)\zeta + \zeta^T R(t) e_i \\ &= e_i^T [2Q^T(t)\eta + 2R(t)\zeta]. \end{aligned}$$

It follows from this that

$$w_\zeta = Q^T(t)\eta + R(t)\zeta. \tag{4.14}$$

Hence the Euler–Lagrange equation for $J_2[\eta]$ is given by the *generalized self–adjoint vector difference equation*

$$\Delta \left[R(t)\Delta\eta(t-1) + Q^T(t)\eta(t)\right] - \left[Q(t)\Delta\eta(t-1) + P(t)\eta(t)\right] = 0$$

and the proof of the theorem is complete. □

Remark 4.15 *We call the difference equation defined by* (4.12) *the generalized self–adjoint difference equation because if* $Q(t) = 0$, *then equation* (4.12) *reduces to the two term self–adjoint difference equation*

$$\Delta\left[R(t)\Delta\eta(t-1)\right] - P(t)\eta(t) = 0. \tag{4.15}$$

Note that if all the partial derivatives $f_{y_i r_j}(t, y, r) = 0$, $1 \le i, j \le n$, *then* $Q(t) = 0$ *and* (4.12) *becomes the self–adjoint equation* (4.15).

We now write equation (4.12) as a three term equation. Expanding out the first term in (4.12) we get that

$$\begin{aligned} R(t+1)\Delta\eta(t) + Q^T(t+1)\eta(t+1) - R(t)\Delta\eta(t-1) & \\ -Q^T(t)\eta(t) - Q(t)\Delta\eta(t-1) - P(t)\eta(t) &= 0. \end{aligned}$$

Expand the remaining forward differences to obtain

$$\begin{aligned} R(t+1)\eta(t+1) - R(t+1)\eta(t) + Q^T(t+1)\eta(t+1) - R(t)\eta(t) & \\ + R(t)\eta(t-1) - Q^T(t)\eta(t) - Q(t)\eta(t) + Q(t)\eta(t-1) - P(t)\eta(t) &= 0. \end{aligned}$$

Regroup this as

$$\begin{aligned} & \left[R(t+1) + Q^T(t+1)\right]\eta(t+1) \\ - & \left[R(t+1) + R(t) + Q^T(t) + Q(t) + P(t)\right]\eta(t) \\ + & \left[R(t) + Q(t)\right]\eta(t-1) = 0. \end{aligned}$$

Hence we can write this equation in the form

$$-K(t)\eta(t+1) + N(t)\eta(t) - K^T(t-1)\eta(t-1) = 0 \tag{4.16}$$

where

$$K(t) = R(t+1) + Q^T(t+1) \tag{4.17}$$

and

$$N(t) = R(t+1) + R(t) + Q^T(t) + Q(t) + P(t). \tag{4.18}$$

Note that $N(t)$ is symmetric. Also, $Q(t)$ is symmetric if and only if $K(t)$ is symmetric.

Legendre's necessary condition will be defined in terms of the matrix $N(t)$ of (4.18).

4.3 LEGENDRE'S NECESSARY CONDITION

We will use the notation $N \geq 0$ to signify that a real symmetric matrix N is positive semidefinite.

Theorem 4.16 (Legendre's necessary condition) *Assume J subject to y in $\mathcal{F}$ has a local minimum $\{$maximum$\}$ at y_0, then*

$$N(t) = N(t; y_0) \geq 0 \qquad \{\leq 0\}$$

for $t \in [a+1, b+1]$.

Proof: Assume J on $\mathcal{F}$ has a local minimum at y_0. By Theorem 4.13

$$J_2[\eta; y_0] \geq 0$$

for all $\eta \in \mathcal{A}$. That is,

$$\sum_{t=a+1}^{b+2} \{\eta^T(t)P(t; y_0)\eta(t) + \eta^T(t)Q(t; y_0)\Delta\eta(t-1)$$
$$+ \Delta\eta^T(t-1)Q^T(t)\eta(t) + \Delta\eta^T(t-1)R(t)\Delta\eta(t-1)\} \geq 0$$

for all $\eta \in \mathcal{A}$. Fix $s \in [a+1, b+1]$ and let γ be an arbitrary but fixed vector in R^n. Define $\eta(s,t)$ by

$$\eta(s,t) = \begin{cases} \gamma, & t = s, \\ 0, & t \in [a, b+2],\ t \neq s. \end{cases}$$

Note that $\eta(s,t) \in \mathcal{A}$. Hence by the above inequality with $\eta = \eta(s,t)$ we obtain

$$\gamma^T P(s; y_0)\gamma + \gamma^T Q(s; y_0)\gamma + \gamma^T Q^T(s; y_0)\gamma$$
$$+ \gamma^T R(s; y_0)\gamma + \gamma^T R(s+1; y_0)\gamma \ \geq\ 0.$$

That is

$$\gamma^T N(s; y_0)\gamma \geq 0$$

for all $\gamma \in \mathrm{R}^n$; i.e.,

$$N(s; y_0) \geq 0.$$

Since $s \in [a+1, b+1]$ is fixed but arbitrary we get that

$$N(t; y_0) \geq 0$$

for all $t \in [a+1, b+1]$. The proof of the local maximum case is similar. □

4.4 DISCRETE HAMILTONIAN SYSTEMS

Let a continuous function $f(t,y,r)$, called the *Lagrangian,* be given. For each fixed t, f is assumed to be of class C^2 with respect to the components of y and r. In this section it is also desirable to assume $f_r(t,y,r)$ has an inverse with respect to r. Specifically, we assume that there is a vector function $g(t,y,z)$ such that

$$z = f_r(t,y,r) \quad \text{if and only if} \quad r = g(t,y,z) \tag{4.19}$$

where we assume that for each fixed t, $g(t,y,z)$ is of class C^1 with respect to the components of y and z. Then we define the Hamiltonian $H(t,y,z)$ by

$$H(t,y,z) = z^T g(t,y,z) - f(t,y,g(t,y,z)). \tag{4.20}$$

Before we state and prove our main theorem of this section we derive some formulas for $H_y(t,y,z)$ and $H_z(t,y,z)$.

To find $H_y(t,y,z)$ we first calculate $H y_i(t,y,z)$. Note that for $1 \leq i \leq n$,

$$\begin{aligned} H_{y_i}(t,y,z) &= z^T g_{y_i}(t,y,z) - f_{y_i}(t,y,g(t,y,z)) \\ &\quad - f_r^T(t,y,g(t,y,z)) g_{y_i}(t,y,z) \\ &= z^T g_{y_i}(t,y,z) - f_{y_i}(t,y,g(t,y,z)) - z^T g_{y_i}(t,y,z) \\ &= -f_{y_i}(t,y,g(t,y,z)). \end{aligned}$$

It follows that

$$H_y(t,y,z) = -f_y(t,y,g(t,y,z)). \tag{4.21}$$

Similarly to find $H_z(t,y,z)$ we first calculate $H_{z_i}(t,y,z)$. Note that

$$\begin{aligned} H_{z_i}(t,y,z) &= z^T g_{z_i}(t,y,z) + e_i^T g(t,y,z) \\ &\quad - f_r^T(t,y,g(t,y,z)) g_{z_i}(t,y,z) \\ &= z^T g_{z_i}(t,y,z) + e_i^T g(t,y,z) - z^T g_{z_i}(t,y,z) \\ &= e_i^T g(t,y,z) \end{aligned}$$

and it follows that

$$H_z(t,y,z) = g(t,y,z). \tag{4.22}$$

We now state and prove the main result of this section. This result shows how one goes from the Euler–Lagrange equation to a Hamiltonian system.

Theorem 4.17 (Lagrangian to Hamiltonian) *Assume f allows definition of a function g as in (4.19) and H is defined by (4.20). Assume $y(t)$ is defined on $[a, b+2]$ and satisfies the Euler–Lagrange equation (4.3) of page 156 on $[a+1, b+1]$. Define a discrete momentum variable by*

$$z(t) = f_r(t+1, y(t+1), \Delta y(t))$$

for $t \in [a, b+1]$. Then $y(t)$, $z(t)$ is a solution of the Hamiltonian system

$$\begin{aligned} \Delta y(t) &= H_z(t+1, y(t+1), z(t)) \\ \Delta z(t) &= -H_y(t+1, y(t+1), z(t)). \end{aligned}$$

Here the first equation is satisfied for $t \in [a, b+1]$ and the second equation is satisfied for $t \in [a, b]$.

Proof: Assume $y(t)$ satisfies the Euler–Lagrange equation on the interval $[a+1, b+1]$ and set

$$z(t) = f_r(t+1, y(t+1), \Delta y(t))$$

for $t \in [a, b+1]$. From (4.19) we get that

$$\Delta y(t) = g(t+1, y(t+1), z(t)). \tag{4.23}$$

It follows from (4.22) that

$$\Delta y(t) = H_z(t+1, y(t+1), z(t)).$$

Next note that

$$\begin{aligned} \Delta z(t) &= \Delta f_r(t+1, y(t+1), \Delta y(t)) \\ &= f_y(t+1, y(t+1), \Delta y(t)) \end{aligned}$$

from the fact that $y(t)$ satisfies the Euler–Lagrange equation (4.3). Hence, using (4.23), we get that

$$\Delta z(t) = f_y(t+1, y(t+1), g(t+1, y(t+1), z(t))).$$

It follows from (4.21) that

$$\Delta z(t) = -\ H_y(t+1, y(t+1), z(t)).$$

Hence $y(t)$, $z(t)$ is a solution of the desired Hamiltonian system. □

For motivation for the next example, we briefly consider the continuous case. Assume that on some t interval, $y(t)$, $z(t)$ is a solution of the autonomous Hamiltonian system of differential equations

$$\begin{aligned} y' &= H_z(y,z) \\ z' &= -\ H_y(y,z). \end{aligned} \tag{4.24}$$

Then

$$\begin{aligned} \frac{d}{dt}H(y(t),z(t)) &= H_y^T(y(t),z(t))y'(t) + H_z^T(y(t),z(t))z'(t) \\ &= H_y^T(y(t),z(t))H_z(y(t),z(t)) \\ &\quad - H_z^T(y(t),z(t))H_y(y(t),z(t)) \equiv 0. \end{aligned}$$

Hence the Hamiltonian $H(y,z)$ is constant along solutions of (4.24). Because of this we say $H(y,z)$ is a *first integral* for (4.24). Note that we used the chain rule of differentiation to prove that the system (4.24) had a first integral. In the discrete case there is no chain rule so we might not be surprised by the following example.

Example 4.18 *The Hamiltonian need not be a "first sum." Consider*

$$J[y] = \sum_{t=a+1}^{b+2} \frac{1}{2}\{y^2(t) + [\Delta y(t-1)]^2\}.$$

Here

$$f(y,r) = \frac{1}{2}y^2 + \frac{1}{2}r^2.$$

The inverse property (4.19) *for this example is*

$$z = f_r(y,r) = r \text{ if and only if } r = g(y,z) = z.$$

Hence by (4.20) *the Hamiltonian is given by*

$$\begin{aligned} H(y,z) &= z^2 - \left(\frac{1}{2}y^2 + \frac{1}{2}r^2\right) \\ &= \frac{1}{2}z^2 - \frac{1}{2}y^2. \end{aligned}$$

The Euler–Lagrange equation for this example is

$$\Delta^2 y(t-1) - y(t) = 0$$

and the corresponding linear autonomous Hamiltonian system is

$$\begin{aligned}\Delta y(t) &= H_z(y(t+1), z(t)) = z(t) \\ \Delta z(t) &= -H_y(y(t+1), z(t)) = y(t+1).\end{aligned}$$

Let $y(t)$, $z(t)$ be the solution of this Hamiltonian system satisfying the initial conditions $y(0) = 0$, $z(0) = 1$. Then it follows that

$$H(y(1), z(0)) = 0 \neq -\frac{5}{2} = H(y(2), z(1)).$$

Hence our discrete Hamiltonian is not constant along solutions.

4.5 HIGHER ORDER DIFFERENCE EQUATIONS

In this section we consider the self–adjoint $2n$–th order difference equation

$$\ell_{2n}u(t) = \sum_{i=0}^{n} \Delta^i [r_i(t)\Delta^i u(t-i)] = 0 \qquad (4.25)$$

$t \in [a+n, b+n]$, where the coefficient functions $r_i(t)$, $0 \leq i \leq n$, are real valued functions on $[a+n, b+n+i]$ respectively and the leading coefficient $r_n(t) > 0$ on $[a, b+2n]$. Solutions of this equation are defined on $[a, b+2n]$. Recall that in Example 3.14 on page 85 we showed that the scalar equation $\ell_{2n}u(t) = 0$ is equivalent to a symplectic system. In this book we only consider real solutions of equation (4.25).

We define a set of admissible variations by

$$A = \{\eta : [a, b+2n] \to \mathrm{R} \mid \eta(a+i) = 0 = \eta(b+2n-i),\ 0 \leq i \leq n-1\}.$$

It is convenient (but not necessary) for the notation and proofs below to extend the domain of definitions of the coefficient functions $r_i(t)$ and the admissible variations $\eta(t)$ by

$$\begin{aligned}r_i(t) &= r_i(b+n+i), \quad t > b+n+i,\ 0 \leq i \leq n \\ \eta(t) &= 0, \quad t \geq b+2n.\end{aligned}$$

Then we can define the quadratic functional Q, associated with $\ell_{2n}u = 0$, on the set of admissible variations by

$$Q[\eta] = \sum_{t=a+n}^{b+2n} \sum_{i=0}^{n} (-1)^{n+i} r_i(t)[\Delta^i \eta(t-i)]^2.$$

First we prove the following very important relationship between Q and the operator ℓ_{2n}.

Theorem 4.19 *If $\eta \in A$, then*

$$Q[\eta] = (-1)^n \sum_{t=a+n}^{b+n} \eta(t)\ell_{2n}\eta(t).$$

Proof: If we set

$$B_i = \sum_{t=a+n}^{b+2n} r_i(t)[\Delta^i \eta(t-i)]^2, \tag{4.26}$$

then

$$Q[\eta] = \sum_{i=0}^{n} (-1)^{n+i} B_i. \tag{4.27}$$

We first prove by induction on j that

$$B_i = \sum_{t=a+n}^{b+2n} (-1)^j [\Delta^{i-j}\eta(t-i+j)]\Delta^j[r_i(t)\Delta^i\eta(t-i)] \tag{4.28}$$

for $0 \le j \le i$. For $j = 0$ this follows from (4.26). Now assume $0 < j < i$ and (4.28) holds. Using a summation by parts formula we get that

$$\begin{aligned} B_i &= (-1)^j \Delta^j[r_i(t)\Delta^i\eta(t-i)]\Delta^{i-j-1}\eta(t-i+j)\Big|_{a+n}^{b+2n+1} \\ &\quad - \sum_{t=a+n}^{b+2n} (-1)^j [\Delta^{i-j-1}\eta(t-i+j+1)]\Delta^{j+1}[r_i(t)\Delta^i\eta(t-i)] \\ &= \sum_{t=a+n}^{b+2n} (-1)^{j+1} [\Delta^{i-(j+1)}\eta(t-i+j+1)]\Delta^{j+1}[r_i(t)\Delta^i\eta(t-i)]. \end{aligned}$$

Hence, by induction (4.28) holds. Setting $j = i$ in (4.28) we get that

$$B_i = \sum_{t=a+n}^{b+2n} (-1)^i \eta(t)\Delta^i[r_i(t)\Delta^i\eta(t-i)].$$

Since $\eta(b+2n-i) = 0$, $0 \le i \le n-1$,

$$B_i = \sum_{t=a+n}^{b+n} (-1)^i \eta(t)\Delta^i[r_i(t)\Delta^i\eta(t-i)].$$

Hence by (4.27)

$$\begin{aligned} Q[\eta] &= (-1)^n \sum_{t=a+n}^{b+n} \eta(t) \sum_{i=0}^{n} \Delta^i [r_i(t) \Delta^i \eta(t-i)] \\ &= (-1)^n \sum_{t=a+n}^{b+n} \eta(t) \ell_{2n} \eta(t) \end{aligned}$$

which is what we wanted to prove. □

A solution $u(t)$ of $\ell_{2n}u = 0$ is said to have a *generalized zero of order k at a* iff $u(a+i) = 0$, $0 \le i \le k-1$. We say $u(t)$ has a *generalized zero of order k at $t_0 > a$* provided $u(t_0 - 1) \neq 0$ and either

$$u(t_0 + i) = 0$$

$0 \le i \le k-1$ or $u(t_0 + i) = 0$, $0 \le i \le k-2$ and

$$(-1)^k u(t_0 - 1) u(t_0 + k - 1) > 0.$$

Exercise 4.20 *Show that no notrivial solution of $\ell_{2n}u = 0$ has a generalized zero of order $2n$ or more.*

Theorem 4.21 *Assume $u(t)$ is a solution of $\ell_{2n}u = 0$ satisfying the $2n-2$ boundary conditions*

$$u(t_1 + i) = 0, \qquad 0 \le i \le n-2$$

$$u(t_2 + i) = 0, \qquad 0 \le i \le n-2$$

where $a + n \le t_1 + n - 1 \le t_2 \le b + n + 1$, then if we set

$$\eta(t) = \begin{cases} 0, & a \le t \le t_1 + n - 2 \\ u(t), & t_1 \le t \le t_2 + n - 2 \\ 0, & t_2 \le t \le b + 2n \end{cases}$$

it follows that $\eta \in A$ and

$$\begin{aligned} Q[\eta] = {} & (-1)^{n+1} r_n(t_1 + n - 1) u(t_1 - 1) u(t_1 + n - 1) \\ & + (-1)^{n+1} r_n(t_2 - 1) u(t_2 - 1) u(t_2 + n - 1). \end{aligned}$$

Proof: Let $u(t)$, $\eta(t)$ be as in the statement of the theorem. Note that $\eta(a+i) = 0$, $0 \le i \le n-1$ and $\eta(b+n+j) = 0$, $1 \le j \le n$ so $\eta \in A$. By Theorem 4.19

$$\begin{aligned} Q[\eta] &= (-1)^n \sum_{t=a+n}^{b+n} \eta(t)\ell_{2n}\eta(t) \\ &= (-1)^n \sum_{t=t_1+n-1}^{t_2-1} u(t)\ell_{2n}\eta(t). \end{aligned}$$

First we consider the case where $t_2 = t_1 + n$. In this case

$$\begin{aligned} Q[\eta] &= (-1)^n u(t_1+n-1)\ell_{2n}\eta(t_1+n-1) \\ &= (-1)^n u(t_1+n-1)[\ell_{2n}u(t_1+n-1) - r_n(t_1+n-1)u(t_1-1) \\ &\quad -r_n(t_1+2n-1)u(t_1+2n-1)] \\ &= (-1)^{n+1} r_n(t_1+n-1)u(t_1+n-1)u(t_1-1) \\ &\quad +(-1)^{n+1} r_n(t_1+2n-1)u(t_1+n-1)u(t_1+2n-1) \end{aligned}$$

which is what we wanted to prove.

Next consider the case where $t_1 + n < t_2$. In this case we can write

$$\begin{aligned} Q[\eta] = &\; (-1)^n u(t_1+n-1)\ell_{2n}\eta(t_1+n-1) + \textstyle\sum_{t=t_1+n}^{t_2-2} u(t)\ell_{2n}u(t) \\ &+(-1)^n u(t_2-1)\ell_{2n}\eta(t_2-1) \\ = &\; (-1)^n u(t_1+n-1)\ell_{2n}\eta(t_1+n-1) \\ &+(-1)^n u(t_2-1)\ell_{2n}\eta(t_2-1) \\ = &\; (-1)^n u(t_1+n-1)[\ell_{2n}u(t_1+n-1) - r_n(t_1+n-1)u(t_1-1)] \\ &+(-1)^n u(t_2-1)[\ell_{2n}u(t_2-1) - r_n(t_2+n-1)u(t_2+n-1)] \\ = &\; (-1)^{n+1} r_n(t_1+n-1)u(t_1-1)u(t_1+n-1) \\ &+(-1)^{n+1} r_n(t_2+n-1)u(t_2-1)u(t_2+n-1) \end{aligned}$$

which completes the proof of the theorem. □

4.6 DISCONJUGACY AND GENERALIZED ZEROS

The $2n$–th order difference equation $\ell_{2n}u = 0$ is said to be (n, n)–disconjugate on $[a, b + 2n]$ provided no nontrivial solution of $\ell_{2n}u = 0$ has two distinct generalized zeros of order n in $[a, b + n + 1]$.

We now state the following important theorem.

Theorem 4.22 *The quadratic functional Q is positive definite on A if and only if $\ell_{2n}u = 0$ is (n, n)–disconjugate on $[a, b + 2n]$.*

Proof: First we prove by contradiction that Q is positive definite implies $\ell_{2n}u = 0$ is (n, n)–disconjugate on $[a, b + 2n]$. Assume $\ell_{2n}u = 0$ is not (n, n)–disconjugate on $[a, b+2n]$. Then there is a nontrivial solution with two distinct generalized zeros of order n. It follows from this that there are integers t_1, t_2 such that $a + n \le t_1 + n - 1 \le t_2 \le b + n + 1$,

$$u(t_1 + i) = 0, \qquad 0 \le i \le n - 2$$

$$u(t_2 + i) = 0, \qquad 0 \le i \le n - 2$$

$$u(t_1 + n - 1) \neq 0$$

and

$$(-1)^n u(t_1 - 1)u(t_1 + n - 1) \ \ge 0$$

$$(-1)^n u(t_2 - 1)u(t_2 + n - 1) \ \ge 0.$$

With η defined as in Theorem 4.21 we get that

$$\begin{aligned} Q[\eta] &= (-1)^{n+1} r_n(t_1 + n - 1)u(t_1 - 1)u(t_1 + n - 1) \\ &\quad +(-1)^{n+1} r_n(t_2 - 1)u(t_2 - 1)u(t_2 + n - 1) \le 0. \end{aligned}$$

Since

$$\eta(t_1 + n - 1) = u(t_1 + n - 1) \neq 0$$

we have that η is not the trivial function in A. Since Q is positive definite on A we get that

$$Q[\eta] > 0$$

which is our contradiction.

The proof of the converse was given by Martin Bohner in Theorem 9 of his dissertation [27]. We will consider Bohner's methods in Chapter 8. □

We now look at some consequences of this last theorem.

Theorem 4.23 *If*

$$(-1)^{n+i}r_i(t) \geq 0 \tag{4.29}$$

on $[a+n, b+n+i]$ *for* $0 \leq i \leq n-1$, *then* $\ell_{2n}u(t) = 0$ *is* (n,n)*–disconjugate on* $[a, b+2n]$.

Proof: Since

$$Q[\eta] = \sum_{t=a+n}^{b+2n} \sum_{i=0}^{n} (-1)^{n+i} r_i(t)[\Delta^i \eta(t-i)]^2$$

it is easy, by (4.29), to see that $Q[\eta] \geq 0$ for all $\eta \in A$ and that $Q[\eta] = 0$ if $\eta = 0$. It remains to show that $\eta \in A$ and $Q[\eta] = 0$ imply that $\eta = 0$. Assuming $\eta \in A$ and $Q[\eta] = 0$ we get that

$$r_n(t)[\Delta^n \eta(t-n)]^2 = 0$$

for t in $[a+n, b+n]$. Since $\eta \in A$ it follows that η solves the initial value problem

$$\begin{aligned} \Delta^n \eta(t-n) &= 0, \qquad & t &\in [a+n, b+n] \\ \eta(a+i) &= 0, & 0 &\leq i \leq n-1. \end{aligned}$$

It follows that η is the trivial function in A and hence Q is positive definite on A. By Theorem 4.22, $\ell_{2n}u = 0$ is (n,n)–disconjugate on $[a, b+2n]$. □

Consider the $2n$–th order difference equation and operator ℓ_{2n}^p defined by

$$\ell_{2n}^p v(t) = \sum_{i=0}^{n} \Delta^i [p_i(t) \Delta^i v(t-i)] = 0$$

where the coefficient functions $p_i(t)$ are defined on $[a+n, b+n+i]$ and $p_n(t) > 0$ on $[a+n, b+2n]$. (For convenience define $p_i(t) \equiv p_i(b+n+i)$ for $t \geq b+n+i$.) We now get the following comparison theorem.

Theorem 4.24 *If*

$$(-1)^{n+i} r_i(t) \geq (-1)^{n+i} p_i(t)$$

on $[a+n, b+n+i]$ *for* $0 \leq i \leq n$, *and* $\ell_{2n}^{p} v = 0$ *is* (n,n)*–disconjugate on* $[a, b+2n]$, *then* $\ell_{2n} u = 0$ *is* (n,n)*–disconjugate on* $[a, b+2n]$.

Proof: Define Q_p on the set of admissible functions A by

$$Q_p[\eta] = \sum_{t=a+n}^{b+2n} \sum_{i=0}^{n} (-1)^{n+i} p_i(t) [\Delta^i \eta(t-i)]^2.$$

Note that for any $\eta \in A$

$$Q[\eta] \geq Q_p[\eta].$$

Since $\ell_{2n}^{p} v = 0$ is (n,n)–disconjugate on $[a, b+2n]$ it follows that Q_p is positive definite on A. This implies that Q is positive definite on A and hence $\ell_{2n} u = 0$ is (n,n)–disconjugate on $[a, b+2n]$. □

Consider the $2n$–th order difference equation and the operator ℓ_{2n}^{q} defined by

$$\ell_{2n}^{q} y(t) = \sum_{i=0}^{n} \Delta^i [q_i(t) \Delta^i y(t-i)] = 0$$

where the coefficients $q_i(t)$ are defined on $[a+n, b+n+i]$ and $q_n(t) > 0$ on $[a+n, b+2n]$. (For convenience define $q_i(t) \equiv q_i(t+n+i)$ for $t \geq b+n+i$). Then we can prove the following comparison theorem.

Theorem 4.25 *Let*

$$r_i(t) = \lambda p_i(t) + \mu q_i(t)$$

for $t \in [a+n, b+n+i]$, $0 \leq i \leq n$, *where* λ, $\mu \geq 0$, *but not both zero. If* $\ell_{2n}^{p} y(t) = 0$ *and* $\ell_{2n}^{q} y(t) = 0$ *are* (n,n)*–disconjugate on* $[a, b+2n]$, *then* $\ell_{2n} y(t) = 0$ *is* (n,n)*–disconjugate on* $[a, b+2n]$.

Exercise 4.26 *Prove Theorem* 4.25.

Theorem 4.27 (Legendre's Necessary Condition) *If* $\ell_{2n} u = 0$ *is* (n,n)*–disconjugate on* $[a, b+2n]$, *then*

$$\sum_{i=0}^{n} (-1)^{n+i} \sum_{t=t_0}^{t_0+i} r_i(t) \binom{i}{t-t_0}^2 > 0$$

for each $t_0 \in [a+n, b+n]$.

Proof: Fix $t_0 \in [a+n, b+n]$ and define η by

$$\eta(t) = \begin{cases} 1, & t = t_0 \\ 0, & t \neq t_0. \end{cases}$$

Since $t_0 \in [a+n, b+n]$, $\eta \in A$. Also $\eta \neq 0$. Since $\ell_{2n}u = 0$ is (n,n)–disconjugate on $[a, b+2n]$, Q is positive definite on A. It follows that

$$Q[\eta] = \sum_{t=a+n}^{b+2n} \sum_{i=0}^{n} (-1)^{n+i} r_i(t)[\Delta^i \eta(t-i)]^2 > 0.$$

Hence

$$\sum_{i=0}^{n} (-1)^{n+i} \sum_{t=a+n}^{b+2n} r_i(t)[\Delta^i \eta(t-i)]^2 > 0.$$

Using the definition of η we get that

$$\sum_{i=0}^{n} (-1)^{n+i} \sum_{t=t_0}^{t_0+i} r_i(t)[\Delta^i \eta(t-i)]^2 > 0$$

and this implies that

$$\sum_{i=0}^{n} (-1)^{n+i} \sum_{t=t_0}^{t_0+i} r_i(t) \binom{i}{t-t_0}^2 > 0$$

and the proof is complete. □

Exercise 4.28 *Show that if the second order scalar equation* (1.1) *is disconjugate on* $[a, b+2]$, *then* $c(t) < 0$ *for all* $t \in [a+1, b+1]$, *where* $c(t)$ *is given by* (1.3).

Exercise 4.29 *Show that if*

$$\Delta^n[p(t)\Delta^n u(t-n)] + q(t)u(t) = 0$$

is (n,n)*–disconjugate on* $[a, b+2n]$, *then*

$$(-1)^{n+1} q(t) \quad < \quad \sum_{s=t}^{t+n} p(s) \binom{n}{s-t}^2$$

for $t \in [a+n, b+n]$.

In the next theorem assume $p(t) > 0$ on $[a+n, b+2n]$ and $q(t)$ is defined on $[a+n, b+n]$.

Theorem 4.30 *If the $2n$–th order two term difference equation*

$$\Delta^n[p(t)\Delta^n u(t-n)] + q(t)u(t) = 0$$

is (n,n)–disconjugate on $[a, b+2n]$, then

$$(-1)^{n+1}\sum_{t=a+n}^{b+n} q(t) < \sum_{i=0}^{n-1}[p(a+n+i) + p(b+n+1+i)]\binom{n-1}{i}^2.$$

Proof: Define

$$\eta(t) = \begin{cases} 0, & a \le t \le a+n-1 \\ 1, & a+n \le t \le b+n \\ 0, & b+n+1 \le t \le b+2n. \end{cases}$$

Then η is a nontrivial admissible function.

Since $\Delta^n[p(t)\Delta^n u(t-n)] + q(t)u(t) = 0$ is (n,n)–disconjugate on $[a, b+2n]$ we have that the corresponding quadratic functional Q is positive definite on A. Hence

$$Q[\eta] = \sum_{t=a+n}^{b+2n} p(t)[\Delta^n\eta(t-n)]^2 + (-1)^n \sum_{t=a+n}^{b+n} q(t)\eta^2(t) > 0.$$

Using the definition of η we get that

$$\sum_{t=a+n}^{a+2n-1} p(t)[\Delta^n\eta(t-n)]^2 + \sum_{t=b+n+1}^{b+2n} p(t)[\Delta^n\eta(t-n)]^2$$
$$> (-1)^{n+1}\sum_{t=a+n}^{b+n} q(t).$$

We can rewrite this as

$$\textstyle\sum_{i=0}^{n-1} p(a+n+i)[\Delta^n\eta(a+i-n)]^2$$
$$+\textstyle\sum_{i=0}^{n-1} p(b+n+1+i)[\Delta^n\eta(b+1+i)]^2 > (-1)^{n+1}\sum_{t=a+n}^{b+n} q(t).$$

But

$$\begin{aligned}\Delta^n \eta(a+i-n) &= \binom{n}{n} - \binom{n}{n-1} + \cdots + (-1)^i \binom{n}{n-i} \\ &= (-1)^i \binom{n-1}{n-i-1} \\ &= (-1)^i \binom{n-1}{i}.\end{aligned}$$

Hence

$$[\Delta^n \eta(a+i-n)]^2 = \binom{n-1}{i}^2.$$

Similarly,

$$[\Delta^n \eta(b+1+i)]^2 = \binom{n-1}{i}^2$$

and this gives the desired result. □

Exercise 4.31 *Show that if the fourth order difference equation*

$$\Delta^2 \left[p(t)\Delta^2 u(t-2)\right] - \Delta\left[q(t)\Delta u(t-1)\right] + r(t)u(t) = 0,$$

where $p(t) > 0$ *on* $[a+2, b+4]$, $q(t)$ *is defined on* $[a+2, b+3]$ *and* $r(t)$ *is defined on* $[a+2, b+2]$, *is disconjugate on* $[a, b+4]$, *then*

$$\sum_{t=a+2}^{b+2} r(t) > q(a+2) + q(b+3) - p(a+2) - p(a+3) - p(b+3) - p(b+4).$$

If r is a real number, let r_- be the negative part of r which is defined by

$$r_- = \max\{0, -r\}.$$

We now state and prove a result of Peterson and Ridenhour [123]. We use the notation

$$t^{(r)} = t(t-1)(t-2)\cdots(t-r+1)$$

for r a positive integer. This notation was used on page 20 of Kelley and Peterson [88].

Theorem 4.32 *If*

$$\left(\frac{b+5-a}{2}\right)^{(3)} \sum_{t=a+2}^{b+2} r_-(t) < 24, \tag{4.30}$$

then $\Delta^4 y(t-2) + r(t)y(t) = 0$ *is* $(2,2)$*–disconjugate on* $[a, b+4]$.

Proof: Assume $\Delta^4 y(t-2)+r(t)y(t)=0$ is not (2,2)–disconjugate on $[a,b+4]$, then there are integers t_1, t_2 and a nontrivial solution $u(t)$ such that $a \leq t_1-1 < t_1 < t_2 \leq b+3$ and

$$u(t_1-1)u(t_1+1) \geq 0, \qquad y(t_1+1) \neq 0$$

$$u(t_1) = 0 = u(t_2),$$

$$u(t_2-1)u(t_2+1) \geq 0, \qquad y(t_2-1) \neq 0.$$

Define η on $[a,b+4]$ by

$$\eta(t) = \begin{cases} 0, & a \leq t \leq t_1-1, \\ u(t), & t_1 \leq t \leq t_2, \\ 0, & t_2+1 \leq t \leq b+4\ . \end{cases}$$

Note $\eta \in A$ and $\eta(t) \not\equiv 0$ on $[a,b+2]$.

For the difference equation $\Delta^4 y(t-2)+r(t)y(t)=0$, the quadratic functional Q is given by

$$Q[\eta] = \sum_{t=a+2}^{b+4} \{[\Delta^2\eta(t-2)]^2 + r(t)\eta^2(t)\}.$$

By Theorem 4.21

$$Q[\eta] = -u(t_1-1)u(t_1+1) - u(t_2-1)u(t_2+1) \leq 0.$$

But we will now use the inequality (4.30) to show that $Q[\eta] > 0$ which would be a contradiction. To see this first pick $t_0 \in [t_1+1, t_2-1]$ so that

$$\begin{aligned} |u(t_0)| &= \max\{|u(t)| : t_1 \leq t \leq t_2\} \\ &= \max\{|\eta(t)| : a \leq t \leq b+4\}. \end{aligned}$$

Define $v(t)$ on $[a, t_0+2]$ to be the solution of the BVP

$$\begin{aligned} &\Delta^4 v(t-2) = 0, \quad t \in [a+2, t_0] \\ &v(a) = \Delta v(a) = 0 \\ &v(t_0) = \eta(t_0) \\ &\Delta v(t_0-1) = 0. \end{aligned}$$

It follows that $v(t)$ is of the form

$$v(t) = \alpha(t-a)^{(2)} + \beta(t-a)^{(3)}.$$

The last two boundary conditions for $v(t)$ gives us the system of equations

$$\alpha(t_0 - a)^{(2)} + B(t_0 - a)^{(3)} = \eta(t_0)$$
$$2\alpha(t_0 - 1 - a)^{(1)} + 3B(t_0 - 1 - a)^{(2)} = 0.$$

Solving for β we get that

$$\beta = \frac{-2\eta(t_0)}{(t_0 - a)^{(3)}}.$$

Since $\Delta^3 v(t) = 3!\beta$ we get that

$$\Delta^3 v(t_0 - 1) = \frac{-12\eta(t_0)}{(t_0 - a)^{(3)}}. \tag{4.31}$$

From the inequality

$$\sum_{t=a+2}^{t_0} [\Delta^2\eta(t-2) - \Delta^2 v(t-2)]^2 \geq 0,$$

one has

$$\sum_{t=a+2}^{t_0} [\Delta^2\eta(t-2)]^2 \geq \sum_{t=a+2}^{t_0} \Delta^2 v(t-2)[2\Delta^2\eta(t-2) - \Delta^2 v(t-2)].$$

Using a summation by parts formula yields

$$\begin{aligned}
\sum_{t=a+2}^{t_0} [\Delta^2\eta(t-1)]^2 \quad &\geq \quad \{\Delta^2 v(t-2)[2\Delta\eta(t-2) - \Delta v(t-2)]\}_{a+2}^{t_0+1} \\
&\quad - \sum_{t=a+2}^{t_0} \Delta^3 v(t-2)[2\Delta\eta(t-1) - \Delta v(t-1)] \\
&= \quad 2\Delta^2 v(t_0 - 1)\Delta\eta(t_0 - 1) \\
&\quad - \sum_{t=a+2}^{t_0} \Delta^3 v(t-2)[2\Delta\eta(t-1) - \Delta v(t-1)].
\end{aligned}$$

Considering the two cases $\eta(t_0) > 0$ and $\eta(t_0) < 0$ separately it can be shown that

$$\Delta^2 v(t_0 - 1)\Delta\eta(t_0 - 1) \leq 0.$$

Hence

$$\sum_{t=a+2}^{t_0} [\Delta^2\eta(t-2)]^2 \geq -\sum_{t=a+2}^{t_0} \Delta^3 v(t-2)[2\Delta\eta(t-1) - \Delta v(t-1)].$$

Using a summation by parts formula we get that

$$\begin{aligned}\sum_{t=a+2}^{t_0} [\Delta^2\eta(t-2)]^2 &\geq -\Delta^3 v(t-2)[2\eta(t-1) - v(t-1)]\Big|_{a+2}^{t_0+1} \\ &\quad + \sum_{t=a+2}^{t_0} \Delta^4 v(t-2)[2\eta(t) - v(t)] \\ &= -\Delta^3 v(t_0-1)\eta(t_0).\end{aligned}$$

Hence by (4.31) we get that

$$\sum_{t=a+2}^{t_0} [\Delta^2\eta(t-2)]^2 \geq \frac{12}{(t_0-a)^{(3)}}\eta^2(t_0). \tag{4.32}$$

Next define $w(t)$ on $[t_0-2, b+4]$ to be the solution of the boundary value problem

$$\begin{aligned}&\Delta^4 w(t-2) = 0, \qquad t \in [t_0, b+2] \\ &w(t_0) = \eta(t_0) \\ &\Delta w(t_0-1) = 0 \\ &w(b+3) = \Delta w(b+3) = 0.\end{aligned}$$

It can be shown that

$$\Delta^3 w(t_0-1) = \frac{12\eta(t_0)}{(b+5-t_0)^{(3)}}. \tag{4.33}$$

The inequality

$$\sum_{t=t_0+1}^{b+4} [\Delta^2\eta(t-2) - \Delta^2 w(t-2)]^2 \geq 0$$

implies that

$$\sum_{t=t_0+1}^{b+4} [\Delta^2\eta(t-2)]^2 \geq \sum_{t=t_0+1}^{b+4} \{\Delta^2 w(t-2)[2\Delta^2\eta(t-2) - \Delta^2 w(t-2)]\}.$$

By summation by parts we get that

$$\begin{aligned}
\sum_{t=t_0+1}^{b+4} [\Delta^2\eta(t-2)]^2 &\geq \Delta^2 w(t-2)[2\Delta\eta(t-2) - \Delta w(t-2)]\Big|_{t_0+1}^{b+5} \\
&\quad - \sum_{t=t_0+1}^{b+4} \{\Delta^3 w(t-2)[2\Delta\eta(t-2) - \Delta w(t-1)]\} \\
&= -2\Delta^2 w(t_0-1)\Delta\eta(t_0-1) \\
&\quad - \sum_{t=t_0+1}^{b+4} \{\Delta^3 w(t-2)[2\Delta\eta(t-1) - \Delta w(t-1)]\} \\
&\geq - \sum_{t=t_0+1}^{b+4} \{\Delta^3 w(t-2)[2\Delta\eta(t-1) - \Delta w(t-1)]\}.
\end{aligned}$$

Using summation by parts again we get that

$$\begin{aligned}
\sum_{t=t_0+1}^{b+4} \left[\Delta^2\eta(t-2)\right]^2 &\geq -\Delta^3 w(t-2)\left[2\eta(t-1) - w(t-1)\right]\Big|_{t_0+1}^{b+5} \\
&\quad + \sum_{t=t_0}^{b+4} \Delta^4 w(t-2)\left[2\eta(t) - w(t)\right] \\
&= \Delta^3 w(t_0-1)\eta(t_0-1).
\end{aligned}$$

Hence by (4.33)

$$\sum_{t=t_0+1}^{b+4} \left[\Delta^2\eta(t-2)\right]^2 \geq \frac{12\eta^2(t_0)}{(b+5-t_0)^{(3)}}. \tag{4.34}$$

From (4.32) and (4.34) we obtain

$$\begin{aligned}
\sum_{t=a+2}^{b+4} \left[\Delta^2\eta(t-2)\right]^2 &= \sum_{t=a+2}^{t_0} \left[\Delta^2\eta(t-2)\right]^2 + \sum_{t=t_0+1}^{b+4} \left[\Delta^2\eta(t-2)\right]^2 \\
&\geq \left[\frac{12}{(t_0-a)^{(3)}} + \frac{12}{(b+s-t_0)^{(3)}}\right]\eta^2(t_0).
\end{aligned}$$

Since the minimum of

$$\frac{12}{(t_0-a)^{(3)}} + \frac{12}{(b+s-t_0)^{(3)}}$$

as a function of a continuous variable t_0 occurs at the midpoint $t_0 = \frac{a+b+5}{2}$ we get that

$$\sum_{t=a+2}^{b+4} \left[\Delta^2\eta(t-2)\right]^2 \geq 24/(\frac{b+5-a}{2})^{(3)}.$$

It follows that

$$\begin{aligned} Q[\eta] &= \sum_{t=a+2}^{b+4} \{\left[\Delta^2\eta(t-2)\right]^2 + r(t)\eta^2(t)\} \\ &\geq \left[\frac{24}{(\frac{b+5-a}{2})^{(3)}} - \sum_{t=a+2}^{b+2} r_-(t)\right]\eta^2(t_0) > 0 \end{aligned}$$

by (4.30) which is the contradiction that we sought. □

We next give an example that shows that the constant 24 in Theorem 4.32 is the smallest constant for which that theorem is true.

Example 4.33 *Consider the difference equation*

$$\Delta^4 y(t-2) + r(t)y(t) = 0, \quad t \in [2,5],$$

where $r(t)$ is defined on $[2,5]$ as follows:

$$\begin{aligned} r(2) &= r(5) = 0 \\ r(3) &= r(4) = -\frac{1}{2}. \end{aligned}$$

Here $a = 0$, $b = 3$, so

$$\left(\frac{b+5-a}{2}\right)^{(3)} \sum_{t=a+2}^{b+2} r(t) = 4^{(3)} \cdot 1 = 24.$$

Let $y(t)$ be the solution of the initial value problem

$$\begin{aligned} &\Delta^4 y(t-2) + r(t)y(t) = 0 \\ &y(0) = y(1) = 0 \\ &y(2) = 1, \ y(3) = 2. \end{aligned}$$

It follows that $y(4) = 2$, $y(5) = 1$, $y(6) = y(7) = 0$. Hence

$$\Delta^4 y(t-2) + r(t)y(t) = 0$$

is not $(2,2)$–disconjugate on $[0,7]$ and Theorem 4.32 is sharp in the sense described in the paragraph proceeding this example.

Example 4.34 *Show for the difference equation*

$$\Delta^4 y(t-2) - 2y(t) = 0, \qquad t \in [2,3],$$

that the inequality (4.30) *is an equality and this difference equation is not* $(2,2)$*-disconjugate on* $[0,5]$.

4.7 VARIABLE STEP VARIATIONAL PROBLEMS

The previous discussion had uniform step size of 1. Because of that restriction, there was no way to let the step size go to zero so as to compare with continuous problems. In order to remedy that severe restriction, we now formulate a variable step size discrete model of the continuous variational problem

$$J[y] = \int_a^b F(t, y(t), y'(t))dt \tag{4.35}$$

on a class of functions $y(t)$ which satisfy fixed end conditions $y(a) = c$ and $y(b) = d$. We assume throughout that $F(t, y, r)$ is continuous on $[a, b] \times R^d \times R^d$ for some positive integer d, although often the domain of F might be some subset of this region. A classical variational problem is the problem of minimal lateral surface of revolution.

Example 4.35 (Mimimal Surface of Revolution) *Set*

$$F(t, y, r) = 2\pi y\sqrt{1 + r^2}.$$

For nonnegative $y(t)$, *the lateral surface area* A *obtained by rotating the curve* $y = y(t)$ *about the* t *axis is represented by the integral*

$$A = \int_a^b F(t, y(t), y'(t))dt.$$

The Euler–Lagrange equation

$$F_y(t, y(t), y'(t)) = (F_r(t, y(t), y'(t)))'$$

for this functional is

$$\sqrt{1 + (y')^2} = \frac{d}{dt}\left[\frac{yy'}{\sqrt{1 + (y')^2}}\right].$$

In order to formulate a discrete variational problem that corresponds reasonably well with continuous variational problems, it is convenient to consider a function f of four variables as follows. Suppose that $f(s,t,y,r)$ is a continuous function defined on $[a,b] \times [a,b] \times R^d \times R^d$ which has continuous second order partials in the variables y and r. Let Π be a partition of the interval $[a,b]$ into N subintervals

$$a = t_0 < t_1 < \ldots < t_N = b.$$

Consider a discrete variational problem on this partition defined by

$$J[y] = \sum_{n=1}^{N} f\Big(t_{n-1}, t_n, y(t_n), \frac{\Delta y(t_{n-1})}{\Delta t_{n-1}}\Big)\Delta t_{n-1} \tag{4.36}$$

on the class of functions $\mathcal{F}$ consisting of R^d valued $y(t)$ defined at $t_0, \ldots, t_N$ which satisfy the fixed end conditions $y(t_0) = y_0 = c$ and $y(t_N) = y_N = d$. We will also use subscript notation y_n for y evaluated at t_n.

The associated continuous variational problem (4.35) has integrand

$$F(t,y,r) = f(t,t,y,r).$$

Theorem 4.36 (Variable Step Euler–Lagrange Equation) *The discrete Euler–Lagrange equation for this fixed endpoint discrete variational problem is*

$$f_y\Big(t_{n-1}, t_n, y_n, \frac{\Delta y_{n-1}}{\Delta t_{n-1}}\Big) = \frac{\Delta f_r(t_{n-1}, t_n, y_n, \frac{\Delta y_{n-1}}{\Delta t_{n-1}})}{\Delta t_{n-1}}. \tag{4.37}$$

If J has a local extremum at $\hat{y} \in \mathcal{F}$, then the Euler–Lagrange equation is satisfied along $\hat{y}$ for $n = 1, 2, \ldots, N-1$.

Proof: Modify the proof of Theorem 4.1 on page 154 as follows. Assume J has a local extremum on $\mathcal{F}$ at $\hat{y}$. Fix $\eta(t_n)$ as an admissible variation on the partition Π with zero end conditions. Define a function $\phi : \mathrm{R} \to \mathrm{R}$ by

$$\begin{aligned}\phi(\varepsilon) &= J[\hat{y} + \varepsilon\eta] \\ &= \sum_{n=1}^{N} f\Big(t_{n-1}, t_n, \hat{y}(t_n) + \varepsilon\eta(t_n), \frac{\Delta\left(\hat{y}(t_{n-1}) + \varepsilon\eta(t_{n-1})\right)}{\Delta t_{n-1}}\Big)\Delta t_{n-1}.\end{aligned}$$

We are now assuming that ϕ has a local extremum at the interior point $\varepsilon = 0$ and hence $\phi'(0) = 0$. Denote the column vector of partials (f_{y^i}) by f_y and similarly for f_r. Calculation of the derivative of $\phi(\varepsilon)$ gives

$$\phi'(\varepsilon) = \sum_{n=1}^{N} \left[\langle f_y, \eta_n \rangle + \langle f_r, \frac{\Delta\eta_{n-1}}{\Delta t_{n-1}} \rangle \right] \Delta t_{n-1} \tag{4.38}$$

where $\langle u, v \rangle = v^T u$ is the usual inner product on R^d. Recall that η_n denotes the d–vector valued function η evaluated at the point t_n. Define the first variation J_1 by $J_1[\eta] = \phi'(0)$. Then

$$J_1[\eta] \equiv \sum_{n=1}^{N} \{ \langle f_y, \eta_n \rangle \Delta t_{n-1} + \langle f_r, \Delta\eta_{n-1} \rangle \} = 0. \tag{4.39}$$

The vector summation by parts formula

$$\sum_{n=1}^{N} \langle u_{n-1}, \Delta v_{n-1} \rangle = \langle u_n, v_n \rangle \Big|_0^N - \sum_{n=1}^{N} \langle \Delta u_{n-1}, v_n \rangle \tag{4.40}$$

applied to the second term gives

$$\sum_{n=1}^{N} \langle f_y \Delta t_{n-1} - \Delta f_r, \eta_n \rangle = 0.$$

Since $\eta_N = 0$ and the values of the d–vector η_n are arbitrary for $n = 1, \ldots,$ $N-1$, the vector Euler–Lagrange equation holds along $\hat{y}$ at the stated values.

□

Exercise 4.37 (Quadratic Functionals and Jacobi Equations) *Suppose that f is a quadratic function (such integrands arise in the second variation)*

$$f(s, t, y, r) = \frac{1}{2}(y^T[Py + Qr] + r^T[Q^T y + Rr]),$$

where P, Q, R are $d \times d$ matrix functions of s, t with P and R symmetric. Show that the Euler Lagrange equation for this f is the discrete Jacobi equation

$$Py_n + Q\frac{\Delta y_{n-1}}{\Delta t_{n-1}} = \frac{\Delta \left(Q^T y_n + R\frac{\Delta y_{n-1}}{\Delta t_{n-1}} \right)}{\Delta t_{n-1}}, \tag{4.41}$$

where P, Q, R are evaluated at (t_{n-1}, t_n).

For the associated discrete Hamiltonian for $f(s,t,y,r)$ we must modify the notation of equation (4.19) on page 165, which defines g, to accommodate the way in which the discrete momentum variable will be defined when step sizes are not 1. Assume that there exists a vector function g such that

$$f_r(s,t,u,v) = w \qquad \text{if and only if} \qquad v = g(s,t,u,w). \tag{4.42}$$

In this case, the Hamiltonian $H(s,t,u,w)$ is the real valued function defined for scalar s, scalar t, and vector arguments u, w by

$$H(s,t,u,w) = \langle w, g(s,t,u,w)\rangle - f(s,t,u,g(s,t,u,w)). \tag{4.43}$$

(The corresponding continuous Hamiltonian for F is $H(t,u,w) = H(t,t,u,w)$.) A *discrete vector momentum variable* z corresponding to a discrete solution y is defined by

$$z_{n-1} = f_r\Big(t_{n-1}, t_n, y_n, \frac{\Delta y_{n-1}}{\Delta t_{n-1}}\Big) \qquad \text{for} \qquad n = 1,2,\ldots,N. \tag{4.44}$$

The *discrete Hamiltonian system* is given by

$$\begin{aligned} \frac{\Delta y_{n-1}}{\Delta t_{n-1}} &= H_w(t_{n-1}, t_n, y_n, z_{n-1}), \qquad \text{for} \quad n = 1,2,\ldots,N, \\ \frac{\Delta z_{n-1}}{\Delta t_{n-1}} &= -H_u(t_{n-1}, t_n, y_n, z_{n-1}), \qquad \text{for} \quad n = 1,2,\ldots,N-1. \end{aligned} \tag{4.45}$$

Exercise 4.38 (The Discrete Legendre Transformation) *Show that if y is such that the discrete Euler–Lagrange equation* (4.37) *is satisfied for $n = 1,\ldots,N-1$, and a discrete momentum variable is defined by* (4.44), *then the pair y, z satisfies the discrete Hamiltonian system* (4.45).

We remark that a discrete Hamiltonian system can be converted [6] to a discrete Euler–Lagrange equation if the equation $H_w(s,t,y,z) = r$ defines z as a function of s, t, y, r, say $z = \gamma(s,t,y,r)$. If that is possible and f is defined by

$$f(s,t,y,r) = \langle r, \gamma(s,t,y,r)\rangle - H(s,t,y,\gamma(s,t,y,r)) \tag{4.46}$$

then solutions of the discrete Hamiltonian system yield solutions of the discrete Euler–Lagrange equation.

Exercise 4.39 (Quadratic Hamiltonians) *Suppose that a Hamiltonian is defined by a quadratic form*

$$H(s,t,u,w) = \frac{1}{2}\begin{bmatrix} u^T & w^T \end{bmatrix} \begin{bmatrix} -C & A^T \\ A & B \end{bmatrix} \begin{bmatrix} u \\ w \end{bmatrix} \tag{4.47}$$

where A, B, and C are $d \times d$ matrix valued functions of s,t with B and C symmetric. Show that the associated Hamiltonian system is the discrete linear Hamiltonian system

$$\begin{aligned} \frac{\Delta y_{n-1}}{\Delta t_{n-1}} &= A(t_{n-1},t_n)y_n + B(t_{n-1},t_n)z_{n-1} \\ \frac{\Delta z_{n-1}}{\Delta t_{n-1}} &= C(t_{n-1},t_n)y_n - A^T(t_{n-1},t_n)z_{n-1}. \end{aligned} \tag{4.48}$$

Exercise 4.40 (Jacobi Equation as a System) *Define the momentum variable $z = f_r$ along a solution from* (4.44) *and* (4.41) *as*

$$z_{n-1} = Q^T y_n + R\frac{\Delta y_{n-1}}{\Delta t_{n-1}} \qquad \text{for} \qquad n = 1,2,\ldots,N, \tag{4.49}$$

with Q and R evaluated at (t_{n-1},t_n). Then the associated discrete Hamiltonian system (4.45) *is of the form of a discrete linear Hamiltonian system* (4.48) *with*

$$\begin{aligned} A &= -R^{-1}Q^T, \qquad B = R^{-1}, \\ C &= P - QR^{-1}Q^T. \end{aligned} \tag{4.50}$$

Exercise 4.41 (Associated Symplectic System) *Let $s = t_{n-1}$, $t = t_n$ and $h = t - s$. Show that the matrix*

$$\left[\left(\frac{1}{h}\right) I - A(s,t)\right] \tag{4.51}$$

is singular if and only if $1/h$ is an eigenvalue of $A(s,t)$. The stepsize criterion is as follows: Assume that given s we may choose t such that

$$\frac{1}{t-s} > \rho(A(s,t)), \tag{4.52}$$

where $\rho(A)$ denotes the spectral radius of A. If this criterion is satisfied, then the system (4.48) *can be solved for y_n, z_n from given values of y_{n-1}, z_{n-1}. Then show that the resulting system has the form*

$$\begin{bmatrix} y(t) \\ z(t) \end{bmatrix} = M(s,t) \begin{bmatrix} y(s) \\ z(s) \end{bmatrix} \tag{4.53}$$

with $M(s,t)$ the symplectic matrix with block entries

$$\begin{aligned} E &= [I - hA]^{-1}, \qquad F = hEB, \\ G &= hCE, \qquad H = (E^T)^{-1} + h^2 CEB. \end{aligned} \tag{4.54}$$

Exercise 4.42 (Conversion to Stepsize of 1) *Suppose that*

$$\hat{M}(n-1,n) = M(t_{n-1}, t_n), \qquad \hat{y}(n) = y(t_n), \quad \text{and} \quad \hat{z}(n) = z(t_n).$$

Show that this converts the linear symplectic system (4.53) *to the form*

$$\begin{bmatrix} \hat{y}(n) \\ \hat{z}(n) \end{bmatrix} = \hat{M}(n-1,n) \begin{bmatrix} \hat{y}(n-1) \\ \hat{z}(n-1) \end{bmatrix} \tag{4.55}$$

which now has stepsize of 1. Compare this system with the systems arising in equation (3.125) *on page* 144. *(Replace t here by $t+1$ to get those systems.)*

Consequently, any definition made for step size 1 can be reinterpreted for variable steps.

Exercise 4.43 (Convergence to Continuous System) *Show that if in the above Exercises* 4.39 *and* 4.41 *one lets s be fixed, assumes $y(s)$ and $z(s)$ are given, lets t go to s, and assumes that there exist $\hat{A}(s)$, $\hat{B}(s)$, and $\hat{C}(s)$ such that $A(s,t) \to \hat{A}(s)$, $B(s,t) \to \hat{B}(s)$, and $C(s,t) \to \hat{C}(s)$, as $t \to s$, then*

$$\frac{y(t) - y(s)}{t - s} \text{ has a limit as } t \to s \text{ of } y'(s)$$

$$\frac{z(t) - z(s)}{t - s} \text{ has a limit as } t \to s \text{ of } z'(s)$$

and

$$y'(s) = \hat{A}(s)y(s) + \hat{B}(s)z(s) \tag{4.56}$$
$$z'(s) = \hat{C}(s)y(s) - \hat{A}^T(s)z(s). \tag{4.57}$$

In particular, if A, B, C are functions of s alone, then the values $y(t), z(t)$ of the solution of the discrete linear Hamiltonian system with initial values $y(s)$, $z(s)$ give difference quotients which converge to the values of the $y'(s)$ and $z'(s)$ for the corresponding linear differential system as $t \to s$. Show this by showing that

$$\frac{1}{h}\left(\begin{bmatrix} y(t) \\ z(t) \end{bmatrix} - \begin{bmatrix} y(s) \\ z(s) \end{bmatrix}\right) = \frac{1}{h}(M(s,t) - I)\begin{bmatrix} y(s) \\ z(s) \end{bmatrix} \to \begin{bmatrix} \hat{A}(s) & \hat{B}(s) \\ \hat{C}(s) & -\hat{A}^T(s) \end{bmatrix}\begin{bmatrix} y(s) \\ z(s) \end{bmatrix}$$

as $t \to s$. Notice that t could be less than s so the limit is two sided.

Exercise 4.44 (Discrete Riccati Converges to Continuous Riccati) *If $h = t - s$, A, B, C are functions of s alone, and $Y(t)$ and $Z(t)$ are $d \times d$ matrix valued functions which satisfy a system*

$$\begin{bmatrix} Y(t) \\ Z(t) \end{bmatrix} = M(s,t)\begin{bmatrix} Y(s) \\ Z(s) \end{bmatrix} \tag{4.58}$$

with $M(s,t)$ the symplectic matrix with block entries

$$\begin{aligned} E(s,t) &= [I - hA(s)]^{-1}, & F(s,t) &= hE(s,t)B(s), \\ G(s,t) &= hC(s)E(s,t), & H(s,t) &= (E^T(s,t))^{-1} + h^2C(s)E(s,t)B(s), \end{aligned} \tag{4.59}$$

and if $Y(s)$ is nonsingular for t close to s, then show that $W(t) = Z(t)Y^{-1}(t)$ satisfies the associated discrete Riccati equation determined by

$$W(t) = (G(s,t) + H(s,t)W(s))\,(E(s,t) + F(s,t)W(s))^{-1} \tag{4.60}$$

which yields the continuous Riccati equation

$$W'(s) = C(s) - A^T(s)W(s) - W(s)A(s) - W(s)B(s)W(s) \tag{4.61}$$

as $t \to s$.

4.8 DISCRETE HAMILTONIAN SYSTEMS YIELD SYMPLECTIC INTEGRATORS

Let us return to the general nonlinear discrete Hamiltonian system (4.45) on page 186. Suppose that $y = y_{n-1}$ and $z = z_{n-1}$ are functional values at t_{n-1}.

Let t_n be fixed and distinct from t_{n-1}. Introduce the labels $p = y_n$ and $q = z_n$ and set $h = t_n - t_{n-1}$. Then system (4.45) may be written as

$$p = y + hH_w, \qquad q = z - hH_u \tag{4.62}$$

where H_w and H_u are evaluated at (t_{n-1}, t_n, p, z). Note that p is implicit and q is explicit.

Theorem 4.45 (Symplectic Maps) *Suppose t_{n-1} and t_n are fixed, $h = t_n - t_{n-1}$, and $\mathcal{U}$ is an open set in $R^d \times R^d$. Assume that Ψ is a C' map from $\mathcal{U}$ to $\mathcal{U}$ such that $(p, q) = \Psi(y, z)$ satisfies system (4.62) on $\mathcal{U}$. Then Ψ is symplectic on $\mathcal{U}$, i.e., $M \equiv \Psi'(y, z)$ implies $M^T J M = J$ where*

$$M \equiv \Psi'(y,z) = \begin{bmatrix} \nabla_y^T p & \nabla_z^T p \\ \nabla_y^T q & \nabla_z^T q \end{bmatrix} \equiv \begin{bmatrix} \mathcal{E} & \mathcal{F} \\ \mathcal{G} & \mathcal{H} \end{bmatrix} \quad \text{and} \quad J = \begin{bmatrix} 0 & I \\ -I & 0 \end{bmatrix}.$$

Here, $\nabla_y^T = \left[\frac{\partial}{\partial y_1} \quad \cdots \quad \frac{\partial}{\partial y_n}\right]$ applied to a vector means it is applied to each component.

Proof: For $H = H(t_{n-1}, t_n, u, w)$ for $u, w \in R^d$, introduce the notation

$$H_{uu} = \left(H_{u_i u_j}\right), \qquad H_{wu} = \left(H_{w_i u_j}\right), \qquad \text{and} \qquad H_{ww} = \left(H_{w_i w_j}\right). \tag{4.63}$$

Thus these second derivative matrices are of order $d \times d$. The chain rule for differentiation applied to

$$p = y + hH_w(t_{n-1}, t_n, p, z), \qquad q = z - hH_u(t_{n-1}, t_n, p, z))$$

gives

$$\mathcal{E} \equiv \left[\tfrac{\partial p_i}{\partial y_j}\right] = I + hH_{wu}\mathcal{E} \;\Rightarrow \qquad I = (I - hH_{wu})\mathcal{E},$$

$$\mathcal{F} \equiv \left[\tfrac{\partial p_i}{\partial z_j}\right] = hH_{wu}\mathcal{F} + hH_{ww} \;\Rightarrow\; (I - hH_{wu})\mathcal{F} = hH_{ww},$$

$$\mathcal{G} \equiv \left[\tfrac{\partial q_i}{\partial y_j}\right] = -hH_{uu}\mathcal{E}, \qquad \mathcal{H} \equiv \left[\tfrac{\partial q_i}{\partial z_j}\right] = I - hH_{uw} - hH_{uu}\mathcal{F}.$$

Also, $H_{uw}^T = H_{wu}$ whereas H_{uu} and H_{ww} are symmetric. In order to show that M is symplectic we observe the following: Premultiply $\mathcal{G} = -hH_{uu}\mathcal{E}$ by $\mathcal{E}^T$ and observe that the matrix $\mathcal{E}^T\mathcal{G} = -h\mathcal{E}^T H_{uu}\mathcal{E}$ is symmetric. Since

$$\mathcal{F}^T\mathcal{H} = \mathcal{F}^T(I - hH_{uw}) - h\mathcal{F}^T H_{uu}\mathcal{F}$$

and because the first term of the right hand side is the transpose of the matrix $(I - hH_{wu})\mathcal{F} = hH_{ww}$, both terms are symmetric and $\mathcal{F}^T\mathcal{H}$ is symmetric. As

$$\mathcal{E}^T\mathcal{H} - \mathcal{G}^T\mathcal{F} = \mathcal{E}^T(I - hH_{uw}) - h\mathcal{E}^T H_{uu}\mathcal{F} + h\mathcal{E}^T H_{uu}\mathcal{F} = \left([I - hH_{wu}]\mathcal{E}\right)^T = I,$$

we conclude that $M = \Psi'$ is symplectic from the characterization of symplectic matrices given in Chapter 3. □

4.9 EXISTENCE AND UNIQUENESS OF LOCAL SOLUTIONS OF DISCRETE HAMILTONIAN SYSTEMS

For vector x we use the infinity norm $|x|_\infty = \max |x_i|$ and for vector a and $\delta > 0$, we let the closed neighborhood

$$\mathcal{N}_\delta(a) = \{x : |x - a|_\infty \leq \delta\}$$

denote the solid box (including the boundary) of side 2δ centered at a. Consider vectors $x \in R^m$ and $y \in R^n$ represented, respectively, by $m \times 1$ and $n \times 1$ column vectors and a vector function $f(x, y) = \big(f_i(x, y)\big)$, expressed as an $n \times 1$ column vector valued function. We now give conditions under which a vector equation $f(x, y) = 0$ determines y as a function of x. We let the gradient in the y variables ∇_y be a column n–vector. Then ∇_y^T is a row vector.

Theorem 4.46 (Implicit Function Theorem) *Let f be a continuous function from an open set S in $R^m \times R^n$ to R^n such that the $n \times n$ partial derivative matrix*

$$f_y = \nabla_y^T f = \left[\tfrac{\partial f_i}{\partial y_j}\right]$$

exists and has continuous entries on S. Assume that $(a, b) \in S$ is such that

$$f(a, b) = 0 \qquad \text{and} \qquad f_y(a, b) \quad \text{is nonsingular.}$$

Then there exists positive real numbers δ and γ and a continuous $n \times 1$ function $\Psi(x)$ from $\mathcal{N}_\delta(a)$ to $\mathcal{N}_\gamma(b)$ such that

$$\Psi(a) = b, \qquad f(x, \Psi(x)) = 0 \quad \text{on} \quad \mathcal{N}_\delta(b)$$

and if $(x, y) \in \mathcal{N}_\delta(a) \times \mathcal{N}_\gamma(b)$ is such that $f(x, y) = 0$, then $y = \Psi(x)$. If, in addition, for some index k the partial derivative $\partial f / \partial x_k$ exists and is continuous, then $\partial \Psi / \partial x_k$ exists, is given by

$$\frac{\partial \Psi}{\partial x_k} = -\left[f_y\right]^{-1} \frac{\partial f}{\partial x_k} \tag{4.64}$$

and is continuous.

We can readily establish this Implicit Function Theorem once we have the following lemma. We will be using the fact that the definition of continuity can be stated in terms of closed neighborhoods instead of open neighborhoods.

Lemma 4.47 *Let f and S be as in the* **IFT**. *Let ρ be such that*

$$\mathcal{N}_\rho(a,b) = \mathcal{N}_\rho(a) \times \mathcal{N}_\rho(b)$$

is contained in S. Then there exists a positive real number γ with $\gamma \le \rho$ such that the following conditions hold.

(a) *The matrix*

$$A(x, y^1, y^2, \ldots, y^n) = \begin{bmatrix} \nabla_y^T f_1(x, y^1) \\ \nabla_y^T f_2(x, y^2) \\ \cdots \\ \nabla_y^T f_n(x, y^n) \end{bmatrix}$$

is nonsingular on $\mathcal{N}_\gamma(a) \times \mathcal{N}_\gamma(b) \times \cdots \times \mathcal{N}_\gamma(b)$.

(b) *If (x, y) and (x, z) in $\mathcal{N}_\gamma(a, b)$ are such that*

$$f(x,y) = 0 = f(x,z), \qquad \text{then} \quad y = z.$$

(c) *Let*

$$\phi(x,y) \equiv |f(x,y)|_2^2 = \sum f_i^2(x,y).$$

Then for each positive ε there exists a $\delta \in (0, \gamma]$ such that

$$x \in \mathcal{N}_\delta(a), \ y \text{ and } z \ \text{ in } \ \mathcal{N}_\gamma(b) \ \text{ with } \ ||y - z||_\infty \le \delta,$$

imply

$$|\phi(x,y) - \phi(a,z)| \le \varepsilon.$$

(d) *There exists a positive real number μ such that*

$$||A^{-1}||_\infty \le \mu$$

on $\mathcal{N}_\gamma(a) \times \mathcal{N}_\gamma(b) \times \cdots \times \mathcal{N}_\gamma(b)$.

Proof of Lemma: **(a)** Since A has continuous entries and $\det A$ converges to $\det A(a, b, \ldots, b) = \det f_y(a, b) \ne 0$ as $(x, y^1, \ldots, y^n) \to (a, b, \ldots b)$, there exists $\gamma \le \rho$ such that $\det A \ne 0$ in the closed γ neighborhood of $(a, b, \ldots, b)$.

(b) Use the mean value theorem on each component f_i of f for the existence of vectors y^i between y and z such that

$$0 = f(x,z) - f(x,y) = A(x, y^1, y^2, \ldots, y^n)(z - y).$$

Nonsingularity of A implies that $y = z$.

(c) Since $\phi(x, y)$ is continuous on a compact set, it is uniformly continuous on $\mathcal{N}_\gamma(a, b)$.

(d) The matrix ∞–norm is the maximum absolute row sum [143]. Since $||A^{-1}||_\infty$ is continuous on a bounded and closed set, there exists a positive constant μ such that $||A^{-1}||_\infty \leq \mu$ on the given box. We have established the Lemma. □

Proof of the **IFT***:* The proof in the scalar case depends upon showing that the function $f^2(x, y)$, for x fixed, has a relative minimum at an interior point $\hat{y}$. Then differentiation gives $0 = 2f(x, \hat{y})f_y(x, \hat{y})$ and since the partial f_y is nonzero, there must exist a $\hat{y}$ at which $f(x, \hat{y}) = 0$. The proof is somewhat involved in higher dimensions, but most of the work has been done in the Lemma. For γ as in the Lemma, let

$$3\varepsilon = \min_{|y-b|=\gamma} \phi(a, y) = \min_{|y-b|=\gamma} |f(a, y)|_2^2,$$

where $|y - b| = |y - b|_\infty$. By **(b)** of the Lemma and $f(a, b) = 0$ we know that $\phi(a, y)$ is positive on the boundary of the box $\mathcal{N}_\gamma(b)$. From compactness, we know that the minimum exists and ε is positive. Then for δ as in **(c)** and $x \in \mathcal{N}_\delta(a)$, and $|y - b| = \gamma$

$$\phi(x, y) = \phi(x, y) - \phi(a, y) + \phi(a, y) \geq -\varepsilon + 3\varepsilon = 2\varepsilon.$$

Note that $\phi(a, b) = 0$. Then use part (c) of the Lemma again with $z = b$ for the observation that $x \in \mathcal{N}_\delta(a)$, and $y \in \mathcal{N}_\delta(b)$ imply $\phi(x, y) \leq \varepsilon$. By compactness of the region and continuity of ϕ, for each such x, the minimum of $\phi(x, y)$ is not on the boundary and there must exist a relative minimum of $\phi(x, y)$ at some point $\hat{y}$ of the interior. More precisely, for each fixed $x \in \mathcal{N}_\delta(a)$, $\phi(x, y)$ has a relative minimum at some $\hat{y}$ interior to $\mathcal{N}_\gamma(b)$ and we have

$$\phi'(\hat{y}) = (\nabla\phi)(\hat{y}) = 2f_y(x, \hat{y})f(x, \hat{y}) = 0.$$

Nonsingularity of f_y implies that $f(x, \hat{y}) = 0$. Part **(b)** of the Lemma tells us that $\hat{y} \equiv \Psi(x)$ is unique in this box. For continuity of $\Psi(x)$, let x and $x + \Delta x$ be in $\mathcal{N}_\delta(a)$, let $y = \Psi(x)$ and $y + \Delta y = \Psi(x + \Delta x)$. Then by the mean value theorem on each component, there exist y^j between y and $y + \Delta y$ such that

$$\begin{aligned} 0 &= f(x + \Delta x, y + \Delta y) - f(x, y) \\ &= f(x + \Delta x, y + \Delta y) - f(x + \Delta x, y) \\ &\quad + f(x + \Delta x, y) - f(x, y) \\ &= A(x + \Delta x, y^1, \ldots, y^n)(\Delta y) + f(x + \Delta x, y) - f(x, y) \end{aligned}$$

for A as in the Lemma. Thus

$$\Delta y = -A^{-1}\big[f(x+\Delta x, y) - f(x,y)\big]. \tag{4.65}$$

Part **(d)** of the Lemma, $|A^{-1}z|_\infty \le ||A^{-1}||_\infty |z|_\infty$, and continuity of f imply

$$|\Delta y|_\infty \le \mu |f(x+\Delta x, y) - f(x,y)|_\infty \to 0$$

as $\Delta x \to 0$. Thus Ψ is continuous. If $\partial f/\partial x_k$ exists, then divide equation (4.65) by Δx_k, set all other $\Delta x_i = 0$, for the limit stated in the **IFT**. Continuity of the partial of f and the form of the partial of Ψ gives continuity of the partial of Ψ. □

We now apply the **IFT** to discrete Hamiltonian systems.

Theorem 4.48 (Local Existence and Uniqueness of Solutions) *Assume that on an open set S of R^{2+2d} the Hamiltonian $H(s,t,u,w)$ is continuous and has continuous partials of the first two orders with respect to all the components of u and w. Let the point $P_0(t_{n-1}, t_{n-1}, y_{n-1}, z_{n-1})$ be in S. Then there exist positive real numbers δ and γ and a 2d–vector valued continuous function $\Psi(s,t,y,z)$ defined on*

$$\mathcal{N}_\delta(P_0) \equiv \{P(s,t,y,z) : |P - P_0|_\infty \le \delta\}$$

with range in

$$\mathcal{N}_\gamma((y_{n-1}, z_{n-1})) \equiv \{(p,q) : |p - y_{n-1}|_\infty \le \gamma, \quad |q - z_{n-1}|_\infty \le \gamma\}$$

such that the following conditions hold:

1. *$\Psi(t_{n-1}, t_{n-1}, y_{n-1}, z_{n-1}) = (y_{n-1}, z_{n-1})$.*
2. *For each $P(s,t,y,z) \in \mathcal{N}_\delta(P_0)$, the point $(p,q) = \Psi(s,t,y,z)$ satisfies the discrete Hamiltonian system* (4.62) *on page* 190.
3. *For each $P(s,t,y,z) \in \mathcal{N}_\delta(P_0)$, the system* (4.62) *has one and only one solution $(p,q) \in \mathcal{N}_\gamma((y_{n-1}, z_{n-1}))$, namely, $(p,q) = \Psi(s,t,y,z)$.*
4. *For fixed t_n such that $h = t_n - t_{n-1}$ satisfies $|h| < \delta$, then for*

   $$|y - y_{n-1}|_\infty < \delta \quad and \quad |z - z_{n-1}|_\infty < \delta,$$

   the function $\Psi(y,z) \equiv \Psi(t_{n-1}, t_n, y, z)$ has continuous partials with respect to the components of y and z and the derivative matrix $\Psi'(y,z)$ is symplectic. Furthermore, for $(y,z) = (y_{n-1}, z_{n-1})$ the point $(p,q) = (y_n, z_n) = \Psi(y_{n-1}, z_{n-1})$ satisfies system (4.62).

It is important to point out that condition 4 gives continuous dependence of solutions upon initial conditions.

Proof: In our application of the above implicit function theorem, x is the vector $\vec{x} = (s,t,y,z)$ with $m = 2+2d$ and y is the vector $\vec{y} = (p,q)$ with $n = 2d$. We use this vector notation to remind us that we are using the **IFT** at points

$$(\vec{x},\vec{y}) \equiv ((s,t,y,z),(p,q)).$$

The function f is defined from the system (4.62) by

$$\begin{aligned} f_i(\vec{x},\vec{y}) &= p_i - y_i - (t-s)H_{w_i}(s,t,p,z) \quad \text{for } i = 1,\ldots d, \\ f_{d+i}(\vec{x},\vec{y}) &= q_i - z_i + (t-s)H_{u_i}(s,t,p,z) \quad \text{for } i = 1,\ldots d. \end{aligned} \tag{4.66}$$

Also, $a = (t_{n-1}, t_{n-1}, y_{n-1}, z_{n-1})$ and $b = (y_{n-1}, z_{n-1})$ satisfy $f(a,b) = 0$. The matrix of partial derivatives $[\partial f_i/\partial y_j]$ is computed from

$$\begin{aligned} \partial f_i/\partial p_j &= \delta_{i,j} - hH_{w_i u_j}, & \partial f_i/\partial q_j &= 0, \\ \partial f_{d+i}/\partial p_j &= hH_{u_i u_j}, & \partial f_{d+i}/\partial q_j &= \delta_{i,j}, \end{aligned}$$

where $h = t - s$. The Jacobian matrix is

$$\begin{bmatrix} I - hH_{wu} & 0 \\ hH_{uu} & I \end{bmatrix}, \tag{4.67}$$

which is continuous. Furthermore, at the point (a,b) we have $h = 0$ and the Jacobian matrix is nonsingular since it is the $2d \times 2d$ identity matrix. Condition 1 says that $\Psi(a) = b$. Condition 2, which actually gives continuity of Ψ as a function of initial conditions, is the condition $f(\vec{x}, \Psi(\vec{x})) = 0$. Condition 3 is the uniqueness of $\vec{y}$ in the restricted box and condition 4 is a consequence of the existence an continuity of certain partials. Since the current x of the implicit function theorem has $m = 2+2d$ we may start k at 3 in equation (4.64) for existence and continuity of the $2d \times 2d$ derivative matrix $\Psi'(y,z)$. The symplectic property of $\Psi'(y,z)$ follows from Theorem 4.45.

Note that h can be negative. From (4.67) the *stepsize criterion* becomes

$$\frac{1}{|h|} > \rho(H_{wu}(t_{n-1}, t_n, y_{n-1}, z_{n-1})) \tag{4.68}$$

where $h = t_n - t_{n-1}$. For autonomous Hamiltonians, $H = H(u,w)$, this criterion gives an *a priori* upper bound for $|h|$.

A solution $(p,q) = \Psi(t_{n-1}, t_n, y_{n-1}, z_{n-1})$ is called *spurious* if as t_n goes to t_{n-1}, the solution (p,q) is defined for every t_n near t_{n-1}, but does not converge to (y_{n-1}, z_{n-1}).

Exercise 4.49 (Spurious Solutions) *For the discrete minimal surface of revolution problem of Example* 4.35, *we have autonomous* f *and* H *given by*

$$f(y,r) = y\sqrt{1+r^2}$$

and

$$H(u,w) = -\sqrt{u^2 - w^2} \quad on \quad \mathcal{U} = \{(u,w) \mid u > 0,\ |w| < u\}.$$

Investigate existence and uniqueness of y_n *from a given* (y_{n-1}, z_{n-1}) *such that*

$$z_{n-1} = \frac{y_n \Delta y_{n-1}}{\sqrt{h^2 + (\Delta y_{n-1})^2}}.$$

Write this equation in the form

$$z_{n-1}\sqrt{h^2 + u^2} = u^2 + y_{n-1}u$$

for $u \equiv y_n - y_{n-1}$. *Plot the two curves* $v = z_{n-1}\sqrt{h^2+u^2}$, $v = u^2 + y_{n-1}u$, *and search for intersection points between the hyperbola and the parabola. Of special note is the case where* $z_{n-1} < 0$ *and* $|z_{n-1}| < y_{n-1}$ *for various* $h > 0$. *Show that in this case there is a critical value of* $h = h_{\text{crit}}$ *such that for large* h *the curves do not intersect and there is no solution; for the critical value of* h *the curves kiss and there is exactly one solution. For* $0 < h < h_{\text{crit}}$ *there are two solutions, one of which is spurious. Note how the boxes of the implicit function theorem determine the correct solution; namely, the unique choice of* (y_n, z_n) *which converges to the base point* (y_{n-1}, z_{n-1}) *as* $h \to 0$.

We remark that for a given h which satisfies the stepsize criterion, the iteration to determine y_n implicitly which starts with initial try at $y_n = y_{n-1}$ puts one on the nonspurious solution branch.

4.10 NOTES

Our discrete Hamiltonians are asymptotically constant as the step size goes to zero, if the numerical scheme converges and if the original variational problem is autonomous. Mickens [101] has an interesting alternative approach to discrete Hamiltonians. For many specific cases he is able to provide discrete Hamiltonians which are constant along solutions. However, there is a trade–off between preserving the Hamiltonian versus having a symplectic flow. Our approach is able to preserve the symplectic structure. This gives some desirable numerical features. References on symplectic integrators can be found in Sanz–Serna [137].

The proof that we have given of the implicit function theorem is closely related to proofs given by McShane and Botts [99] and in the unpublished book of W. T. Reid [135].

We have not considered discrete variational problems with constraints nor have we considered discrete variational problems associated with double integral problems. Those areas have applications to economics and to numerical solutions of partial differential equations, respectively.

Related work on discrete variational theory is that of Betty Harmsen [72, 73, 74]. Bohner [27]–[31] has considered more general boundary conditons.

5

SYMMETRIC THREE TERM RECURRENCE RELATIONS

5.1 A DISCRETE REID ROUNDABOUT THEOREM

In this chapter we will study the three term vector difference equation

$$\mathcal{L}u(t) = -K(t)u(t+1) + N(t)u(t) - K^*(t-1)u(t-1) = 0. \tag{5.1}$$

We assume $N(t)$ is an $n \times n$ Hermitian matrix function defined on the discrete interval $[a+1, b+1]$ and $K(t)$ is an $n \times n$ nonsingular matrix function defined on $[a, b+1]$. Solutions $u(t)$ of (5.1) are defined on $[a, b+2]$.

In Chapter 4 we saw that the Jacobi equation of the functional J in Chapter 4 along some $y_0(t)$ leads to an equation (see equation (4.15)) of the form (5.1). See Examples 3.17 and 3.27 for results related to equation (5.1).

A quadratic form $\mathcal{J}$ associated with (5.1) is defined by

$$\mathcal{J}[\eta] = \sum_{t=a+1}^{b+1} \{\eta^*(t)N(t)\eta(t) - \eta^*(t-1)K(t-1)\eta(t) - \eta^*(t)K^*(t-1)\eta(t-1)\},$$

where $\eta \in \mathcal{F}$, where

$$\mathcal{F} \equiv \{\eta : [a, b+2] \to \mathrm{C}^n \text{ such that } \eta(a) = 0 = \eta(b+2)\}.$$

Quadratic functionals of this form arise as second variations in the discrete calculus of variations. Hence we will consider $n \times n$ real valued matrix functions

$N(t)$ and $K(t)$ and we will usually make the following assumptions:

$$\begin{aligned} &N(t) \textit{ is symmetric on } [a+1, b+2] \\ &K(t) \textit{ is nonsingular on } [a+1, b+1] \\ &\mathcal{J}[\eta] \textit{ has domain } \mathcal{F}. \end{aligned} \tag{5.2}$$

Recall, that in Chapter 4, instead of the above class of complex variations $\mathcal{F}$, we had defined a set of *admissible variations* by

$$\mathcal{A} = \{\eta : [a, b+2] \to \mathrm{R}^n \text{ such that } \eta(a) = \eta(b+2) = 0\}.$$

Note that even if the coefficients in (5.1) are real, it is of interest to consider complex solutions and complex $\eta \in \mathcal{F}$. Indeed, scalar second order recurrences with constant coefficients are studied by means of the characteristic equation, which can have complex roots.

Since the complex results contain the real results by replacing conjugate transpose * by transpose ^{T} it is simplest to just do everything for the complex case. However, the concept of a "prepared vector solution" is not needed if we restrict to real vector solutions of systems with real coefficients.

For problems with variational origins, all coefficients and all solutions are real valued. Consequently, we will present the section on discrete Jacobi conditions only in the real case. Otherwise, this entire chapter allows complex solutions. If one relaxes the assumption of real coefficients to complex coefficients, then hypotheses of symmetric coefficients must be replaced by hypotheses of Hermitian coefficients. We have used the notation for conjugate transpose of * on matrices even if they are real. Then the results are easily reinterpreted for the complex case if desired.

Exercise 5.1 *Assume* (5.2) *holds. Show that* $\mathcal{J}[\eta]$ *can be written in the form*

$$\begin{aligned} \mathcal{J}[\eta] = \textstyle\sum_{t=a+1}^{b+1}\{\eta^*(t)C(t)\eta(t) \quad &+\Delta\eta^*(t-1)K(t-1)\eta(t) \\ &+\eta^*(t)K^*(t-1)\Delta\eta(t-1)\} \end{aligned}$$

for $\eta \in \mathcal{A}$, *where* $C(t)$ *is the real symmetric matrix*

$$C(t) = N(t) - K(t-1) - K^*(t-1),$$

$t \in [a+1, b+1]$. *Use this to show that the Euler–Lagrange equation for* $\mathcal{J}[\eta]$ *is*

$$-K(t)\eta(t+1) + N(t)\eta(t) - K^*(t-1)\eta(t-1) = 0.$$

Exercise 5.2 *Show that the self–adjoint vector difference equation*

$$\Delta\left[P(t)\Delta y(t-1)\right] + Q(t)y(t) = 0$$

with $P(t)$ *and* $Q(t)$ *Hermitian on* $[a+1, b+2]$ *and* $[a+1, b+1]$, *respectively, has associated quadratic form*

$$\mathcal{J}[\eta] = \sum_{t=a+1}^{b+2} \Delta\eta^*(t-1)P(t)\Delta\eta(t-1) - \sum_{t=a+1}^{b+1} \eta^*(t)Q(t)\eta(t).$$

We will consider the three term matrix difference equation

$$\mathcal{L}U(t) \equiv -K(t)U(t+1) + N(t)U(t) - K^*(t-1)U(t-1) = 0 \qquad (5.3)$$

corresponding to (5.1) where $U(t)$ is an $n \times m$ matrix function on $[a, b+2]$.

Theorem 5.3 (Lagrange Identity) *Assume* $U(t)$, $V(t)$ *are* $n \times m$ *matrix functions defined on* $[a, b+2]$, *then*

$$\{\mathcal{L}U(t)\}^* V(t) - U^*(t)\mathcal{L}V(t) = \Delta\{U(t); V(t)\} \qquad (5.4)$$

for $t \in [a+1, b+1]$, *where* $\{U(t); V(t)\}$, *called the Lagrange bracket of* $U(t)$ *and* $V(t)$, *is defined by*

$$\{U(t); V(t)\} = U^*(t-1)K(t-1)V(t) - U^*(t)K^*(t-1)V(t-1)$$

for $t \in [a+1, b+2]$.

Formula (5.4) is called the *Lagrange identity* for $\mathcal{L}$.

Proof: Assume $U(t)$, $V(t)$ are $n \times m$ matrix functions defined on $[a, b+2]$ and consider for $t \in [a+1, b+1]$

$$\{\mathcal{L}U(t)\}^*V(t) - U^*(t)\mathcal{L}V(t) =$$

$$- U^*(t+1)K^*(t)V(t) + U^*(t)N(t)V(t)$$

$$
\begin{aligned}
& - U^*(t-1)K(t-1)V(t) + U^*(t)K(t)V(t+1) \\
& - U^*(t)N(t)V(t) + U^*(t)K^*(t-1)V(t-1) \\
= \; & \{U^*(t)K(t)V(t+1) - U^*(t+1)K^*(t)V(t)\} \\
& - \{U^*(t-1)K(t-1)V(t) - U^*(t)K^*(t-1)V(t-1)\} \\
= \; & \Delta\{U(t); V(t)\}.
\end{aligned}
$$

□

Exercise 5.4 *Show that if $U(t)$, $V(t)$ are $n \times m$ matrix functions on $[a, b+2]$, then*

$$\{U(t); V(t)\}^* = -\{V(t); U(t)\}$$

for $t \in [a+1, b+2]$.

Corollary 5.5 *If $U(t)$, $V(t)$ are $n \times m$ matrix solutions of $\mathcal{L}U(t) = 0$, then*

$$\{U(t); V(t)\} \equiv C$$

for $t \in [a+1, b+2]$, where C is an $m \times m$ constant matrix.

Proof: Since $U(t)$, $V(t)$ are solutions of $\mathcal{L}U(t) = 0$ we have from the Lagrange identity for $\mathcal{L}$ that

$$\Delta\{U(t); V(t)\} = 0$$

for $t \in [a+1, b+1]$. This implies the desired result. □

If $U(t)$, $V(t)$ are solutions of $\mathcal{L}U(t) = 0$ such that

$$\{U(t); V(t)\} = 0$$

for $t \in [a+1, b+2]$, then we say they are a *prepared pair* of solutions of $\mathcal{L}U(t) = 0$. (Or we might say that the solutions U and V are *conjoined*.) If

$$\{U(t); U(t)\} = 0,$$

then we say $U(t)$ is a *prepared solution* of $\mathcal{L}U(t) = 0$. It follows that if $U(t)$, $V(t)$ are solutions of $\mathcal{L}U(t) = 0$, then $U(t)$, $V(t)$ is a prepared pair of solutions iff

$$U^*(t-1)K(t-1)V(t) \equiv U^*(t)K^*(t-1)V(t-1)$$

for all $t \in [a+1, b+2]$ iff

$$U^*(t_0-1)K(t_0-1)V(t_0) = U^*(t_0)K^*(t_0-1)V(t_0-1)$$

for some $t_0 \in [a+1, b+2]$. As a special case we know that a solution $U(t)$ of $\mathcal{L}U(t) = 0$ is a prepared solution iff

$$U^*(t-1)K(t-1)U(t) \equiv U^*(t)K^*(t-1)U(t-1)$$

for all $t \in [a+1, b+2]$ iff

$$U^*(t_0-1)K(t_0-1)U(t_0) = U^*(t_0)K^*(t_0-1)U(t_0-1)$$

for some $t_0 \in [a+1, b+2]$. Equivalently a solution $U(t)$ of $\mathcal{L}U(t) = 0$ is a prepared solution of $\mathcal{L}U(t) = 0$ iff $U^*(t-1)K(t-1)U(t)$ is Hermitian for all $t \in [a+1, b+2]$ iff $U^*(t_0-1)K(t_0-1)U(t_0)$ is Hermitian for some $t_0 \in [a+1, b+2]$. Hence to insure a prepared solution $U(t)$ we just need to specify initial conditions $U(t_0)$, $U(t_0+1)$ such that $U^*(t_0)K(t_0)U(t_0+1)$ is Hermitian. Finally a solution $u(t)$ of the vector equation is a prepared solution iff

$$u^*(t_0-1)K(t_0-1)u(t_0)$$

is real for some $t_0 \in [a+1, b+2]$.

Exercise 5.6 *Show that if $U(t)$, $V(t)$ are $n \times m$ prepared solutions of $\mathcal{L}U(t) = 0$ such that $\{U; V\}$ is Hermitian, then $Y(t) = U(t) + V(t)$ is a prepared solution of $\mathcal{L}U(t) = 0$.*

Exercise 5.7 *Show that if $U(t)$ is an $n \times m$ prepared solution of $\mathcal{L}U(t) = 0$ and Γ is an $m \times p$ constant matrix, then $Y(t) = U(t)\Gamma$ is an $n \times p$ prepared solution of $\mathcal{L}U(t) = 0$.*

Exercise 5.8 *Show that if $U(t)$ is a prepared solution of* (5.3), *then any two linear combinations $u(t)$, $v(t)$ of the columns of $U(t)$ is a prepared pair of solutions of* (5.1). *In particular any two columns of $U(t)$ is a prepared pair of solutions of* (5.1).

Theorem 5.9 (Discrete Legendre-Clebsch Transformation) *Let $U(t)$ be an $n \times n$ prepared matrix solution of $\mathcal{L}U(t) = 0$. Assume η, $\psi \in \mathcal{F}$ are related by $\eta(t) = U(t)\psi(t)$ for $t \in [a, b+2]$, then*

$$\mathcal{J}[\eta] = \sum_{t=a+1}^{b+2} [\Delta\psi(t-1)]^* U^*(t-1)K(t-1)U(t)\Delta\psi(t-1).$$

Proof: Assume $U(t)$ is a prepared solution of $\mathcal{L}U(t) = 0$ and η, $\psi \in \mathcal{F}$ are related by the equation $\eta(t) = U(t)\psi(t)$. Then

$$\begin{aligned}
\mathcal{J}[\eta] &= \sum_{t=a+1}^{b+1} \{\eta^*(t)N(t)\eta(t) - \eta^*(t-1)K(t-1)\eta(t) \\
&\qquad -\eta^*(t)K^*(t-1)\eta(t-1)\} \\
&= \sum_{t=a+1}^{b+1} \{\psi^*(t)U^*(t)N(t)U(t)\psi(t) \\
&\qquad -\psi^*(t-1)U^*(t-1)K(t-1)U(t)\psi(t) \\
&\qquad -\psi^*(t)U^*(t)K^*(t-1)U(t-1)\psi(t-1)\}.
\end{aligned}$$

Since $U(t)$ is a solution of $\mathcal{L}U(t) = 0$,

$$N(t)U(t) = K(t)U(t+1) + K^*(t-1)U(t-1),$$

and hence

$$\begin{aligned}
\mathcal{J}[\eta] &= \sum_{t=a+1}^{b+1} \{\psi^*(t)U^*(t)K(t)U(t+1)\psi(t) \\
&\qquad +\psi^*(t)U^*(t)K^*(t-1)U(t-1)\psi(t) \\
&\qquad -\psi^*(t-1)U^*(t-1)K(t-1)U(t)\psi(t) \\
&\qquad -\psi^*(t)U^*(t)K^*(t-1)U(t-1)\psi(t-1)\}.
\end{aligned}$$

In the first term under the sum we replace t by $t-1$, use the fact that $\psi(a)=0$ and separate the last term from the first sum to obtain

$$\begin{aligned}
\mathcal{J}[\eta] &= \psi^*(b+1)U^*(b+1)K(b+1)U(b+2)\psi(b+1) \\
&\quad + \sum_{t=a+1}^{b+1} \{\psi^*(t-1)U^*(t-1)K(t-1)U(t)\psi(t-1) \\
&\quad +\psi^*(t)U^*(t)K^*(t-1)U(t-1)\psi(t) \\
&\quad -\psi^*(t-1)U^*(t-1)K(t-1)U(t)\psi(t) \\
&\quad -\psi^*(t)U^*(t)K^*(t-1)U(t-1)\psi(t-1)\}.
\end{aligned}$$

Since $U(t)$ is a prepared solution

$$\begin{aligned}
\mathcal{J}[\eta] = &\sum_{t=a+1}^{b+1} \{\psi^*(t-1)U^*(t-1)K(t-1)U(t)\psi(t-1) \\
&+\psi^*(t)U^*(t-1)K(t-1)U(t)\psi(t) \\
&-\psi^*(t-1)U^*(t-1)K(t-1)U(t)\psi(t) \\
&-\psi^*(t)U^*(t-1)K(t-1)U(t)\psi(t-1)\} \\
&+\psi^*(b+1)U^*(b+1)K(b+1)U(b+2)\psi(b+1).
\end{aligned}$$

Finally we have

$$\begin{aligned}
\mathcal{J}[\eta] &= \sum_{t=a+1}^{b+1} \Delta\psi^*(t-1)U^*(t-1)K(t-1)U(t)\Delta\psi(t-1) \\
&\quad +\psi^*(b+1)U^*(b+1)K(b+1)U(b+2)\psi(b+1) \\
&= \sum_{t=a+1}^{b+2} \Delta\psi^*(t-1)U^*(t-1)K(t-1)U(t)\Delta\psi(t-1)
\end{aligned}$$

which is what we wanted to prove. □

Compare the next result to Lemma 1.37 on page 29 and Theorem 4.19 on page 169.

Theorem 5.10 *If $\eta \in \mathcal{F}$, then*

$$\mathcal{J}[\eta] = \sum_{t=a+1}^{b+1} \eta^*(t)\mathcal{L}\eta(t).$$

Proof: By definition

$$\begin{aligned}\mathcal{J}[\eta] &= \textstyle\sum_{t=a+1}^{b+1}\{\eta^*(t)N(t)\eta(t) - \eta^*(t-1)K(t-1)\eta(t) \\ &\quad -\eta^*(t)K^*(t-1)\eta(t-1)\}.\end{aligned}$$

In the second term if we replace t by $t+1$ and use $\eta(a) = \eta(b+2) = 0$ we have

$$\begin{aligned}\mathcal{J}[\eta] &= \sum_{t=a+1}^{b+1} \{\eta^*(t)[N(t)\eta(t) - K(t)\eta(t+1) - K^*(t-1)\eta(t-1)]\} \\ &= \sum_{t=a+1}^{b+1} \eta^*(t)\mathcal{L}\eta(t).\end{aligned}$$

□

Compare the next result to Lemma 1.38 on page 30 and Theorem 4.21 on page 170.

Corollary 5.11 *Assume $u(t)$ is a prepared solution of $\mathcal{L}u(t) = 0$ and α, β are integers satisfying $a \le \alpha < \beta < b+2$. If*

$$\eta(t) = \begin{cases} 0, & a \le t \le \alpha \\ u(t), & \alpha+1 \le t \le \beta \\ 0, & \beta+1 \le t \le b+2, \end{cases}$$

then $\eta \in \mathcal{F}$ and

$$\mathcal{J}[\eta] = u^*(\alpha)K(\alpha)u(\alpha+1) + u^*(\beta)K(\beta)u(\beta+1).$$

Proof: Obviously $\eta \in \mathcal{F}$ so by Theorem 5.10

$$\begin{aligned}\mathcal{J}[\eta] &= \textstyle\sum_{t=a+1}^{b+1} \eta^*(t)\mathcal{L}\eta(t) \\ &= \textstyle\sum_{t=\alpha+1}^{\beta} u^*(t)\mathcal{L}\eta(t).\end{aligned} \tag{5.5}$$

First consider the case where $\alpha + 1 = \beta$. In this case

$$\begin{aligned}\mathcal{J}[\eta] &= u^*(\alpha+1)\mathcal{L}\eta(\alpha+1)\\ &= u^*(\alpha+1)N(\alpha+1)u(\alpha+1).\end{aligned}$$

Since $u(t)$ is a solution we have

$$\mathcal{J}[\eta] = u^*(\alpha+1)K(\alpha+1)u(\alpha+2) + u^*(\alpha+1)K^*(\alpha)u(\alpha).$$

Use the fact that $u(t)$ is prepared and that $\beta = \alpha + 1$ for

$$\mathcal{J}[\eta] = u^*(\alpha)K(\alpha)u(\alpha+1) + u^*(\beta)K(\beta)u(\beta+1),$$

which is what we wanted to prove.

Next consider the case where $\alpha + 1 < \beta$. From (5.5) we see that

$$\begin{aligned}\mathcal{J}[\eta] &= u^*(\alpha+1)\mathcal{L}\eta(\alpha+1) + \sum_{t=\alpha+2}^{\beta-1} u^*(t)\mathcal{L}u(t) + u^*(\beta)\mathcal{L}\eta(\beta)\\ &= u^*(\alpha+1)\mathcal{L}\eta(\alpha+1) + u^*(\beta)\mathcal{L}\eta(\beta)\\ &= u^*(\alpha+1)[\mathcal{L}u(\alpha+1) + K^*(\alpha)u(\alpha)]\\ &\quad + u^*(\beta)[\mathcal{L}u(\beta) + K(\beta)u(\beta+1)]\\ &= u^*(\alpha+1)K^*(\alpha)u(\alpha) + u^*(\beta)K(\beta)u(\beta+1).\end{aligned}$$

Since $u(t)$ is a prepared solution we have

$$\mathcal{J}[\eta] = u^*(\alpha)K(\alpha)u(\alpha+1) + u^*(\beta)K(\beta)u(\beta+1)$$

which concludes the proof of this corollary. □

With the conclusion of Corollary 5.11 in mind we make the following definition. We say a prepared solution $u(t)$ of (5.1) has a *generalized zero at* a only if $u(a) = 0$ and $u(t)$ has a *generalized zero at* $t_0 > a$ provided $u(t_0 - 1) \neq 0$ and

$$u^*(t_0-1)K(t_0-1)u(t_0) \leq 0.$$

Of course this last inequality would be true if $u(t_0) = 0$, but this inequality could also hold if $u(t_0 - 1) \neq 0$ and $u(t_0) \neq 0$. We say that (5.1) is *disconjugate* on $[a, b+2]$ provided no nontrivial prepared solution has two generalized zeros in $[a, b+2]$.

Exercise 5.12 *Show that when Corollary* 5.11 *is applied to the self-adjoint vector difference equation*

$$\Delta [P(t)\Delta y(t-1)] + Q(t)y(t) = 0$$

with $P(t)$ *and* $Q(t)$ *Hermitian, the conclusion of that corollary becomes*

$$\mathcal{J}[\eta] = u^*(\alpha)P(\alpha+1)u(\alpha+1) + u^*(\beta)P(\beta+1)u(\beta+1).$$

Because of this a generalized zero of a prepared solution of this self-adjoint difference equation is defined similar to above in terms of the expression

$$u^*(t_0-1)P(t_0)u(t_0).$$

Theorem 5.13 (Discrete Reid Roundabout Theorem) *Suppose we have the assumptions* (5.2). *Then the following conditions are equivalent:*
(i) $\mathcal{J}[\eta]$ *is positive definite on* $\mathcal{F}$.
(ii) The recurrence relation (5.1) *is disconjugate on* $[a, b+2]$.
(iii) If $U(t)$ *is the* $n \times n$ *matrix solution of the* **IVP** $\mathcal{L}U(t) = 0$, $U(a) = 0$, $U(a+1) = I$, *then*

$$U^*(t-1)K(t-1)U(t) > 0 \quad on \quad [a+2, b+2].$$

(iv) If $V(t)$ *is the* $n \times n$ *matrix solution of the* **IVP** $\mathcal{L}V(t) = 0$, $V(b+1) = I$, $V(b+2) = 0$, *then*

$$V^*(t-1)K(t-1)V(t) > 0 \quad on \quad [a+1, b+1].$$

(v) There is a prepared solution $Y(t)$ *of* $\mathcal{L}U(t) = 0$ *with*

$$Y^*(t-1)K(t-1)Y(t) > 0 \quad on \quad [a+1, b+2].$$

Proof: First we show that (i) implies (ii). To see this assume $\mathcal{J}[\eta]$ is positive definite on $\mathcal{F}$ but the difference equation (5.1) is not disconjugate on $[a, b+2]$. Then there is a nontrivial prepared solution $u(t)$ of (5.1) with at least two generalized zeros in $[a, b+2]$. This implies there exist integers t_1, t_2 such that $a \le t_1 < t_2 < b+2$,

$$u(t_1+1) \neq 0, \qquad u(t_2) \neq 0,$$

and

$$u^*(t_1)K(t_1)u(t_1+1) \le 0, \qquad u^*(t_2)K(t_2)u(t_2+1) \le 0.$$

Define η by

$$\eta(t) = \begin{cases} 0, & a \le t \le t_1 \\ u(t), & t_1 + 1 \le t \le t_2 \\ 0, & t_2 + 1 \le t \le b + 2. \end{cases}$$

By Corollary 5.11, $\eta \in \mathcal{F}$ and

$$\mathcal{J}[\eta] = u^*(t_1)K(t_1)u(t_1 + 1) + u^*(t_2)K(t_2)u(t_2 + 1) \le 0.$$

But $u(t_1+1) \ne 0$ implies η is a nontrivial element of $\mathcal{F}$. It follows that $\mathcal{J}[\eta] > 0$ which is a contradiction. Hence (i) implies (ii).

Next we show that (ii) implies (iii) and (ii) implies (iv). Assume (ii) holds, that is, assume (5.1) is disconjugate on $[a, b+2]$. Let $U(t)$ be the $n \times n$ matrix solution of the **IVP**

$$\mathcal{L}U(t) = 0$$

$$U(a) = 0, \qquad U(a + 1) = I.$$

Let $\alpha \ne 0$ be an arbitrary but fixed complex n vector. Set $u(t) = U(t)\alpha$. Then $u(t)$ is a nontrivial prepared solution of (5.1) with $u(a) = 0$. Since $u(t)$ has a generalized zero at a it can have no other generalized zeros in $[a, b + 2]$ since we are assuming (5.1) is disconjugate on $[a, b + 2]$. Hence

$$u^*(t)K(t)u(t + 1) > 0$$

for $t \in [a + 1, b + 1]$. But this implies that

$$\alpha^*U^*(t)K(t)U(t + 1)\alpha > 0$$

for $t \in [a + 1, b + 1]$. Since $\alpha \ne 0$ in C^n is arbitrary it follows that

$$U^*(t)K(t)U(t + 1) > 0$$

for $t \in [a+1, b+1]$ and hence (iii) holds. The proof of (ii) implies (iv) is similar and is left as an exercise.

Next we show that each of (iii), (iv), and (v) implies (i). To see that (iii) implies (i), assume the $n \times n$ matrix solution $U(t)$ of the **IVP**

$$\mathcal{L}U(t) = 0, \qquad U(a) = 0, \quad U(a + 1) = I$$

satisfies

$$U^*(t - 1)K(t - 1)U(t) > 0$$

on $[a+2, b+2]$. Let $\eta \in \mathcal{F}$, since $U(t)$ is nonsingular on $[a+1, b+2]$ we can define $\psi \in \mathcal{F}$ by

$$\psi(t) = \begin{cases} 0, & t = a \\ U^{-1}(t)\eta(t), & a+1 \le t \le b+2. \end{cases}$$

It follows that $\eta(t) = U(t)\psi(t)$ for all $a \le t \le b+2$ so by the Legendre-Clebsch transformation, Theorem 5.9 on page 204, we deduce that

$$\mathcal{J}[\eta] = \sum_{t=a+1}^{b+2} [\Delta\psi(t-1)]^* U^*(t-1)K(t-1)U(t)\Delta\psi(t-1) \ge 0.$$

Since $\eta(t) \equiv 0$ on $[a, b+2]$ iff $\Delta\psi(t) \equiv 0$ on $[a, b+1]$ we have $\mathcal{J}[\eta] = 0$ iff $\eta = 0$. Hence $\mathcal{J}[\eta]$ is positive definite on $\mathcal{F}$ and (i) holds. Similarly (iv) implies (i) and (v) implies (i).

To complete the proof of this theorem we now show that (i) implies (v). Assume (i) holds and let $U(t)$, $V(t)$ be $n \times n$ matrix solutions of $\mathcal{L}U(t) = 0$ satisfying the initial conditions

$$U(a) = 0, \quad U(a+1) = I$$

and

$$V(b+1) = I, \quad V(b+2) = 0$$

respectively. Earlier we showed that (i) implies (iii) and (i) implies (iv). Hence we have that

$$U^*(t-1)K(t-1)U(t) > 0 \tag{5.6}$$

on $[a+2, b+2]$ and

$$V^*(t-1)K(t-1)V(t) > 0 \tag{5.7}$$

on $[a+1, b+1]$.

We want to choose an $n \times n$ constant matrix Γ so that if we set

$$V_1(t) = V(t)\Gamma, \tag{5.8}$$

then $V_1(t)$ is a solution of $\mathcal{L}U(t) = 0$ and

$$\{U; V_1\} = -I. \tag{5.9}$$

Let $t = b+1$ in (5.8) for $V_1(b+1) = \Gamma$. Let $t = b+2$ in (5.9) to get

$$U^*(b+1)K(b+1)V_1(b+2) - U^*(b+2)K^*(b+1)V_1(b+1) = -I.$$

Since $V_1(b+2) = 0$ and $V_1(b+1) = \Gamma$ this leads to the equation

$$U^*(b+2)K^*(b+1)\Gamma = I.$$

From (5.6) we know that $U(t)$ is nonsingular on $[a+1, b+2]$. Solving for Γ we have

$$\Gamma = [U^*(b+2)K^*(b+1)]^{-1}.$$

It follows that if Γ is given by this last equation then (5.9) holds. Furthermore,

$$V_1^*(t-1)K(t-1)V_1(t) = \Gamma^* V^*(t-1)K(t-1)V(t)\Gamma.$$

Since Γ is nonsingular and (5.7) holds we have

$$V_1^*(t-1)K(t-1)V_1(t) > 0 \tag{5.10}$$

on $[a+1, b+1]$.

Set

$$Y(t) = U(t) + V_1(t)$$

for $t \in [a, b+2]$. Since $U(t)$ and $V_1(t)$ are prepared solutions of $\mathcal{L}U(t) = 0$ and, by (5.9), $\{U; V_1\}$ is Hermitian, it follows from Exercise 5.6 that $Y(t)$ is a prepared solution of $\mathcal{L}U(t) = 0$.

It remains to show that

$$Y^*(t-1)K(t-1)Y(t) > 0 \tag{5.11}$$

for $t \in [a+1, b+2]$.

To see that (5.11) holds for $t = a+1$, first let $t = a+1$ in (5.9) in order to obtain

$$U^*(a)K(a)V_1(a+1) - U^*(a+1)K^*(a)V_1(a) = -I.$$

Using $U(a) = 0$, $U(a+1) = I$ and solving for $V_1(a)$ we have

$$V_1(a) = [K^*(a)]^{-1}.$$

It follows that

$$\begin{aligned} Y^*(a)K(a)Y(a+1) &= V_1^*(a)K(a)[I + V_1(a+1)] \\ &= I + V_1^*(a)K(a)V_1(a+1) > 0 \end{aligned}$$

by (5.10). Hence (5.11) holds for $t = a+1$.

Assume $a+1 < t_0 \leq b+2$. Then it remains to show that (5.11) holds for $t = t_0$. To see this let γ be an arbitrary nonzero vector in C^n and define

$$\eta(t) = \begin{cases} U(t)\gamma, & a \leq t \leq t_0 - 1 \\ -V_1(t)\gamma, & t_0 \leq t \leq b+2. \end{cases}$$

Since $\eta \in \mathcal{F}$ we have by Theorem 5.10 on page 206 that

$$\mathcal{J}[\eta] = \sum_{t=a+1}^{b+1} \eta^*(t)\mathcal{L}\eta(t).$$

Since $\eta(t)$ is a solution of $\mathcal{L}u(t) = 0$ on $[a, t_0 - 1]$ and on $[t_0, b+2]$ (not on $[a, b+2]$) this last equation simplifies to give us

$$\begin{aligned}
\mathcal{J}[\eta] &= \eta^*(t_0 - 1)\mathcal{L}\eta(t_0 - 1) + \eta^*(t_0)\mathcal{L}\eta(t_0) \\
&= \eta^*(t_0 - 1)\{-K(t_0 - 1)\eta(t_0) + N(t_0 - 1)\eta(t_0 - 1) \\
&\quad -K^*(t_0 - 2)\eta(t_0 - 2)\} \\
&\quad +\eta^*(t_0)\{-K(t_0)\eta(t_0 + 1) + N(t_0)\eta(t_0) - K^*(t_0 - 1)\eta(t_0 - 1)\} \\
&= \eta^*(t_0 - 1)\{K(t_0 - 1)V_1(t_0)\gamma + \mathcal{L}[U(t)\gamma](t_0 - 1) \\
&\quad +K(t_0 - 1)U(t_0)\gamma\} \\
&\quad +\eta^*(t_0)\{\mathcal{L}[-V_1(t)\gamma](t_0) - K^*(t_0 - 1)V_1(t_0 - 1)\gamma \\
&\quad -K^*(t_0 - 1)U(t_0 - 1)\gamma\} \\
&= \eta^*(t_0 - 1)K(t_0 - 1)Y(t_0)\gamma - \eta^*(t_0)K^*(t_0 - 1)Y(t_0 - 1)\gamma \\
&= \gamma^*U^*(t_0 - 1)K(t_0 - 1)Y(t_0)\gamma + \gamma^*V_1^*(t_0)K^*(t_0 - 1)Y(t_0 - 1)\gamma \\
&= \gamma^*Y^*(t_0 - 1)K(t_0 - 1)Y(t_0)\gamma - \gamma^*V_1^*(t_0 - 1)K(t_0 - 1)Y(t_0)\gamma \\
&\quad +\gamma^*V_1^*(t_0)K^*(t_0 - 1)Y(t_0 - 1)\gamma \\
&= \gamma^*Y^*(t_0 - 1)K(t_0 - 1)Y(t_0)\gamma - \gamma^*\{V_1; Y\}\gamma.
\end{aligned}$$

But

$$\{V_1; Y\} = \{V_1; U + V_1\} = \{V_1; U\} + \{V_1; V_1\} = -\{U; V_1\}^* = I.$$

So from above

$$\gamma^* Y^*(t_0 - 1)K(t_0 - 1)Y(t_0)\gamma = \mathcal{J}[\eta] + \|\gamma\|^2 > 0$$

for all $\gamma \neq 0$ in C^n. This implies that

$$Y^*(t_0 - 1)K(t_0 - 1)Y(t_0) > 0.$$

Hence (5.11) holds for $t \in [a+1, b+2]$ and the proof of the Reid Roundabout Theorem is complete. □

Exercise 5.14 *Prove that in Theorem* 5.13 *on page* 208, *condition (ii) implies condition (iv).*

5.2 DISCRETE LEGENDRE CONDITIONS

Let $d = b - a + 1$ and define a Hermitian $d \times d$ block tridiagonal matrix Q and a $d \times 1$ block column vector z by

$$Q = \begin{bmatrix} N(a+1) & -K(a+1) & 0 & \cdots & 0 \\ -K^*(a+1) & N(a+2) & -K(a+2) & \ddots & \vdots \\ 0 & \ddots & \ddots & \ddots & 0 \\ \vdots & \ddots & \ddots & \ddots & -K(b) \\ 0 & \cdots & 0 & -K^*(b) & N(b+1) \end{bmatrix}$$

and, if $\eta \in \mathcal{F}$,

$$z = \begin{bmatrix} \eta(a+1) \\ \vdots \\ \eta(b+1) \end{bmatrix}$$

respectively. Then we can write $\mathcal{J}[\eta]$ in the form

$$\mathcal{J}[\eta] = z^* Q z. \tag{5.12}$$

We say $\mathcal{J}[\eta]$ is *positive semidefinite* on $\mathcal{F}$ provided $\mathcal{J}[\eta] \geq 0$ for all $\eta \in \mathcal{F}$. If further $\mathcal{J}[\eta] = 0$ iff $\eta = 0$, then we say $\mathcal{J}[\eta]$ is *positive definite* on $\mathcal{F}$.

Now it is an easy matter to prove the following theorem.

Theorem 5.15 *$\mathcal{J}[\eta]$ is positive definite {semidefinite} on $\mathcal{F}$ iff the matrix Q is positive definite {semidefinite}. If $\mathcal{J}[\eta]$ is positive semidefinite on $\mathcal{F}$, then the Legendre condition*

$$N(t) \geq 0$$

holds on $[a+1, b+1]$. If $\mathcal{J}[\eta]$ is positive definite on $\mathcal{F}$, then the strict Legendre condition

$$N(t) > 0$$

holds on $[a+1, b+1]$.

Proof: The first sentence follows from equation (5.12). Assume $\mathcal{J}[\eta]$ is positive definite {semidefinite} on $\mathcal{F}$. Let $t_0 \in [a+1, b+1]$ and for $t \in [a, b+2]$ set

$$\eta(t) = \begin{cases} 0, & t \neq t_0 \\ \gamma, & t = t_0, \end{cases}$$

where γ is an arbitrary column n-vector.

It follows from (5.12) that

$$\mathcal{J}[\eta] = \gamma^* N(t_0)\gamma$$

and consequently $N(t_0)$ is positive definite {semidefinite}. Since t_0 is an arbitrary point in $[a+1, b+1]$, the theorem is true. □

Corollary 5.16 *Equation* (5.1) *is disconjugate on $[a, b+2]$ iff Q is positive definite.*

Proof: This result follows from Theorems 5.15 and 5.13. □

Remark 5.17 *It follows from Theorem* 7.2.5 *of* [4] *that $\mathcal{J}[\eta]$ is positive definite iff any nested sequence of n principal minors of Q (not just the leading principal minors) are positive.*

Exercise 5.18 *Use Remark* 5.17 *to show that*

$$-y(t+1) + 2y(t) - y(t-1) = 0$$

is disconjugate on $(-\infty, \infty)$.

Exercise 5.19 *Show that the strict Legendre condition holds for the three term vector difference equation*

$$-y(t+1)+\begin{bmatrix} 3 & -2 & 3 \\ -2 & 2 & -1 \\ 3 & -1 & 5 \end{bmatrix} y(t)-y(t-1)=0.$$

5.3 A STURMIAN COMPARISON THEOREM

In the next theorem we will have a comparison theorem for (5.1) and the difference equation

$$-K_1(t)u(t+1)+N_1(t)u(t)-K_1^*(t-1)u(t-1)=0 \tag{5.13}$$

where we assume $K_1(t)$ is nonsingular on $[a,b+1]$ and $N_1(t)$ is Hermitian on $[a+1,b+1]$. Define the Hermitian block tridiagonal matrix Q_1 by

$$Q_1=\begin{bmatrix} N_1(a+1) & -K_1(a+1) & 0 & \cdots & 0 \\ -K_1^*(a+1) & N_1(a+2) & -K_1(a+2) & \ddots & \vdots \\ 0 & \ddots & \ddots & \ddots & 0 \\ \vdots & \ddots & \ddots & \ddots & -K_1(b) \\ 0 & \cdots & 0 & -K_1^*(b) & N_1(b+1) \end{bmatrix}.$$

Now we state and prove a very important comparison theorem.

Theorem 5.20 (Sturm Comparison Theorem) *If* (5.1) *is disconjugate on* $[a,b+2]$ *and* $Q_1\geq Q$, *then* (5.13) *is disconjugate on* $[a,b+2]$.

Proof: Assume (5.1) is disconjugate on $[a,b+2]$. By Corollary 5.16 we know that Q is positive definite. Since $Q_1\geq Q$, Q_1 is also positive definite and, again by Corollary 5.16, equation (5.13) is disconjugate on $[a,b+2]$. □

Now also consider the difference equation

$$-K_2(t)u(t+1)+N_2(t)u(t)-K_2^*(t-1)u(t-1)=0 \tag{5.14}$$

where we assume $K_2(t)$ is nonsingular on $[a,b+1]$ and $N_2(t)$ is Hermitian on $[a+1,b+1]$. Then we have the following result.

Theorem 5.21 *If* (5.13) *and* (5.14) *are disconjugate on* $[a, b+2]$ *and we define*

$$K(t) = \lambda K_1(t) + \mu K_2(t) \qquad \text{for } t \in [a, b+1],$$

and

$$N(t) = \lambda N_1(t) + \mu N_2(t) \qquad \text{for } t \in [a+1, b+1],$$

where $\lambda \geq 0$, $\mu \geq 0$, $\lambda + \mu > 0$, *then it follows that* (5.1) *is disconjugate on* $[a, b+2]$.

Proof: Let the matrix Q_2 be defined in terms of the coefficients of (5.14) like Q_1 was defined in terms of the coefficients of (5.13). Since we are assuming that (5.13) and (5.14) are disconjugate on $[a, b+2]$ we know from Corollary 5.16 that Q_1 and Q_2 are positive definite. Since

$$Q = \lambda Q_1 + \mu Q_2$$

it follows that Q is positive definite. By Corollary 5.16 we conclude that equation (5.1) is disconjugate on $[a, b+2]$. □

The above theorem implies that the space of equations (in terms of the coefficient functions $K(t)$, $N(t)$) of the form (5.1) which are disconjugate on $[a, b+2]$ is a convex set.

5.4 PREPARED BASES

If $U(t)$ is an $n \times p$ matrix solution of (5.3), then we define the $2n \times p$ matrix $\mathcal{U}(t)$ by

$$\mathcal{U}(t) = \begin{bmatrix} U(t-1) \\ U(t) \end{bmatrix}$$

for $t \in [a+1, b+2]$. Then we define

$$\mathcal{K}(t) = \begin{bmatrix} 0 & K(t) \\ -K^*(t) & 0 \end{bmatrix}$$

for $t \in [a, b+1]$. It follows that

$$\{U; U\} = \mathcal{U}^*(t)\mathcal{K}(t-1)\mathcal{U}(t) \tag{5.15}$$

for $t \in [a+1, b+2]$.

Theorem 5.22 *Assume* $U(t)$ *is an* $n \times p$ *matrix solution of* (5.3), *then* $\mathcal{U}(t)$ *has constant rank on* $[a+1, b+2]$. *If, furthermore,* $U(t)$ *is a prepared solution of* (5.3) *and* rank $\mathcal{U}(t) = p$, *then* $p \leq n$.

Proof: Assume $U(t)$ is an $n \times p$ matrix solution of (5.3). Fix $t_0 \in [a+1, b+1]$ and assume

$$\mathcal{U}(t_0)c = 0$$

where $c \in C^p$. It follows that

$$U(t_0 - 1)c = 0, \qquad U(t_0)c = 0.$$

Use equation (5.3) for

$$U(t_0 + 1)c = 0.$$

It follows that

$$\mathcal{U}(t_0 + 1)c = 0.$$

Hence we have shown that $\mathcal{U}(t_0)c = 0$ implies $\mathcal{U}(t_0+1)c = 0$. The proof of the converse $\mathcal{U}(t_0+1)c = 0$ implies $\mathcal{U}(t_0)c = 0$ is similar. But then we know that $\mathcal{U}(t_0)$ and $\mathcal{U}(t_0+1)$ have the same nullity and hence the same rank. But $t_0 \in [a+1, b+1]$ is arbitrary so $\mathcal{U}(t)$ has constant rank on $[a+1, b+2]$.

Next assume $U(t)$ is an $n \times p$ prepared matrix solution of (5.3) such that

$$\text{rank}\, \mathcal{U}(t) = p.$$

Since $U(t)$ is a prepared solution, $\{U; U\} = 0$ for $t \in [a+1, b+2]$. Hence, by (5.15)

$$\mathcal{U}^*(t)\mathcal{K}(t-1)\mathcal{U}(t) = 0 \tag{5.16}$$

for $t \in [a+1, b+2]$. Since $\mathcal{K}(t-1)$ is nonsingular,

$$\text{rank}\, \{\mathcal{K}(t-1)\mathcal{U}(t)\} = p.$$

It then follows from (5.16) that the nullity of $\mathcal{U}^*(t)$ is at least p. Use

$$\text{rank}\, \mathcal{U}^*(t) + \text{nullity}\, \mathcal{U}^*(t) = 2n$$

and, hence,

$$2p = p + p \leq p + \text{nullity}\, \mathcal{U}^*(t) = 2n,$$

i.e., $p \leq n$. □

If $U(t)$ is an $n \times n$ prepared solution of $\mathcal{L}U(t) = 0$ such that

$$\text{rank}\, \mathcal{U}(t) = n$$

on $[a+1, b+2]$, then we say that $U(t)$ is a *prepared basis*. From above rank $\mathcal{U}(t) = n$ on $[a+1, b+2]$ iff there is a $t_0 \in [a+1, b+2]$ such that rank $\mathcal{U}(t_0) = n$.

Theorem 5.23 *Assume $U(t)$, $V(t)$ are $n \times n$ prepared solutions of $\mathcal{L}U(t) = 0$. If $\{U; V\}$ is nonsingular then $U(t)$ and $V(t)$ are prepared bases.*

Proof: Assume $U(t)$, $V(t)$ are $n \times n$ prepared solutions of $\mathcal{L}U(t) = 0$ such that $\{U; V\}$ is nonsingular on $[a+1, b+2]$. Note that for $t \in [a+1, b+2]$

$$\begin{aligned} \{U; V\} &= U^*(t-1)K(t-1)V(t) - U^*(t)K^*(t-1)V(t-1) \\ &= \mathcal{U}^*(t)\mathcal{K}(t-1)\mathcal{V}(t). \end{aligned}$$

Fix $t_0 \in [a+1, b+2]$. Assume $c \in C^n$ is such that $\mathcal{V}(t_0)c = 0$. Then from the above equation $\{U; V\}c = 0$. Since $\{U; V\}$ is nonsingular we have $c = 0$. Hence rank $\mathcal{V}(t_0) = n$ and because of constant rank, (see Theorem 5.22)

$$\text{rank}\, \mathcal{V}(t) \equiv n$$

for $t \in [a+1, b+2]$ and consequently $V(t)$ is a prepared basis. Similarly $c^*\mathcal{U}^*(t) = 0$ implies $c^*\{U; V\} = 0$ implies $c = 0$. Hence rank $\mathcal{U}(t_0) = n$ which implies that $U(t)$ is a prepared basis. □

5.5 AN ASSOCIATED BILINEAR FORM

Assume $\phi, \theta : [a, b+2] \to C^n$ and define the *bilinear form* $B(\phi, \theta)$ by

$$B(\phi, \theta) = \sum_{t=a+1}^{b+1} \{\theta^*(t)N(t)\phi(t) - \theta^*(t-1)K(t-1)\phi(t) - \theta^*(t)K^*(t-1)\phi(t-1)\}.$$

Actually, this form is a *sesquilinear form* since

$$B(\phi, \lambda\theta) = \bar{\lambda}B(\phi, \theta)$$

for complex λ so the form is only "one and a half" linear. However, we will use B nonetheless because we have already used an S.

Theorem 5.24 *If* $\phi,\ \theta : [a, b+2] \to C^n$, *then*

$$B(\phi,\theta) = \left[\sum_{t=a+1}^{b+1} \theta^*(t)\mathcal{L}\phi(t)\right] + \theta^*(b+1)K(b+1)\phi(b+2) - \theta^*(a)K(a)\phi(a+1).$$

Proof: By definition

$$B(\phi,\theta) = \sum_{t=a+1}^{b+1} \{\theta^*(t)N(t)\phi(t) - \theta^*(t-1)K(t-1)\phi(t) - \theta^*(t)K^*(t-1)\phi(t-1)\}.$$

Replace t by $t+1$ in the second term under the sum in order to obtain

$$\begin{aligned} B(\phi,\theta) &= -\sum_{t=a}^{b} \theta^*(t)K(t)\phi(t+1) \\ &\quad + \sum_{t=a+1}^{b+1} \{\theta^*(t)N(t)\phi(t) - \theta^*(t)K^*(t-1)\phi(t-1)\} \\ &= -\theta^*(a)K(a)\phi(a+1) + \theta^*(b+1)K(b+1)\phi(b+2) \\ &\quad + \sum_{t=a+1}^{b+1} \theta^*(t)[N(t)\phi(t) - K(t)\phi(t+1) - K^*(t-1)\phi(t-1)] \\ &= -\theta^*(a)K(a)\phi(a+1) + \theta^*(b+1)K(b+1)\phi(b+2) \\ &\quad + \sum_{t=a+1}^{b+1} \theta^*(t)\mathcal{L}\phi(t) \end{aligned}$$

which is the desired result. □

Exercise 5.25 *Show that*

$$B(\phi,\theta) = [B(\theta,\phi)]^*.$$

Theorem 5.26 *Assume* $u(t)$, $v(t)$ *is a prepared pair of solutions of (5.1). Assume* t_1, $t_2 \in [a+1, b+2]$, *where it is not true that* $t_1 = t_2 = a+1$, *and* *define*

$$\phi(t) = \begin{cases} u(t), & a \le t \le t_1 - 1 \\ 0, & t_1 \le t \le b+2 \end{cases}$$

$$\theta(t) = \begin{cases} v(t), & a \le t \le t_2 - 1 \\ 0, & t_2 \le t \le b+2. \end{cases}$$

Then

$$B(\phi,\theta) = \begin{cases} -v^*(a)K(a)u(a+1), & t_1 \neq t_2 \\ -v^*(a)K(a)u(a+1) + v^*(t_1-1)K(t_1-1)u(t_1), & t_1 = t_2. \end{cases}$$

Proof: We will prove this theorem in the case where $a+1 \le t_1 \le t_2 \le b+2$ and $t_2 \neq a+1$. The proof for the case $a+1 \le t_2 \le t_1 \le b+2$ and $t_1 \neq a+1$, is analogous.

Since $\phi(b+2) = 0$ we have by Theorem 5.24

$$B(\phi,\theta) = \left[\sum_{t=a+1}^{b+1} \theta^*(t)\mathcal{L}\phi(t)\right] - \theta^*(a)K(a)\phi(a+1). \tag{5.17}$$

If $t_1 = t_2 > a+1$, then from (5.17) we have

$$\begin{aligned} B(\phi,\theta) &= v^*(t_1-1)\mathcal{L}\phi(t_1-1) - v^*(a)K(a)u(a+1) \\ &= v^*(t_1-1)[\mathcal{L}u(t_1-1) + K(t_1-1)u(t_1)] - v^*(a)K(a)u(a+1) \\ &= v^*(t_1-1)K(t_1-1)u(t_1) - v^*(a)K(a)u(a+1) \end{aligned}$$

which is the desired result. Next assume $a+1 \le t_1 < t_2 \le b+2$. Then from (5.17)

$$\begin{aligned} B(\phi,\theta) &= v^*(t_1-1)\mathcal{L}\phi(t_1-1) + v^*(t_1)\mathcal{L}\phi(t_1) \\ &\quad -v^*(a)K(a)u(a+1) \\ &= v^*(t_1-1)[\mathcal{L}u(t_1-1) + K(t_1-1)u(t_1)] \\ &\quad +v^*(t_1)[-K^*(t_1-1)u(t_1-1)] - v^*(a)K(a)u(a+1) \\ &= v^*(t_1-1)K(t_1-1)u(t_1) - v^*(t_1)K^*(t_1-1)u(t_1-1) \\ &\quad -v^*(a)K(a)u(a+1) \\ &= -v^*(a)K(a)u(a+1) \end{aligned}$$

since $u(t)$, $v(t)$ is a prepared pair of solutions. □

Exercise 5.27 *Show that if* $u(t)$ *is a solution of* (5.1) *on* $[a, b+2]$ *and* v *maps* $[a, b+2]$ *to* C^n, *then*

$$B(u, v) = u^*(b+1)K(b+1)v(b+2) - u^*(a)K(a)v(a+1).$$

A set $\{u_1, \ldots, u_m\}$ of solutions of (5.1) is said to be *mutually prepared* provided

$$\{u_i; u_j\} = 0$$

for $1 \le i, j \le m$.

Exercise 5.28 *Show that if* $U(t)$ *is an* $n \times m$ *prepared solution of* $\mathcal{L}U(t) = 0$ *and*

$$u_i(t) = U(t)d_i,$$

$1 \le i \le p$, *where the* d_i *are constant* $m \times 1$ *vectors, then* $\{u_1(t), \ldots, u_p(t)\}$ *is a mutually prepared set of solutions of (5.1).*

We say that a prepared basis $U(t)$ has a *generalized zero* at a only if $U(a)$ is singular. We say $U(t)$ has a *generalized zero* at $t_0 \in [a+1, b+2]$ provided either $U(t_0)$ is singular or

$$U^*(t_0 - 1)K(t_0 - 1)U(t_0)$$

is nonsingular and not positive definite. As some motivation for this definition consider the following exercise.

Exercise 5.29 *Show that if a prepared basis* $U(t)$ *has a generalized zero at* t_0, *then there is an* $n \times 1$ *vector* γ_0 *such that* $u_0(t) \equiv U(t)\gamma_0$ *is a nontrivial prepared solution of* (5.1) *with a generalized zero at* t_0.

5.6 A DISCRETE STURM SEPARATION THEOREM

Theorem 5.30 (Discrete Matrix Sturm Separation Theorem) *Suppose equation* (5.1) *is disconjugate on* $[a, b+2]$. *If* $U(t)$ *is a prepared basis, then* $U(t)$ *has at most* n *generalized zeros in* $[a+1, b+2]$ *(hence at most* $n+1$ *generalized zeros in* $[a, b+2]$*).*

Proof: Assume $U(t)$ has $m > n$ generalized zeros at $t_1 < t_2 < \cdots < t_m$ in $[a+1, b+2]$. By Exercise 5.29 there exist nonzero $n \times 1$ vectors γ_i, $1 \le i \le m$ such that

$$u_i(t) \equiv U(t)\gamma_i$$

is a nontrivial prepared solution of (5.1) with a generalized zero at t_i. By Exercise 5.28 $\{u_1(t), \ldots, u_m(t)\}$ is a mutually prepared set of solutions of (5.1).

Define

$$\eta_i(t) = \begin{cases} u_i(t), & a \le t \le t_i - 1 \\ 0, & t_i \le t \le b+2. \end{cases}$$

Since $m > n$, then the m vectors

$$\{\eta_i(a)\}_{i=1}^m = \{u_i(a)\}_{i=1}^m$$

in C^n have to be linearly dependent. Hence there exist constants c_i, $1 \le i \le m$, not all zero, such that

$$\sum_{i=1}^m c_i \eta_i(a) = \sum_{i=1}^m c_i u_i(a) = 0. \tag{5.18}$$

Define η on $[a, b+2]$ by

$$\eta(t) = \sum_{i=1}^m c_i \eta_i(t).$$

Since $\eta(a) = \eta(b+2) = 0$, $\eta \in \mathcal{F}$. Consider

$$\begin{aligned} \mathcal{J}[\eta] &= B[\eta, \eta] \\ &= B\Big[\sum_{i=1}^m c_i \eta_i(t), \sum_{j=1}^m c_j \eta_j(t)\Big] \\ &= \sum_{i=1}^m \sum_{j=1}^m c_i \bar{c}_j B[\eta_i(t), \eta_j(t)]. \end{aligned}$$

Use Theorem 5.26 on page 219 to obtain

$$\begin{aligned}\mathcal{J}[\eta] &= -\sum_{i=1}^{m}\sum_{j=1}^{m} c_i\bar{c}_j u_j^*(a)K(a)u_i(a+1)\\ &+\sum_{i=1}^{m}|c_i|^2 u_i^*(t_i-1)K(t_i-1)u_i(t_i)\\ &\leq -\sum_{i=1}^{m}\sum_{j=1}^{m} c_i\bar{c}_j u_j^*(a)K(a)u_i(a+1)\end{aligned}$$

since $u_i(t)$ has a generalized zero at t_i, $1 \leq i \leq m$.

Hence

$$\mathcal{J}[\eta] \leq -\sum_{i=1}^{m}\left\{c_i\left[\sum_{j=1}^{m} c_j u_j(a)\right]^* K(a)u_i(a+1)\right\}.$$

Since the inside sum is zero by (5.18) we know that

$$\mathcal{J}[\eta] \leq 0.$$

But (5.1) is disconjugate on $[a, b+2]$ implies that $\mathcal{J}$ is positive definite on $\mathcal{F}$. Hence

$$\eta(t) = \sum_{i=1}^{m} c_i\eta_i(t) \equiv 0$$

on $[a, b+2]$. Let p be the largest integer $1 \leq p \leq m$ such that $c_p \neq 0$. Then

$$\eta(t) = \sum_{i=1}^{p} c_i\eta_i(t) = 0$$

on $[a, b+2]$. Hence

$$0 = \eta(t_p-1) = \sum_{i=1}^{p} c_i\eta_i(t_p-1) = c_p u_p(t_p-1).$$

This implies that

$$u_p(t_p-1) = 0.$$

But $u_p(t)$ has a generalized zero at t_p. Since $u_p(t_p-1) = 0$ we must have that $u_p(t_p) = 0$. But $u_p(t_p-1) = u_p(t_p) = 0$ implies $u_p(t)$ is the trivial solution which is a contradiction. □

The next two corollaries follow easily from the Sturm separation theorem.

Corollary 5.31 *If* (5.1) *is disconjugate on* $[a, \infty)$ *and if* $U(t)$ *is a prepared basis, then there is an integer* $t_0 \in [a+1, \infty)$ *such that*

$$D(t) \equiv U^*(t)K(t)U(t+1) > 0$$

for $t \geq t_0$. *In particular every prepared basis of* (5.1) *is eventually nonsingular.*

Corollary 5.32 *If one prepared basis has a finite number of generalized zeros in* $[a, \infty)$, *then every prepared basis has a finite number of generalized zeros in* $[a, \infty)$.

We now state and prove another corollary that follows from the Sturm separation theorem.

Corollary 5.33 *Assume* $W(t)$ *is positive definite on* $[a+1, \infty)$, *then there is an extended real number* μ *such that the recurrence relation*

$$-K(t)u(t+1) + N(t)u(t) - K^*(t-1)u(t-1) = \lambda W(t)u(t) \tag{5.19}$$

satisfies the property that every prepared basis of (5.19) *has a finite number of generalized zeros in* $[a, \infty)$ *if* $\lambda < \mu$ *whereas if* $\lambda > \mu$ *every prepared basis has infinitely many generalized zeros in* $[a, \infty)$.

Proof: By Corollary 5.32 we have that for each fixed λ every prepared basis of (5.19) either has finitely many generalized zeros in $[a, \infty)$ or every prepared basis of (5.19) has infinitely many generalized zeros in $[a, \infty)$. Assume there is a λ_0 such that every prepared basis of (5.19) for $\lambda = \lambda_0$ has finitely many generalized zeros in $[a, \infty)$. We will show that if $\lambda_1 < \lambda_0$, then every prepared basis of (5.19) for $\lambda = \lambda_1$ has finitely many generalized zeros in $[a, \infty)$. Assume $U_{\lambda_0}(t)$ is a prepared basis for (5.19) for $\lambda = \lambda_0$. Since $U_{\lambda_0}(t)$ has a finite number of generalized zeros there is a $t_0 \geq a$ such that

$$U_{\lambda_0}(t-1)K(t-1)U_{\lambda_0}(t) > 0$$

on $[t_0 + 1, \infty)$. Let $\mathcal{J}_\lambda[\eta]$ be the quadratic form corresponding to (5.19) like $\mathcal{J}[\eta]$ is the quadratic form corresponding to (5.1) where a is replaced by t_0 and b is replaced by t_1 where $t_1 \geq t_0$. By the Reid Roundabout theorem $\mathcal{J}_{\lambda_0}[\eta]$ is positive definite on $\mathcal{F}$ where in the definition of $\mathcal{F}$ we replace a and b by t_0 and t_1 respectively. Let Q_λ correspond to (5.19) as Q does to (5.1) with a and b replaced by t_0 and t_1 respectively. Then

$$Q_\lambda = Q - \lambda \hat{W}$$

where $\hat{W}$ is the block diagonal matrix

$$\hat{W} = \begin{bmatrix} W(t_0+1) & & & \\ & W(t_0+2) & & \\ & & \ddots & \\ & & & W(t_1+1) \end{bmatrix}.$$

Since $W(t)$ is positive definite on $[a+1,\infty)$, $\hat{W}$ is a positive definite matrix. By Theorem 5.15 on page 214, Q_{λ_0} is positive definite. Since $\lambda_1 < \lambda_0$

$$Q_{\lambda_1} = Q - \lambda_1 \hat{W} \geq Q - \lambda_0 \hat{W} = Q_{\lambda_0}.$$

It follows that Q_{λ_1} is positive definite. But then by Theorem 5.15 we have that $\mathcal{J}_{\lambda_1}[\eta]$ is positive definite on $\mathcal{F}$. By Theorem 5.13, the difference equation (5.19) is disconjugate on $[t_0, t_1+2]$. Since $t_1 \geq t_0$ is arbitrary, (5.19) for $\lambda = \lambda_1$ is disconjugate on $[t_0,\infty)$. Hence any prepared basis of (5.19) has only finitely many generalized zeros in $[a,\infty)$ by Corollary 5.31.

If there is no λ such that (5.19) has the property that every prepared basis of (5.19) has finitely many generalized zeros in $[a,\infty)$, then $\mu = -\infty$. Otherwise let

$$\mu = \sup\{\lambda : (5.19) \textit{ has the property that every prepared basis } of (5.19) \textit{ has finitely many generalized zeros in } [a,\infty)\}.$$

If $\mu = \infty$, then for all λ the difference equation (5.19) has the property that every prepared basis of (5.19) has a finite number of generalized zeros in $[a,\infty)$.

□

A special case of Corollary (5.33) would be when $K(t)$ is positive definite on $[a,\infty)$. In this case Smith [142] proved that μ is the infimum of the essential spectrum of an associated self-adjoint operator. Corollary 5.33 in this general setting was proven by Ahlbrandt [6].

5.7 DISCRETE JACOBI CONDITIONS

Recall that $\mathcal{A}$ is the class

$$\mathcal{A} = \{\eta : [a, b+2] \to \mathrm{R}^n \text{ such that } \eta(a) = \eta(b+2) = 0\}.$$

For the purposes of this section, we will restrict our admissible variations to this real class instead of the class $\mathcal{F}$ because our variational problems will be restricted to real vectors. The following theorem could be restated for complex solutions if $\mathcal{A}$ is replaced by $\mathcal{F}$, solutions u are complex valued which are prepared and all transposes are replaced by conjugate transposes.

Theorem 5.34 (Jacobi) *Assume* (5.2) *holds and that $u(t)$ is an $n \times 1$ real vector solution of* (5.1) *such that*

$$u^T(a)K(a)u(a+1) \leq 0$$

and $u(a+1) \neq 0$. Define $d(t) = u^T(t-1)K(t-1)u(t)$ for $t \in [a+1, b+2]$. If $\mathcal{J}[\eta]$ is positive semidefinite on $\mathcal{A}$, then the Jacobi condition

$$\begin{aligned} d(t) &> 0 \text{ on } [a+2, b+1] \\ d(b+2) &\geq 0 \end{aligned}$$

holds. If $\mathcal{J}[\eta]$ is positive definite on $\mathcal{A}$, then the strengthened Jacobi condition

$$d(t) > 0 \text{ on } [a+2, b+2]$$

holds.

Proof: Assume $a+2 \leq t_0 \leq b+2$ and set

$$\eta(t) = \begin{cases} 0, & t = a, \\ u(t), & a+1 \leq t \leq t_0 - 1, \\ 0, & t_0 \leq t \leq b+2. \end{cases}$$

By Corollary 5.11 on page 206, $\eta \in \mathcal{A}$ and

$$\mathcal{J}[\eta] = u^T(a)K(a)u(a+1) + u^T(t_0-1)K(t_0-1)u(t_0).$$

Therefore, $d(t_0) = \mathcal{J}[\eta] - d(a+1) \geq \mathcal{J}[\eta]$. Hence if $\mathcal{J}[\eta]$ is positive semidefinite on $\mathcal{A}$, then $d(t_0) \geq 0$. Since t_0 is an arbitrary point of $[a+2, b+2]$ we have

$$d(t) \geq 0 \quad \text{on} \quad [a+2, b+2].$$

Now assume there is a $t_1 \in [a+2, b+1]$ such that

$$d(t_1) = 0.$$

Define $\hat{\eta}$ on $[a, b+2]$ by

$$\hat{\eta}(t) = \begin{cases} 0, & t = a \\ u(t), & a+1 \le t \le t_1 - 1 \\ 0, & t_1 \le t \le b+2. \end{cases}$$

By Corollary 5.11 on page 206, $\hat{\eta} \in \mathcal{A}$ and

$$\begin{aligned} \mathcal{J}[\hat{\eta}] &= u^T(a)K(a)u(a+1) + u^T(t_1-1)K(t_1-1)u(t_1) \\ &= d(a+1) + d(t_1) \\ &= d(a+1) \le 0. \end{aligned}$$

Since we are assuming $\mathcal{J}[\eta]$ is positive semidefinite

$$\mathcal{J}[\hat{\eta}] = 0.$$

Thus $\mathcal{J}$ has a global minimum on $\mathcal{A}$ at $\hat{\eta}$. Hence by Theorem 4.3 on page 156 $\hat{\eta}$ is a solution of the Euler-Lagrange equation for $\mathcal{J}[\eta]$. Therefore by Exercise 5.1, $\hat{\eta}(t)$ is a solution of

$$-K(t)\eta(t+1) + N(t)\eta(t) - K^T(t-1)\eta(t-1) = 0.$$

But $\hat{\eta}$ satisfies the initial conditions $\hat{\eta}(b+1) = \hat{\eta}(b+2) = 0$. Hence it follows that

$$\hat{\eta}(t) \equiv 0 \quad \text{on} \quad [a, b+2].$$

But $\hat{\eta}(a+1) = u(a+1) \ne 0$ which is a contradiction. Therefore $d(t_1) > 0$. Since $t_1 \in [a+2, b+1]$ is arbitrary we have

$$d(t) > 0 \quad \text{on} \quad [a+2, b+1].$$

Finally assume $\mathcal{J}[\eta]$ is positive definite on $\mathcal{A}$. It remains to show that $d(b+2) > 0$. To see this define

$$\eta_0(t) = \begin{cases} 0, & t = a \\ u(t), & a+1 \le t \le b+1 \\ 0, & t = b+2. \end{cases}$$

Then, by Corollary 5.11 on page 206, $\eta_0 \in \mathcal{A}$ and

$$\mathcal{J}[\eta_0] = d(a+1) + d(b+2).$$

Since $\eta_0 \ne 0$,

$$d(b+2) = \mathcal{J}[\eta_0] - d(a+1) \ge \mathcal{J}[\eta_0] > 0.$$

□

5.8 REDUCTION OF ORDER

We next prove a reduction of order theorem for the matrix version $\mathcal{L}U(t) = 0$ of equation (5.1), i.e.,

$$\mathcal{L}U(t) = -K(t)U(t+1) + N(t)U(t) - K^*(t-1)U(t-1) = 0. \tag{5.20}$$

Theorem 5.35 (Reduction of Order Theorem) *Assume $U_0(t)$ is an $n \times n$ prepared solution of $\mathcal{L}U(t) = 0$ with $U_0(t)$ nonsingular on $[a, \infty)$. Let*

$$D_0(t) = U_0^*(t)K(t)U_0(t+1). \tag{5.21}$$

Then $U(t)$ is an $n \times m$ matrix solution of $\mathcal{L}U(t) = 0$ iff $U(t)$ satisfies a first order difference equation

$$\Delta[U_0^{-1}(t)U(t)] = D_0^{-1}(t)Q, \qquad \text{for } t \geq a, \tag{5.22}$$

with Q a constant $n \times m$ matrix iff $U(t)$ is of the form

$$U(t) = U_0(t)[P + \mathcal{S}_0(t)Q] \tag{5.23}$$

where P and Q are constant $n \times m$ matrices and

$$\mathcal{S}_0(t) = \sum_{s=a}^{t-1} D_0^{-1}(s) \qquad \text{for } t \geq a. \tag{5.24}$$

Furthermore,

$$P = U_0^{-1}(a)U(a) \tag{5.25}$$

$$Q = \{U_0; U\} \tag{5.26}$$

and $U(t)$ is a prepared solution iff $P^*Q = Q^*P$.

We are using the convention that when a summation index runs from a to $a-1$, then the summation is defined to be zero; hence $\mathcal{S}_0(a) = 0$.

Because the matrix D arises in a different manner than the matrix D of Chapter 3, the matrix D^{-1} of this chapter is the matrix D of Chapter 3.

Proof: First assume $U(t)$ is an $n \times m$ matrix solution of $\mathcal{L}U(t) = 0$. We will show that $U(t)$ is of the form (5.23) where $\mathcal{S}_0(t)$, P and Q are given by (5.24), (5.25), and (5.26) respectively. To see this define the $n \times m$ matrix Q by

$$Q = \{U_0; U\}.$$

By the definition of the Lagrange bracket we have that

$$U_0^*(t-1)K(t-1)U(t) - U_0^*(t)K^*(t-1)U(t-1) = Q$$

for $t \geq a+1$. Replacing t by $t+1$ gives

$$U_0^*(t)K(t)U(t+1) - U_0^*(t+1)K^*(t)U(t) = Q$$

for $t \geq a$. Multiplying by $U_0^{-1}(t+1)K^{-1}(t)[U_0^*(t)]^{-1}$ we have that

$$U_0^{-1}(t+1)U(t+1) - U_0^{-1}(t+1)K^{-1}(t)[U_0^*(t)]^{-1}U_0^*(t+1)K^*(t)U(t)$$

$$= U_0^{-1}(t+1)K^{-1}(t)[U_0^*(t)]^{-1}Q$$

for $t \geq a$. Hence

$$U_0^{-1}(t+1)U(t+1) - [U_0^*(t)K(t)U_0(t+1)]^{-1}U_0^*(t+1)K^*(t)U(t)$$

$$= [U_0^*(t)K(t)U_0(t+1)]^{-1}Q.$$

Using $U_0(t)$ is a prepared solution and the definition of $D_0(t)$ we obtain

$$U_0^{-1}(t+1)U(t+1) - [U_0^*(t+1)K^*(t)U_0(t)]^{-1}U_0^*(t+1)K^*(t)U(t) = D_0^{-1}(t)Q.$$

Simplifying the second term we get that

$$U_0^{-1}(t+1)U(t+1) - U_0^{-1}(t)U(t) = D_0^{-1}(t)Q.$$

Hence

$$\Delta[U_0^{-1}(t)U(t)] = D_0^{-1}(t)Q$$

for $t \geq a$. Summing both sides from a to $t-1$ we obtain

$$U_0^{-1}(t)U(t) - U_0^{-1}(a)U(a) = \sum_{s=a}^{t-1} D_0^{-1}(s)Q = \mathcal{S}_0(t)Q$$

by (5.24). Solving for $U(t)$ we get (5.23) provided P is given by (5.25).

Conversely, we will show that if $U(t)$ is given by (5.23), then $U(t)$ is a solution of $\mathcal{L}U(t) = 0$ on $[a, b+2]$. From (5.23) we have that

$$U(t+1) = U_0(t+1)\left[P + \mathcal{S}_0(t+1)Q\right]$$

for $t \geq a+1$. But from (5.24)

$$\begin{aligned} \mathcal{S}_0(t+1) &= \mathcal{S}_0(t) + D_0^{-1}(t) \\ &= \mathcal{S}_0(t) + \left[U_0^*(t)K(t)U_0(t+1)\right]^{-1}. \end{aligned}$$

Hence

$$U(t+1) = U_0(t+1)\left[P + \mathcal{S}_0(t)Q\right] + K^{-1}(t)\left[U_0^*(t)\right]^{-1}Q.$$

Multiplying by $-K(t)$ we obtain

$$-K(t)U(t+1) = -K(t)U_0(t+1)[P + \mathcal{S}_0(t)Q] - [U_0^*(t)]^{-1}Q.$$

Use the fact that $U_0(t)$ is a solution of $\mathcal{L}U(t) = 0$ for the computation

$$\begin{aligned} -K(t)U(t+1) = &- N(t)U_0(t)[P + \mathcal{S}_0(t)Q] \\ &+ K^*(t-1)U_0(t-1)[P + \mathcal{S}_0(t)Q] - [U_0^*(t)]^{-1}Q \\ = &- N(t)U(t) + K^*(t-1)U_0(t-1)\{P + \mathcal{S}_0(t-1)Q \\ &+ [U_0^*(t-1)K(t-1)U_0(t)]^{-1}Q\} - [U_0^*(t)]^{-1}Q \\ = &- N(t)U(t) + K^*(t-1)U(t-1) + K^*(t-1)U_0(t-1) \\ &\cdot [U_0^*(t-1)K(t-1)U_0(t)]^{-1}Q - [U_0^*(t)]^{-1}Q. \end{aligned}$$

Since $U_0(t)$ is a prepared solution we can write this as

$$\begin{aligned} -K(t)U(t+1) &= -N(t)U(t) + K^*(t-1)U(t-1) + K^*(t-1)U_0(t-1) \cdot \\ &\quad [U_0^*(t)K^*(t-1)U_0(t-1)]^{-1}Q - [U_0^*(t)]^{-1}Q. \end{aligned}$$

Simplifying we obtain

$$\begin{aligned} -K(t)U(t+1) &= - N(t)U(t) + K^*(t-1)U(t-1) \\ &\quad + [U_0^*(t)]^{-1}Q - [U_0^*(t)]^{-1}Q. \end{aligned}$$

Because the last two terms cancel, $U(t)$ is a solution of $\mathcal{L}U(t) = 0$ on $[a, b+2]$. Letting $t = a$ in (5.23) and solving for P gives (5.25).

To show that (5.26) holds, use the constancy of the bracket function (see Corollary 5.5 on page 202) and the definition of the bracket function given in Theorem 5.3 on page 201 for

$$\begin{aligned}\{U_0(t);U(t)\} &= \{U_0(a+1);U(a+1)\}\\ &= U_0^*(a)K(a)U(a+1) - U_0^*(a+1)K^*(a)U(a).\end{aligned}$$

But (5.23) and (5.24) give

$$\begin{aligned}\{U_0;U\} &= U_0^*(a)K(a)U_0(a+1)\{P+[U_0^*(a)K(a)U_0(a+1)]^{-1}Q\}\\ &\quad -U_0^*(a+1)K^*(a)U_0(a)P\\ &= U_0^*(a)K(a)U_0(a+1)P+Q-U_0^*(a+1)K^*(a)U_0(a)P = Q\end{aligned}$$

because $U_0(t)$ is a prepared solution. Hence (5.26) holds.

It only remains to show that $U(t)$ is a prepared solution iff $P^*Q = Q^*P$. This is left as the next exercise. □

Exercise 5.36 *Prove that $U(t)$ in Theorem 5.35 is a prepared solution of $\mathcal{L}U(t) = 0$ iff P^*Q is Hermitian.*

Exercise 5.37 *Show that $u(t) = 2^t$ is a solution of each of the following and use formula (5.23) to find a general solution of each of the following:*

(a) $-\frac{1}{6^t}u(t+1) + \frac{5}{6^t}u(t) - \frac{1}{6^{t-1}}u(t-1) = 0, \quad t \geq 1.$

(b) $-\frac{(t-1)!}{2^t}u(t+1) + \frac{(2t-1)(t-2)!}{2^t}u(t) - \frac{(t-2)!}{2^{t-1}}u(t-1) = 0, \quad t \geq 2.$

5.9 BACKWARDS REDUCTION OF ORDER

We next look at what we call the *backwards reduction of order theorem.*

Theorem 5.38 (Backwards Reduction of Order Theorem) *Suppose $U_0(t)$ is an $n \times n$ prepared matrix solution of $\mathcal{L}U(t) = 0$ on $(-\infty, a]$ with $U_0(t)$ nonsingular on $(-\infty, a]$. Then $U(t)$ is an $n \times m$ matrix solution of $\mathcal{L}U(t) = 0$ iff $U(t)$ is of the form*

$$U(t) = U_0(t)[P - \bar{S}_0(t)Q] \qquad \text{for } t \leq a, \tag{5.27}$$

where

$$\bar{S}_0(t) = \sum_{s=t}^{a-1} D_0^{-1}(s)$$

for $t \leq a$, where $D_0(t)$, P, and Q are as in Theorem 5.35. Furthermore $U(t)$ is a prepared solution iff $P^*Q = Q^*P$.

Proof: If U is a solution, then set $Q \equiv \{U_0; U\}(t)$ as in the proof of the forward reduction of order theorem. We will leave the proof that U can be written in the form (5.27) as an exercise. We will now show the converse, namely, that U of the form (5.27) is a solution. Assume

$$U(t) = U_0(t)[P - \bar{S}_0(t)Q]$$

for $t \leq a$, where P and Q are $n \times m$ constant matrices. We now show that $U(t)$ is a solution of $\mathcal{L}U(t) = 0$ on $(-\infty, a]$. For $t \leq a - 1$, consider

$$\begin{aligned} -K(t)U(t+1) &= -K(t)U_0(t+1)[P - \bar{S}_0(t+1)Q] \\ &= -K(t)U_0(t+1)\{[P - \bar{S}_0(t)Q] + D_0^{-1}(t)Q\} \\ &= -K(t)U_0(t+1)\left[P - \bar{S}_0(t)Q\right] \\ &\quad -K(t)U_0(t+1)[U_0^*(t)K(t)U_0(t+1)]^{-1}Q. \end{aligned}$$

Use the fact that U_0 is a solution in the first term and simplify the second term to obtain

$$\begin{aligned} -K(t)U(t+1) &= [-N(t)U_0(t) + K^*(t-1)U_0(t-1)]\left[P - \bar{S}_0(t)Q\right] \\ &\quad -(U_0^*(t))^{-1}Q \\ &= [-N(t)U_0(t) + K^*(t-1)U_0(t-1)]\{[P - \bar{S}_0(t-1)Q] \\ &\quad +D_0^{-1}(t-1)Q\} - (U_0^*(t))^{-1}Q \\ &= -N(t)U(t) + K^*(t-1)U(t-1) - (U_0^*(t))^{-1}Q \\ &\quad +K^*(t-1)U_0(t-1)[U_0^*(t-1)K(t-1)U_0(t)]^{-1}Q. \end{aligned}$$

Using the assumption that $U_0(t)$ is a prepared solution, we have

$$\begin{aligned} -K(t)U(t+1) &= -N(t)U(t) + K^*(t-1)U(t-1) - (U_0^*(t))^{-1}Q \\ &\quad +K^*(t-1)U_0(t-1)[U_0^*(t)K^*(t-1)U_0(t-1)]^{-1}Q \\ &= -N(t)U(t) + K^*(t-1)U(t-1) \\ &\quad -(U_0^*(t))^{-1}Q + (U_0^*(t))^{-1}Q. \end{aligned}$$

Since the last two terms cancel we have the desired result that $U(t)$ is a solution of $\mathcal{L}U(t) = 0$ on $(-\infty, a]$. Letting $t = a$ in (5.27) gives that $P = U_0^{-1}(a)U(a)$ which is the desired result, namely, equation (5.25) on page 228. To see that equation (5.26) holds, note that

$$\begin{aligned} \{U_0; U\}(t) &= \{U_0; U\}(a-1) \\ &= U_0^*(a-1)K(a-1)U(a) - U_0^*(a)K^*(a-1)U(a-1) \\ &= U_0^*(a-1)K(a-1)U_0(a)P \\ &\quad -U_0^*(a)K^*(a-1)U_0(a-1)[P - D_0^{-1}(a-1)Q] \\ &= [U_0^*(a-1)K(a-1)U_0(a) - U_0^*(a)K^*(a-1)U_0(a-1)]P \\ &\quad +U_0^*(a)K^*(a-1)U_0(a-1)D_0^{-1}(a-1)Q. \end{aligned}$$

Since $U_0(t)$ is a prepared solution

$$\{U_0; U\}(t) = D_0(a-1)D_0^{-1}(a-1)Q = Q.$$

Hence (5.26) holds. The remainder of the first part of the proof is similar to the corresponding part of the proof of the forward reduction of order theorem, Theorem 5.35, and will be left as an exercise. □

Exercise 5.39 *Complete the proof of Theorem* 5.38.

5.10 DOMINANT AND RECESSIVE SOLUTIONS

We are particularly interested in the existence of two types of solutions of $\mathcal{L}U(t) = 0$ on the infinite interval $[a, \infty)$. These are the dominant solutions at ∞ and the recessive solutions at ∞. In general, computation of the recessive solution of a three term recurrence is a difficult problem. It is useful to motivate this study by the continuous example of $u'' - u = 0$ which has general solution of the form $u = C_1 e^t + C_2 e^{-t}$. As $t \to \infty$, the solution $u_1(t) = e^{-t}$ is recessive in the sense that $u_1(t)/u(t) \to 0$ as $t \to \infty$ if u is any solution which is linearly independent of u_1. The recessive solution at ∞ is unique up to nonzero constant multiples, but the solutions $\cosh t$ and $\sinh t$ are both dominant and are linearly independent. Note that if one numerically follows a recessive solution to the right, any numerical error will incorporate some part of a dominant solution and soon the recessive behavior is destroyed. In this example the solution $u_1(t) = e^{-t}$ is dominant at $-\infty$ and the solution $u_2(t) = e^t$ is recessive at $-\infty$. Recessive solutions do not always have to go to zero as is seen by the example of the "Euler equation" $4t^2 u'' + u = 0$ where $u(t) = t^{1/2}$ is recessive at ∞. Recessive solutions of difference equations are of particular significance because of the connection between existence of recessive solutions and convergence of associated continued fractions first shown by Pincherle [127]. One major difference between the continuous and discrete theories is that in the continuous theory there exists a recessive solution if and only if the equation is nonoscillatory (or eventually disconjugate in the matrix case). However the Fibonacci recurrence $u_{n+1} = u_n + u_{n-1}$ placed in self–adjoint form by multiplying both sides by $(-1)^n$ has a recessive solution at ∞ but has an oscillatory solution and fails to be disconjugate in any neighborhood of ∞. Thus the question of existence of a recessive solution is somehow a much deeper question for difference equations than for differential equations.

We say that a solution $U_1(t)$ of $\mathcal{L}U(t) = 0$ is a *dominant solution* at ∞ provided $U_1(t)$ is a prepared basis, there is an integer t_0 such that $U_1(t)$ is nonsingular on $[t_0, \infty)$ and if

$$\mathcal{S}_1(t) \equiv \sum_{s=t_0}^{t-1} D_1^{-1}(s),$$

where $D_1(t) = U_1^*(t)K(t)U_1(t+1)$, then the series

$$\sum_{s=t_0}^{\infty} D_1^{-1}(s) = \lim_{t\to\infty} \mathcal{S}_1(t)$$

converges (to a Hermitian matrix with finite entries). We repeat our caveat that the matrix D arises in a different manner than the matrix D of Chapter 3, i.e., the matrix D^{-1} of this chapter is the matrix D of Chapter 3.

Theorem 5.40 *Assume $\mathcal{L}U(t) = 0$ has a dominant solution $U_1(t)$ at ∞. If $U(t)$ is any $n \times m$ matrix solution of $\mathcal{L}U(t) = 0$, then*

$$\lim_{t\to\infty} U_1^{-1}(t)U(t) = C$$

where C is a constant $n \times m$ matrix.

Proof: Since $U_1(t)$ is a dominant solution there is an integer $t_0 \geq a$ such that $U_1(t)$ is nonsingular on $[t_0, \infty)$. By the reduction of order theorem, i.e., Theorem 5.35 on page 228, with a replaced by t_0 and $\mathcal{S}_0$ replaced by $\mathcal{S}_1$, we have

$$U(t) = U_1(t)[P + \mathcal{S}_1(t)Q]$$

where P and Q are the $n \times m$ matrices given by

$$P = U_1^{-1}(t_0)U(t_0) \qquad \text{and} \qquad Q = \{U_1; U\}.$$

It follows that

$$U_1^{-1}(t)U(t) = P + \mathcal{S}_1(t)Q \qquad \text{for } t \geq t_0.$$

Since $U_1(t)$ is a dominant solution at ∞, $\lim_{t\to\infty} \mathcal{S}_1(t)$ exists and consequently

$$\lim_{t\to\infty} U_1^{-1}(t)U(t)$$

exists and the proof is complete. □

A solution $U_0(t)$ is said to be *recessive at* ∞ provided $U_0(t)$ is a prepared basis and whenever $U(t)$ is an $n \times n$ matrix solution of $\mathcal{L}U(t) = 0$ such that $\{U; U_0\}$ is nonsingular, then $U(t)$ is nonsingular for all sufficiently large t and

$$\lim_{t\to\infty} U^{-1}(t)U_0(t) = 0.$$

Exercise 5.41 *Suppose that $U_0(t)$ is a solution of $\mathcal{L}U(t) = 0$ which is recessive at ∞. Show that if C is a nonsingular $n \times n$ constant matrix, then the solution $U(t) \equiv U_0(t)C$ is also recessive at ∞.*

Exercise 5.42 *Suppose that $U_0(t)$ is a solution of $\mathcal{L}U(t) = 0$ which is recessive at ∞. Show that if $U_1(t)$ is a prepared solution of $\mathcal{L}U(t) = 0$ such that $\{U_1; U_0\}$ is nonsingular, then $U_1(t)$ is a dominant solution at ∞.*

We are also interested in two types of solutions of $\mathcal{L}U(t) = 0$ on the infinite discrete interval $(-\infty, a]$. These are dominant solutions at $-\infty$ and recessive solutions at $-\infty$. We say that $U_1(t)$ is a *dominant* solution of $\mathcal{L}U(t) = 0$ at $-\infty$ provided $U_1(t)$ is a prepared basis, there is an integer $t_0 \le a$ such that $U_1(t)$ is nonsingular on $(-\infty, t_0]$ and if for $t \le t_0$,

$$\mathcal{S}_1(t, t_0) \equiv \sum_{s=t}^{t_0-1} D_1^{-1}(s),$$

where

$$D_1(t) = U_1^*(t)K(t)U_1(t+1),$$

then the series

$$\sum_{s=-\infty}^{t_0-1} D_1^{-1}(s) = \lim_{t\to-\infty} \mathcal{S}_1(t, t_0) \equiv \mathcal{S}_1(-\infty, t_0)$$

converges (to a Hermitian matrix with finite entries).

We say that $U_0(t)$ is a *recessive* solution of $\mathcal{L}U(t) = 0$ at $-\infty$ provided $U_0(t)$ is a prepared basis and whenever $U(t)$ is an $n \times n$ matrix solution of $\mathcal{L}U(t) = 0$ such that $\{U; U_0\}$ is nonsingular, then $U(t)$ is nonsingular for all sufficiently small integers t and

$$\lim_{t\to-\infty} U^{-1}(t)U_0(t) = 0.$$

Exercise 5.43 *Show that if $U_0(t)$ is a recessive solution at $-\infty$ and C is a nonsingular $n \times n$ constant matrix, then $U(t) = U_0(t)C$ is also a recessive solution at $-\infty$.*

Exercise 5.44 *Assume $\mathcal{L}U(t) = 0$ has a solution $U_0(t)$ which is recessive at $-\infty$. Show that if $U_1(t)$ is a prepared solution such that $\{U_1; U_0\}$ is nonsingular, then $U_1(t)$ is a dominant solution at $-\infty$.*

Theorem 5.45 *Assume $\mathcal{L}U(t) = 0$ has a solution $U_1(t)$ which is dominant at ∞. For t_0 sufficiently large define $\mathcal{S}_1(t, \infty) \equiv \sum_{s=t}^{\infty} D_1^{-1}(s)$ for $t \ge t_0$. Then for $t \ge t_0$ the solution U_0 defined by*

$$U_0(t) \equiv U_1(t)\mathcal{S}_1(t, \infty)$$

is recessive at ∞ and

$$\{U_1; U_0\} = -I.$$

Proof: Since $U_1(t)$ is dominant at ∞, $U_1(t)$ is a prepared basis and there is an integer $t_0 \geq a$ such that $U_1(t)$ is nonsingular for $t \geq t_0$ and the series

$$\sum_{s=t_0}^{\infty} D_1^{-1}(s)$$

converges to a Hermitian matrix limit. Hence $\mathcal{S}_1(t,\infty)$ defined in the statement of the theorem is a well defined Hermitian matrix for $t \geq t_0$. Define

$$U_0(t) \equiv U_1(t)\mathcal{S}_1(t,\infty)$$

for $t \geq t_0$. Then we can write $U_0(t)$ in the form

$$U_0(t) = U_1(t)\left\{\mathcal{S}_1(t_0,\infty) - \mathcal{S}_1(t)I\right\}.$$

Hence $U_0(t)$ is of the form (5.23) on page 228 with $P = \mathcal{S}_1(t_0,\infty)$ and $Q = -I$. It follows from the reduction of order theorem that $U_0(t)$ is a solution of $\mathcal{L}U(t) = 0$. Also by (5.26)

$$\{U_1; U_0\} = -I.$$

Since $P^*Q = -\mathcal{S}_1(t_0,\infty)$ is Hermitian, $U_0(t)$ is a prepared solution by the reduction of order theorem, Theorem 5.35 on page 228 under the replacement of U_0 by U_1 and U by U_0. From Theorem 5.23 on page 218, $U_0(t)$ is a prepared basis. Let $U(t)$ be an $n \times n$ matrix solution of $\mathcal{L}U(t) = 0$ such that $\{U; U_0\}$ is nonsingular. By the reduction of order theorem, Theorem 5.35, with a replaced by t_0 and U_0 replaced by U_i,

$$U(t) = U_1(t)[P_1 + \mathcal{S}_1(t)Q_1] \quad \text{for } t \geq t_0,$$

where

$$P_1 = U_1^{-1}(t_0)U(t_0) \quad \text{and} \quad Q_1 = \{U_1; U\}.$$

Since

$$\begin{aligned}
\mathcal{S}_1(t) &= \mathcal{S}_1(t_0,\infty) - \mathcal{S}_1(t,\infty), \\
U(t) &= U_1(t)\{P_1 + [\mathcal{S}_1(t_0,\infty) - \mathcal{S}_1(t,\infty)]Q_1\} \\
&= U_1(t)[P_1 + \mathcal{S}_1(t_0,\infty)Q_1] - U_1(t)\mathcal{S}_1(t,\infty)Q_1,
\end{aligned}$$

and we have

$$U(t) = U_1(t)C_1 + U_0(t)C_2, \tag{5.28}$$

where

$$C_1 = P_1 + \mathcal{S}_1(t_0, \infty)Q_1 \qquad \text{and} \qquad C_2 = -Q_1 .$$

To see that $U(t)$ is nonsingular for large t, note that

$$\begin{aligned} \{U; U_0\} &= \{U_1C_1 + U_0C_2; U_0\} \\ &= C_1^*\{U_1; U_0\} + C_2^*\{U_0; U_0\} = -C_1^* . \end{aligned}$$

Since $\{U; U_0\}$ is nonsingular, C_1 is nonsingular. From (5.28)

$$\begin{aligned} \lim_{t\to\infty} U_1^{-1}(t)U(t) &= \lim_{t\to\infty} [C_1 + U_1^{-1}(t)U_0(t)C_2] \\ &= \lim_{t\to\infty} [C_1 + \mathcal{S}_1(t, \infty)C_2] = C_1 \end{aligned}$$

is nonsingular. Hence for large t, $U(t)$ is nonsingular. Finally,

$$\begin{aligned} \lim_{t\to\infty} U^{-1}(t)U_0(t) &= \lim_{t\to\infty} \{[U_1(t)C_1 + U_0(t)C_2]^{-1}U_0(t)\} \\ &= \lim_{t\to\infty} \{[U_1(t)\left(C_1 + U_1^{-1}(t)U_0(t)C_2\right)]^{-1}U_0(t)\} \\ &= \lim_{t\to\infty} \{\left[C_1 + U_1^{-1}(t)U_0(t)C_2\right]^{-1} \cdot U_1^{-1}(t)U_0(t)\} \\ &= \lim_{t\to\infty} \{[C_1 + \mathcal{S}_1(t, \infty)C_2]^{-1} \cdot \mathcal{S}_1(t, \infty)\} = C_1^{-1} \cdot 0 = 0. \end{aligned}$$

Hence $U_0(t)$ is a recessive solution of $\mathcal{L}U(t) = 0$ at ∞ . □

Theorem 5.46 *Assume that $\mathcal{L}U(t) = 0$ has a solution $U_1(t)$ which is dominant at $-\infty$. For t_0 sufficiently small and fixed, define*

$$\mathcal{S}_1(-\infty, t) = \sum_{s=-\infty}^{t-1} D_1^{-1}(s) \qquad \textit{for } t \le t_0.$$

If for $t \le t_0$ we define a solution U_0 by

$$U_0(t) \equiv U_1(t)\mathcal{S}_1(-\infty, t),$$

then $U_0(t)$ is recessive at $-\infty$ and

$$\{U_1; U_0\} = I.$$

Exercise 5.47 *Prove Theorem* 5.46 *by mimicking the proof of Theorem* 5.45.

Theorem 5.48 *Suppose that $U_0(t)$ is a solution of $\mathcal{L}U(t) = 0$ which is recessive at ∞ with $U_0(t_0)$ nonsingular for some $t_0 \in [a, \infty)$. Then $U_0(t)$ is uniquely determined by $U_0(t_0)$ and $\mathcal{L}U(t) = 0$ has a solution $U_1(t)$ which is dominant at ∞.*

Proof: Assume $U_0(t)$ is nonsingular for some $t_0 \in [a, \infty)$. Let $U_1(t)$ be the solution of the **IVP**

$$\begin{aligned} \mathcal{L}U_1(t) &= 0, \\ U_1(t_0) &= 0, \\ U_1(t_0+1) &= I. \end{aligned}$$

Then $U_1(t)$ is a prepared solution and

$$\begin{aligned} \{U_1; U_0\} &= \{U_1; U_0\}(t_0+1) \\ &= U_1^*(t_0)K(t_0)U_0(t_0+1) - U_1^*(t_0+1)K^*(t_0)U_0(t_0) \\ &= -K^*(t_0)U_0(t_0) \end{aligned}$$

is nonsingular. Using Exercise 5.42 on page 235 we know that $U_1(t)$ is a dominant solution of $\mathcal{L}U(t) = 0$ at ∞. Suppose that Γ is an arbitrary but fixed $n \times n$ constant matrix. Let $U(t)$ solve the **IVP**

$$\begin{aligned} \mathcal{L}U(t) &= 0, \\ U(t_0) &= I, \\ U(t_0+1) &= \Gamma. \end{aligned}$$

By Theorem 5.40

$$\lim_{t\to\infty} U_1^{-1}(t)U(t) = \Omega$$

where Ω is an $n \times n$ constant matrix. Note that Ω is independent of the recessive solution $U_0(t)$. By using initial conditions at t_0 and t_0+1 it is easy to see that there are $n \times n$ constant matrices C_1 and C_2 such that

$$U_0(t) = U(t)C_1 + U_1(t)C_2 \tag{5.29}$$

where

$$C_1 = U_0(t_0)$$

is invertible. Multiplying both sides of equation (5.29) by $U_1^{-1}(t)$ we obtain

$$U_1^{-1}(t)U_0(t) = U_1^{-1}(t)U(t)U_0(t_0) + C_2.$$

Let $t \to \infty$ in order to obtain

$$0 = \Omega U_0(t_0) + C_2$$

and therefore

$$C_2 = -\Omega U_0(t_0).$$

Substitute this into equation (5.29) and let $t = t_0 + 1$ for

$$U_0(t_0 + 1) = (\Gamma - \Omega)U_0(t_0).$$

Hence the initial conditions of $U_0(t)$ at t_0, $t_0 + 1$ are determined by $U_0(t_0)$. Thus $U_0(t)$ is uniquely determined by $U_0(t_0)$. □

Theorem 5.49 *Assume $U_0(t)$ is a solution of $\mathcal{L}U(t) = 0$ which is recessive solution at $-\infty$ with $U_0(t_0)$ nonsingular for some $t_0 \in (\infty, a]$. Then $\mathcal{L}U(t) = 0$ has a dominant solution $U_1(t)$ at $-\infty$ and $U_0(t)$ is uniquely determined by $U_0(t)$.*

Exercise 5.50 *Prove Theorem* 5.49 *by a proof similar to the proof of Theorem* 5.48.

Theorem 5.51 *Suppose $U_0(t)$ is recessive at ∞ and $U_1(t)$ is dominant at ∞. Assume $\mathcal{S}_1(t, \infty) \equiv \sum_{s=t}^{\infty} D_1^{-1}(s)$ and $U_0(t)$ are nonsingular for large t. Then there exists a nonsingular $n \times n$ constant matrix C such that*

$$U_0(t) = U_1(t)\mathcal{S}_1(t, \infty)C$$

for large t. Furthermore, $\{U_0; U_1\}$ is nonsingular and

$$\lim_{t \to \infty} U_1^{-1}(t)U_0(t) = 0.$$

Proof: For sufficiently large t define a solution U by

$$U(t) = U_1(t)\mathcal{S}_1(t, \infty), \tag{5.30}$$

where

$$\mathcal{S}_1(t, \infty) \equiv \sum_{s=t}^{\infty} D_1^{-1}(s) = \sum_{s=t}^{\infty} [U_1^*(s)K(s)U_1(s+1)]^{-1}.$$

By Theorem 5.45 $U(t)$ is a recessive solution at ∞ and

$$\{U_1; U\} = -I.$$

Since, for large t, $U_1(t)$ and $\mathcal{S}_1(t,\infty)$ are nonsingular it follows from (5.30) that $U(t)$ is nonsingular for large t. Since $U(t)$ is recessive at ∞

$$\lim_{t\to\infty} U_1^{-1}(t)U(t) = 0.$$

Choose an integer $t_0 \geq a$ sufficiently large so that $U_0(t)$ and $U(t)$ are nonsingular in $[t_0,\infty)$. Define a solution V on $[t_0,\infty)$ by

$$V(t) = U(t)U^{-1}(t_0)U_0(t_0).$$

From Exercise 5.41 we know that $V(t)$ is also a recessive solution of $\mathcal{L}U(t) = 0$ at ∞. Since $U_0(t)$ and $V(t)$ are recessive at ∞ and $U_0(t_0) = V(t_0)$ we conclude from the uniqueness in Theorem 5.48 that $V(t) \equiv U_0(t)$. Hence

$$U_0(t) = U(t)U^{-1}(t_0)U_0(t_0)$$

for $t \geq t_0$. Using (5.30) we get that

$$\begin{aligned} U_0(t) &= U_1(t)\mathcal{S}_1(t,\infty)U^{-1}(t_0)U_0(t_0) \\ &= U_1(t)\mathcal{S}_1(t,\infty)C \end{aligned}$$

where

$$C = U^{-1}(t_0)U_0(t_0)$$

is a nonsingular $n \times n$ constant matrix. □

Theorem 5.52 (Left Counterpart) *Suppose that $U_0(t)$ is recessive at $-\infty$ and $U_1(t)$ is dominant at $-\infty$. Assume $\mathcal{S}_1(-\infty,t) \equiv \sum_{s=-\infty}^{t-1} D_1^{-1}(s)$ and $U_0(t)$ are nonsingular for sufficiently small t. Then*

$$U_0(t) = U_1(t)\mathcal{S}_1(-\infty,t)C$$

in a neighborhood of $-\infty$, where C is an $n\times n$ nonsingular constant matrix and

$$\lim_{t\to-\infty} U_1^{-1}(t)U_0(t) = 0.$$

Exercise 5.53 *Prove Theorem* 5.52 *by a proof similar to the proof of Theorem* 5.51.

The next result relates the convergence of infinite series, the convergence of continued fractions, and the existence of recessive solutions. For related results see Jones and Thron [87, p. 403], Euler [60], Pincherle [127], Gautschi [67], and Ahlbrandt [5, 7, 13, 9, p. 248]. This result is important in the development of continued fractions [9]. We will refer to it as the *connection theorem* because it connects the topics of convergence of continued fractions, convergence of infinite series, existence of recessive solutions at ∞, and ultimately provides a continued fraction representation of the minimal solution of the associated matrix Riccati equation.

Theorem 5.54 (The Connection Theorem) *Let $U(t)$, $V(t)$ be solutions of $\mathcal{L}U(t) = 0$ determined by the initial conditions*

$$\begin{aligned} U(t_0) &= I, \qquad U(t_0+1) = K^{-1}(t_0)\Gamma, \\ V(t_0) &= 0, \qquad V(t_0+1) = K^{-1}(t_0), \end{aligned}$$

where $t_0 \in [a, \infty)$ and Γ is a constant $n \times n$ Hermitian matrix. Then $U(t)$, $V(t)$ are prepared solutions with $\{U; V\} = I$ and the following are equivalent:

(a) $V(t)$ is dominant at ∞.

(b) $V(t)$ is nonsingular for large t and $\lim_{t\to\infty} V^{-1}(t)U(t)$ exists as a Hermitian matrix $\Omega(\Gamma)$ with finite entries.

(c) There exists a solution $U_0(t)$ which is recessive at ∞ and has $U_0(t_0)$ nonsingular.

If (a)–(c) hold then

$$U_0(t_0+1)U_0^{-1}(t_0) = U(t_0+1) - V(t_0+1)\Omega(\Gamma) = -K^{-1}(t_0)\Omega(0). \qquad (5.31)$$

Proof: Since $V(t_0) = 0$, $V(t)$ is a prepared solution. Also

$$\begin{aligned} \{U; U\} &= \{U; U\}(t_0+1) \\ &= U^*(t_0)K(t_0)U(t_0+1) - U^*(t_0+1)K^*(t_0)U(t_0) \\ &= \Gamma - \Gamma^* = 0. \end{aligned}$$

Hence $U(t)$ is a prepared solution. Furthermore

$$\begin{aligned} \{U; V\} &= \{U; V\}(t_0+1) \\ &= U^*(t_0)K(t_0)V(t_0+1) - U^*(t_0+1)K^*(t_0)V(t_0) = I. \end{aligned}$$

Now we show that (a) implies (b). So assume $V(t)$ is a dominant solution of $\mathcal{L}U(t) = 0$ at ∞. Then there is an integer $t_1 \geq a$ such that $V(t)$ is nonsingular for $t \geq t_1$ and the series

$$\mathcal{S}(t_1, \infty) \equiv \sum_{s=t_1}^{\infty} [V^*(s)K(s)V(s+1)]^{-1}$$

converges to a Hermitian matrix $\mathcal{S}(t_1, \infty)$. By the reduction of order theorem, Theorem 5.35 on page 228, with a replaced by t_1,

$$U(t) = V(t)[P + \mathcal{S}(t)Q] \tag{5.32}$$

for $t \geq t_1$, where

$$\begin{aligned} \mathcal{S}(t) &= \sum_{s=t_1}^{t-1} [V^*(s)K(s)V(s+1)]^{-1} \\ P &= V^{-1}(t_1)U(t_1) \\ Q &= \{V; U\} = -\{U; V\}^* = -I. \end{aligned}$$

Since $U(t)$ is a prepared solution we have by the reduction of order theorem that $P^*Q = -P^*$ is Hermitian. Consequently P is Hermitian. Use (5.32) to conclude that

$$\lim_{t\to\infty} V^{-1}(t)U(t) = P + \mathcal{S}(t_1)Q = P - \mathcal{S}(t_1)$$

is Hermitian. Hence (b) holds.

Next we show that (b) implies (c). Assume $V(t)$ is nonsingular for $t \geq t_1$ and

$$\lim_{t\to\infty} V^{-1}(t)U(t) = \Omega \tag{5.33}$$

is Hermitian. From equation (5.32)

$$V^{-1}(t)U(t) = [P + \mathcal{S}(t)Q] = P - \mathcal{S}(t).$$

Letting $t \to \infty$ and using (5.33) we get that

$$\Omega = P - \mathcal{S}(t_1, \infty).$$

Therefore

$$\mathcal{S}(t_1, \infty) = P - \Omega.$$

Define U_0 by

$$U_0(t) = U(t) - V(t)\Omega. \tag{5.34}$$

Using (5.34) we get that

$$\begin{aligned}\{U_0;U_0\} &= \{U-V\Omega;U-V\Omega\}\\ &= \{U;U\}-\{U;V\}\Omega-\Omega^*\{V;U\}+\Omega^*\{V;V\}\Omega\\ &= -\Omega+\Omega^*=0.\end{aligned}$$

Hence $U_0(t)$ is a prepared solution. Since, by (5.34),

$$U_0(t_0)=U(t_0)=I,$$

$U_0(t)$ is a prepared basis. Next assume $U_1(t)$ is an $n\times n$ matrix solution of $\mathcal{L}U(t)=0$ such that $\{U_1;U_0\}$ is nonsingular. Let C_1 and C_2 be $n\times n$ constant matrices such that

$$U_1(t)=V(t)C_1+U_0(t)C_2 \tag{5.35}$$

(C_1 and C_2 can be determined by letting $t=t_0$ and $t=t_0+1$). Consider

$$\begin{aligned}\{U_1;U_0\} &= \{VC_1+U_0C_2;U_0\}\\ &= C_1^*\{V;U_0\}+C_2^*\{U_0;U_0\}\\ &= C_1^*\{V;U_0\}\\ &= C_1^*\{V;U_0\}(t_0+1)\\ &= C_1^*\left[V^*(t_0)K(t_0)U_0(t_0+1)-V^*(t_0+1)K^*(t_0)U_0(t_0)\right]\\ &= -C_1^*.\end{aligned}$$

Hence C_1 is nonsingular. Using (5.34) and (5.33) we have that

$$\lim_{t\to\infty}V^{-1}(t)U_0(t) = \lim_{t\to\infty}[V^{-1}(t)U(t)-\Omega]=0.$$

Therefore, by (5.35)

$$\lim_{t\to\infty}V^{-1}(t)U_1(t) = \lim_{t\to\infty}\left[C_1+V^{-1}(t)U_0(t)C_2\right]=C_1$$

is nonsingular. Thus $U_1(t)$ is nonsingular for large t and for large t

$$\begin{aligned}\lim_{t\to\infty}U_1^{-1}(t)U_0(t) &= \lim_{t\to\infty}\{[V(t)C_1+U_0(t)C_2]^{-1}U_0(t)\}\\ &= \lim_{t\to\infty}\{[V(t)(C_1+V^{-1}(t)U_0(t)C_2)]^{-1}U_0(t)\}\\ &= \lim_{t\to\infty}\{(C_1+V^{-1}(t)U_0(t)C_2)^{-1}V^{-1}(t)U_0(t)\}\\ &= C_1^{-1}\cdot 0=0.\end{aligned}$$

Hence $U_0(t)$ is a recessive solution of $\mathcal{L}U(t)=0$ at ∞.

We next show that (c) implies (a). Assume $U_0(t)$ is a recessive solution at ∞ with $U_0(t_0)$ nonsingular. Then

$$\begin{aligned} \{V; U_0\} &= \{V; U_0\}(t_0+1) \\ &= V^*(t_0)K(t_0)U_0(t_0+1) - V^*(t_0+1)K^*(t_0)U_0(t_0) \\ &= -U_0(t_0) \end{aligned}$$

is nonsingular. Hence by Exercise 5.42, $V(t)$ is a dominant solution of $\mathcal{L}U(t) = 0$ at ∞. To complete the proof of this theorem it remains to show that if (a)–(c) hold, then (5.31) holds. It can be shown by checking initial conditions at t_0, $t_0 + 1$ that

$$U_0(t) = U(t)U_0(t_0) + V(t)C \tag{5.36}$$

for a suitable $n \times n$ constant matrix C. By (b)

$$\lim_{t\to\infty} V^{-1}(t)U(t) = \Omega(\Gamma). \tag{5.37}$$

From (5.36) we get that

$$V^{-1}(t)U_0(t) = V^{-1}(t)U(t)U_0(t_0) + C.$$

Since $U_0(t)$ is a recessive solution

$$\begin{aligned} 0 &= \lim_{t\to\infty} (V^{-1}(t)U(t)U_0(t_0) + C) \\ &= \Omega(\Gamma)U_0(t_0) + C \end{aligned}$$

by (5.37). Hence $C = -\Omega(\Gamma)U_0(t_0)$ and so from (5.36) we obtain

$$U_0(t) = (U(t) - V(t)\Omega(\Gamma))U_0(t_0).$$

Set $t = t_0 + 1$ to obtain

$$U_0(t_0+1)U_0^{-1}(t_0) = U(t_0+1) - V(t_0+1)\Omega(\Gamma)$$

which is the first part of (5.31). To complete the proof of (5.31), let $W(t)$ be the solution of the **IVP**

$$\begin{aligned} \mathcal{L}W(t) &= 0, \\ W(t_0) &= I, \\ W(t_0+1) &= 0. \end{aligned}$$

Checking initial conditions at t_0, $t_0 + 1$ we see that

$$U(t) = W(t) + V(t)\Gamma.$$

Hence

$$V^{-1}(t)U(t) = V^{-1}(t)W(t) + \Gamma.$$

Letting $t \to \infty$ and using the fact that when $\Gamma = 0$, then $U(t)$ and $W(t)$ are the same, we get from (5.37) that

$$\Omega(\Gamma) = \Omega(0) + \Gamma.$$

Therefore

$$\begin{aligned} & U(t_0+1) - V(t_0+1)\Omega(\Gamma) \\ & \quad = K^{-1}(t_0)\Gamma - K^{-1}(t_0)[\Omega(0) + \Gamma] \\ & \quad = -K^{-1}(t_0)\Omega(0). \end{aligned}$$

Hence (5.31) holds and the proof of this theorem is complete. □

The following counterpart of the previous theorem at $-\infty$ is useful for representing the "maximal solution" of the associated Riccati equation.

Theorem 5.55 *Let $U(t)$, $V(t)$ be solutions of $\mathcal{L}U(t) = 0$ determined by the initial conditions*

$$\begin{aligned} U(t_0-1) &= K^{*-1}(t_0-1), \qquad U(t_0) = I \\ V(t_0-1) &= K^{-1}(t_0-1), \qquad V(t_0) = 0, \end{aligned}$$

where Γ is a constant $n \times n$ Hermitian matrix. Then $U(t)$, $V(t)$ are prepared solutions with $\{U; V\} = -I$ and the following are equivalent:

(a) $V(t)$ is dominant at $-\infty$.

(b) $V(t)$ is nonsingular in a neighborhood of $-\infty$ and $\lim_{t \to -\infty} V^{-1}(t)U(t) = \Omega(\Gamma)$ exists as a Hermitian matrix $\Omega(\Gamma)$ with finite entries.

(c) There exists a solution $U_0(t)$ which is recessive at $-\infty$ and has $U_0(t_0)$ nonsingular.

If (a)–(c) hold, then

$$U_0(t_0-1)U_0^{-1}(t_0) = U(t_0-1) - V(t_0-1)\Omega(\Gamma) = -K^{-1}(t_0-1)\Omega(0).$$

Exercise 5.56 *Prove Theorem 5.55 by a proof similar to the proof of Theorem 5.54.*

Assume $h(t)$ is an $n \times 1$ vector function in $[t_0 + 1, \infty)$. We say that the vector difference equation $\mathcal{L}u(t) = h(t)$ has the *unique two point property* on $[t_0, \infty)$ provided given any $t_0 \leq t_1 < t_2$ if $u(t)$, $v(t)$ are solutions of $\mathcal{L}u(t) = h(t)$ with $u(t_1) = v(t_1)$, $u(t_2) = v(t_2)$ then $u(t) \equiv v(t)$ in $[t_0, \infty)$.

Theorem 5.57 *If $\mathcal{L}u(t) = 0$ has the unique two point property on $[t_0, \infty)$, then the* **BVP**

$$\begin{aligned} \mathcal{L}u(t) &= h(t), \\ u(t_1) &= \alpha, \\ u(t_2) &= \beta, \end{aligned}$$

where $t_0 \leq t_1 < t_2$ and $\alpha, \beta \in \mathrm{C}^n$, has a unique solution on $[t_0, \infty)$.

Proof: If $t_2 = t_1 + 1$, then the given **BVP** is an **IVP** and the result is true. So assume $t_1 + 1 < t_2$. Let $U(t, t_1)$, $V(t, t_1)$ by $n \times n$ matrix solutions of $\mathcal{L}U(t) = 0$ determined by the initial conditions

$$\begin{aligned} U(t_1, t_1) &= 0, & U(t_1 + 1, t_1) &= I, \\ V(t_1, t_1) &= I, & V(t_1 + 1, t_1) &= 0. \end{aligned}$$

Then a general solution of $\mathcal{L}u(t) = 0$ is given by

$$u(t) = U(t, t_1)\gamma + V(t, t_1)\delta$$

where γ, $\delta \in \mathrm{C}^n$. Indeed, note that $u(t_1) = \delta$, $u(t_1 + 1) = \gamma$. Hence, by varying γ and $\delta \in \mathrm{C}^n$ we get all initial conditions at t_1, $t_1 + 1$ and hence all solutions.

Because of the unique two point property the homogeneous **BVP**

$$\mathcal{L}u(t) = 0, \quad u(t_1) = 0, \quad u(t_2) = 0 \tag{5.38}$$

has only the trivial solution. Let

$$u(t) = U(t, t_1)\gamma + V(t, t_1)\delta.$$

The boundary condition $u(t_1) = 0$ implies $\delta = 0$. The boundary condition $u(t_2) = 0$ then is equivalent to

$$U(t_2, t_1)\gamma = 0. \tag{5.39}$$

Since the unique solution of (5.38) is the trivial solution, the unique solution of (5.39) is $\gamma = 0$. This implies that $U(t_2, t_1)$ is nonsingular. Next let $w(t)$ be the solution of the **IVP**

$$\begin{aligned} \mathcal{L}w(t) &= h(t), \\ w(t_1) &= 0, \\ w(t_1+1) &= 0. \end{aligned}$$

Then a general solution of $\mathcal{L}u(t) = h(t)$ is given by

$$y(t) = U(t, t_1)\gamma + V(t, t_1)\delta + w(t).$$

We now show that the **BVP**

$$\begin{aligned} \mathcal{L}y(t) &= h(t) \\ y(t_1) &= \alpha, \quad y(t_2) = \beta \end{aligned}$$

has a unique solution. The boundary condition $y(t_1) = \alpha$ implies $\delta = \alpha$. The boundary condition $y(t_2) = \beta$ leads to the equation

$$\beta = U(t_2, t_1)\gamma + V(t_2, t_1)\alpha + w(t_2).$$

Since $U(t_2, t_1)$ is nonsingular we can uniquely solve this for γ. Similarly, a general solution of the homogeneous difference equation is

$$u(t) = U(t, t_1)\gamma + V(t, t_1)\delta.$$

Note that $u(t_1) = 0$ implies $\delta = 0$. Consider the boundary condition $u(t_2) = 0$. That is, consider

$$U(t_2, t_1)\gamma = 0.$$

Since $\gamma = 0$ is the only solution of the vector equation, $U(t_2, t_1)$ is a nonsingular matrix. Next, as noted above,

$$y(t) = U(t, t_1)\gamma + V(t, t_1)\delta + y_p(t)$$

is a general solution of $\mathcal{L}u(t) = h(t)$. The boundary condition $y(t_1) = \alpha$ implies $\delta = \alpha$. The second boundary condition $y(t_2) = \beta$ is equivalent to

$$\beta = U(t_2, t_1)\gamma + V(t_2, t_1)\alpha + y_p(t_2).$$

Since $U(t_2, t_1)$ is nonsingular, we can uniquely solve for γ which implies the result. □

Corollary 5.58 *If $\mathcal{L}u(t) = 0$ has the unique two point property on $[t_0, \infty)$, then the matrix* **BVP**

$$\begin{aligned} \mathcal{L}U(t) &= 0, \\ U(t_1) &= A, \qquad U(t_2) = B, \end{aligned}$$

where A and B are given $n \times m$ constant matrices, has a unique $n \times m$ matrix solution $U(t)$.

Exercise 5.59 *Prove Corollary* 5.58.

The construction used in the following theorem to get a recessive solution was used by Reid [131] in the matrix continuous case. Gautschi [67] credits this construction in the scalar case to J. C. P. Miller. This method was also used by Olver and Sookne [108].

Theorem 5.60 *Assume $\mathcal{L}u(t) = 0$ has the unique two point property on $[t_0, \infty)$ and assume $U_0(t)$ is a solution of $\mathcal{L}U(t) = 0$ which is recessive at ∞ and has $U_0(t_0)$ nonsingular. Let $U(t, s)$ be the solution of the* **BVP**

$$\begin{aligned} \mathcal{L}U(t) &= 0, \\ U(t_0, s) &= I, \qquad U(s, s) = 0. \end{aligned}$$

Then the recessive solution $U_0(t)U_0^{-1}(t_0)$ is uniquely determined by

$$U_0(t)U_0^{-1}(t_0) = \lim_{s \to \infty} U(t, s). \tag{5.40}$$

Proof: Assume $U_0(t)$ is a recessive solution of $\mathcal{L}U(t) = 0$ at ∞ with $U_0(t_0)$ nonsingular. Let $V(t)$ be the solution of the **IVP**

$$\begin{aligned} \mathcal{L}V(t) &= 0, \\ V(t_0) &= 0, \\ V(t_0 + 1) &= K^{-1}(t_0). \end{aligned}$$

By Theorem 5.54 on page 242, $V(t)$ is nonsingular for large t. By checking boundary conditions at $t = t_0$, $t = s$, s large, we get that

$$U(t, s) = -V(t)V^{-1}(s)U_0(s)U_0^{-1}(t_0) + U_0(t)U_0^{-1}(t_0). \tag{5.41}$$

Since $U_0(t)$ is a recessive solution and

$$\begin{aligned}\{V;U_0\} &= \{V;U_0\}(t_0+1)\\ &= V^*(t_0)K(t_0)U_0(t_0+1)-V^*(t_0+1)K^*(t_0)U_0(t_0)\\ &= -U_0(t_0)\end{aligned}$$

is nonsingular we get that

$$\lim_{t\to\infty} V^{-1}(t)U_0(t)=0.$$

Hence from (5.41) we get that

$$\lim_{s\to\infty} U(t,s)=U_0(t)U_0^{-1}(t_0)$$

and the proof is complete. □

Exercise 5.61 *Assume that the scalar difference equation*

$$u(t+1)-5u(t)+6u(t-1)=0$$

has a recessive solution at ∞. *Find* $u(t,s)$ *as in Theorem* 5.60 *and use equation* (5.40) *to find the recessive solution at* ∞ *which has the value one at* $t_0=1$.

Theorem 5.62 *Assume* $\mathcal{L}u(t)=0$ *has the unique two point property on* $(-\infty,t_0)$ *and assume* $U_0(t)$ *is a recessive solution at* $-\infty$ *with* $U_0(t_0)$ *nonsingular. Let* $U(t,s)$, *for* $s<t_0-1$ *be the solution of the* **BVP**

$$\begin{aligned}\mathcal{L}U(t) &= 0,\\ U(s,s) &= 0, \qquad U(t_0,s)=I.\end{aligned}$$

Then the recessive solution $U_0(t)U_0^{-1}(t_0)$ *at* $-\infty$ *is uniquely determined by*

$$U_0(t)U_0^{-1}(t_0)=\lim_{s\to-\infty} U(t,s). \tag{5.42}$$

Exercise 5.63 *Prove Theorem* 5.62 *by a proof similar to the proof of Theorem* 5.60.

Exercise 5.64 *Assume that the scalar difference equation*

$$u(t+1) - 5u(t) + 6u(t-1) = 0$$

has a recessive solution at $-\infty$. *Use formula* (5.42) *to find the recessive solution* $u_0(t)$ *at* $-\infty$ *such that* $u_0(0) = 1$.

Theorem 5.65 *If* $\mathcal{L}u(t) = 0$ *is disconjugate on* $[t_0 - 1, \infty)$, *then* $\mathcal{L}U(t) = 0$ *has a solution* $U_0(t)$ *which is recessive at* ∞ *and has* $U_0(t)$ *nonsingular for* $t \geq t_0$. *Furthermore*

$$U_0^*(t)K(t)U_0(t+1) > 0$$

for $t \geq t_0$.

Proof: Let $U(t)$ be the solution of the **IVP**

$$\begin{aligned} \mathcal{L}U(t) &= 0 \\ U(t_0 - 1) &= 0, \qquad U(t_0) = I. \end{aligned}$$

Note that $U(t)$ is a prepared basis for $\mathcal{L}U(t) = 0$. We claim that $U(t)$ is nonsingular in $[t_0, \infty)$. Assume not. Then there is an integer $t_1 \geq t_0 + 1$ such that $U(t_1)$ is singular. But then there is a nontrivial vector δ such that

$$U(t_1)\delta = 0.$$

Set

$$u(t) = U(t)\delta$$

then $u(t)$ is a nontrivial prepared solution with

$$u(t_0 - 1) = 0, \qquad u(t_1) = 0.$$

This contradicts disconjugacy of $\mathcal{L}u(t) = 0$ on $[t_0 - 1, \infty)$. Hence $U(t)$ is nonsingular in $[t_0, \infty)$. We next claim that

$$U^*(t)K(t)U(t+1) > 0 \tag{5.43}$$

in $[t_0, \infty)$. If not then there is an integer $t_2 \geq t_0$ such that

$$U^*(t_2)K(t_2)U(t_2+1) \not> 0.$$

It follows from Exercise 5.29 that there is a nontrivial vector γ such that

$$v(t) \equiv U(t)\gamma$$

is a nontrivial prepared solution of $\mathcal{L}u(t) = 0$ with a generalized zero at $t_2 + 1$. But also $U(t_0 - 1) = 0$ and this contradicts the assumption that $\mathcal{L}u(t) = 0$ is disconjugate in $[t_0 - 1, \infty)$. Hence (5.43) holds. Define for $t \geq t_0$

$$U_1(t) = U(t)[I + S(t)] \tag{5.44}$$

where

$$S(t) = \sum_{s=t_0}^{t-1} D^{-1}(s)$$

where

$$D(t) = U^*(t)K(t)U(t+1).$$

By the reduction of order theorem $U_1(t)$ is a prepared solution of $\mathcal{L}U(t) = 0$ and

$$\{U; U_1\} = I.$$

By (5.43) $D(t) > 0$ for $t \geq t_0$. It follows that

$$I < I + S(t) < I + S(t+1). \tag{5.45}$$

From (5.44) we get that $U_1(t)$ is nonsingular for $t \geq t_0$. By the reduction of order theorem for $t \geq t_0$

$$U(t) = U_1(t)[P + \mathcal{S}_1(t)Q]$$

where

$$D_1(t) = U_1^*(t)K(t)U_1(t+1) \quad \text{and} \quad \mathcal{S}_1(t) = \sum_{s=t_0}^{t-1} D_1^{-1}(s).$$

Also by the reduction of order theorem

$$\begin{aligned} P &= U_1^{-1}(t_0)U(t_0) = I \\ Q &= \{U_1; U\} = -I \end{aligned}$$

so

$$U(t) = U_1(t)\{I - \mathcal{S}_1(t)\} \tag{5.46}$$

for $t \geq t_0$. Using (5.46) and (5.44) we get that

$$\begin{aligned} I &= [U_1^{-1}(t)U(t)][U^{-1}(t)U_1(t)] \\ &= [I - \mathcal{S}_1(t)][I + S(t)]. \end{aligned}$$

Since the second factor is strictly increasing (see (5.45)) and bounded below by I, the first factor $I - \mathcal{S}_1(t)$ is positive definite, strictly decreasing and has a limit. Since

$$0 < I - \mathcal{S}_1(t+1) < I - \mathcal{S}_1(t) \leq I,$$

it follows that

$$0 \leq I - \mathcal{S}_1(t_0, \infty) < I - \mathcal{S}_1(t) \leq I$$

for $t > t_0$, where

$$\mathcal{S}_1(t_0, \infty) \equiv \sum_{t=t_0}^{\infty} D_1^{-1}(t).$$

It follows that

$$0 \leq \mathcal{S}_1(t) < \mathcal{S}_1(t_0, \infty) \leq I.$$

Note that $U_1(t)$ is a dominant solution at ∞. Set

$$U_0(t) = U_1(t)\mathcal{S}_1(t, \infty).$$

By Theorem 5.45 on page 236, $U_0(t)$ is a recessive solution at ∞. Since

$$U_0(t) = U_1(t)[\mathcal{S}_1(t_0, \infty) - \mathcal{S}_1(t)],$$

$U_1(t)$ is nonsingular for $t \geq t_0$ and $\mathcal{S}_1(t_0, \infty) - \mathcal{S}_1(t) > 0$ for $t \geq t_0$ we get that $U_0(t)$ is nonsingular for $t \geq t_0$. It remains to show that

$$U_0^*(t)K(t)U_0(t+1) > 0$$

in $[t_0, \infty]$. We will prove this immediately following Corollary 5.72 on page 255.

Theorem 5.66 *If $\mathcal{L}u(t) = 0$ is disconjugate on $(-\infty, t_0 - 1]$, then $\mathcal{L}U(t) = 0$ has a solution $U_0(t)$ which is recessive at $-\infty$ and has $U_0(t)$ nonsingular on $(-\infty, t_0]$. Furthermore*

$$U_0^*(t)K(t)U_0(t+1) > 0$$

on $(-\infty, t_0 - 1]$.

Exercise 5.67 *Prove the first statement in Theorem* 5.66 *by a proof similar to the proof of Theorem* 5.65.

Theorem 5.68 *Assume $\mathcal{L}u(t) = 0$ is disconjugate on $[t_0 - 1, \infty)$ and Γ is a Hermitian matrix. Let $U(t), V(t)$ be the solutions of $\mathcal{L}U(t) = 0$ satisfying the initial conditions*

$$\begin{aligned} U(t_0) &= I, \qquad U(t_0+1) = K^{-1}(t_0)\Gamma, \\ V(t_0) &= 0, \qquad V(t_0+1) = K^{-1}(t_0). \end{aligned}$$

Then $V(t)$ is nonsingular for $t \geq t_0 + 1$, $V(t)$ is dominant at ∞, and

$$\lim_{t \to \infty} [V^{-1}(t)U(t)]$$

exists and is a Hermitian matrix.

Proof: By the part of Theorem 5.65 on page 251 that we already proved we get that $\mathcal{L}U(t) = 0$ has a recessive solution $U_0(t)$ at ∞ with $U_0(t)$ nonsingular for $t \geq t_0$. Hence part (c) of Theorem 5.54 on page 242 holds. Thus by part (a) of Theorem 5.54, $V(t)$ is a dominant solution at ∞. By part (b) in Theorem 5.54

$$\lim_{t\to\infty} [V^{-1}(t)U(t)]$$

is a Hermitian matrix. Since $V(t_0) = 0$ and $\mathcal{L}u(t) = 0$ is disconjugate on $[t_0, \infty)$

$$V^*(t)K(t)V(t+1) > 0$$

in $[t_0 + 1, \infty)$. In particular $V(t)$ is nonsingular in $[t_0 + 1, \infty)$. □

Theorem 5.69 *Assume $\mathcal{L}u(t) = 0$ is disconjugate on $(-\infty, t_0 + 1]$ and Γ is a Hermitian matrix. Let $U(t), V(t)$ be the solutions of $\mathcal{L}U(t) = 0$ satisfying the initial conditions*

$$\begin{aligned} U(t_0 - 1) &= K^{*-1}(t_0 - 1)\Gamma, \qquad U(t_0) = I, \\ V(t_0 - 1) &= K^{*-1}(t_0 - 1), \qquad V(t_0) = 0. \end{aligned}$$

Then $V(t)$ is a dominant solution at $-\infty$, $V(t)$ is nonsingular on $(-\infty, t_0 - 1]$ and

$$\lim_{t\to-\infty} [V^{-1}(t)U(t)]$$

exists and is a Hermitian matrix.

Exercise 5.70 *Prove Theorem* 5.69. *Use the ideas of the proof of Theorem* 5.68.

Theorem 5.71 *If $\mathcal{L}u(t) = 0$ is disconjugate on $[t_0, \infty)$, then $\mathcal{L}u(t) = h(t)$ has the unique two point property in $[t_0, \infty)$. In particular every* **BVP** *of the form*

$$\begin{aligned} \mathcal{L}u(t) &= h(t), \\ u(t_1) &= \alpha, \qquad u(t_2) = \beta \end{aligned}$$

where $t_0 \leq t_1 < t_2$ and α, β are given $n+1$ vectors has a unique solution.

Proof: By Theorem 5.57 on page 247 it suffices to show that $\mathcal{L}u(t) = 0$ has the unique two point property in $[t_0, \infty)$. To this end assume $u(t), v(t)$ are solutions of $\mathcal{L}u(t) = 0$ and there are integers s_1, s_2 such that $t_0 \leq s_1 < s_2$ and

$$\begin{aligned} u(s_1) &= v(s_1) \\ u(s_2) &= v(s_2). \end{aligned}$$

If $s_2 = s_1+1$, then u, v satisfy the same initial conditions and hence $u(t) \equiv v(t)$ in $[t_0, \infty)$. Next assume $s_1 + 1 < s_2$. In this case set

$$y(t) = u(t) - v(t).$$

Then $y(t)$ solves the **BVP**

$$\mathcal{L}y(t) = 0, \qquad y(t_1) = 0, \quad y(t_2) = 0.$$

Since $\mathcal{L}u(t) = 0$ is disconjugate and $y(t)$ is a prepared solution with two generalized zeros it follows that $y(t) \equiv 0$ in $[t_0, \infty)$. This implies $u(t) \equiv v(t)$ in $[t_0, \infty)$ and hence the two point property holds. □

Corollary 5.72 (Construction of the Recessive Solution) *Assume the equation $\mathcal{L}u(t) = 0$ is disconjugate on $[t_0, \infty)$. For each point s of (t_0, ∞), let $U(t, s)$ be the solution of the* **BVP** *$\mathcal{L}U(t) = 0$,*

$$U(t_0, s) = I, \quad U(s, s) = 0.$$

Then the recessive solution $U_0(t)$ of $\mathcal{L}U(t) = 0$ with $U_0(t_0) = I$ is given by

$$U_0(t) = \lim_{s\to\infty} U(t, s).$$

Proof: By Theorem 5.71, $\mathcal{L}u(t) = h(t)$ has the unique two point property in $[t_0, \infty)$. The conclusion of the theorem follows from Theorem 5.60 on page 249.

□

Finally we complete the proof of Theorem 5.65 on page 251. From the initial condition $U(s, s) = 0$ and the fact that $\mathcal{L}u(t) = 0$ is disconjugate on $[t_0 - 1, \infty)$ it follows that

$$U^*(t, s)K(t)U(t + 1, s) > 0$$

in $[t_0 - 1, s - 2]$. By Theorem 5.60 on page 249

$$\lim_{s\to\infty} U(t, s) = U_0(t)U_0^{-1}(t_0).$$

It follows that

$$U_0^*(t)K(t)U_0(t + 1) \geq 0$$

for $t \geq t_0 - 1$. But $U_0(t)$ nonsingular for $t \geq t_0$ implies

$$U_0^*(t)K(t)U_0(t + 1) > 0$$

for $t \geq t_0$ and the proof of Theorem 5.65 is complete. □

Corollary 5.73 *Assume $\mathcal{L}u(t) = 0$ is disconjugate on $(-\infty, t_0]$ and let $U(t,s)$, for $s \leq t_0 - 2$, be defined as the solution of the* **BVP**

$$\mathcal{L}U(t) = 0, \qquad U(t_0, s) = I, \quad U(s,s) = 0$$

for $s < t_0$. Then the solution $U_0(t)$ of $\mathcal{L}U(t) = 0$ which is recessive at $-\infty$ and has $U_0(t_0) = I$ is given by

$$U_0(t) = \lim_{s \to -\infty} U(t,s).$$

Exercise 5.74 *Prove Corollary* 5.73.

Exercise 5.75 *Prove the last statement in Theorem* 5.65.

Define for $t \geq t_0 \geq a+1$, the block tridiagonal matrix

$$Q(t) = \begin{bmatrix} N(t_0) & -K(t_0) & 0 & & & 0 \\ -K^*(t_0) & N(t_0+1) & -K(t_0+1) & 0 & & \\ 0 & & & & & \\ & & & \ddots & 0 & \\ & & & & N(t-1) & -K(t-1) \\ 0 & & & 0 & -K^*(t-1) & N(t) \end{bmatrix}.$$

Theorem 5.76 *Assume there is an integer $t_0 \geq a+1$ such that $Q(t)$ is positive definite for $t \geq t_0$. Then $\mathcal{L}U(t) = 0$ has a recessive solution $U_0(t)$ at ∞ with*

$$U_0^*(t)K(t)U_0(t+1) > 0$$

for $t \geq t_0$.

Proof: Assume $t_1 \geq t_0$. Since $Q(t)$ is positive definite we have by Corollary 5.16 on page 214 that $\mathcal{L}u(t) = 0$ is disconjugate on $[t_0 - 1, t_1 + 1]$. Since this is true for each integer $t_1 \geq t_0$, $\mathcal{L}u(t) = 0$ is disconjugate on $[t_0 - 1, \infty]$. Hence by Theorem 5.65 on page 251 $\mathcal{L}U(t) = 0$ has a recessive solution $U_0(t)$ at ∞ with

$$U_0^*(t)K(t)U_0(t+1) > 0$$

in $[t_0, \infty)$. □

For the corresponding result at $-\infty$, define, for $t \leq t_0$, the block tridiagonal matrix

$$\bar{Q}(t) = \begin{bmatrix} N(t) & -K(t) & 0 & & & 0 \\ -K^*(t) & N(t+1) & -K(t+1) & 0 & & \\ 0 & & & & & \\ & & & \ddots & 0 & \\ & & & & N(t_0-1) & -K(t_0-1) \\ 0 & & & 0 & -K^*(t_0-1) & N(t_0) \end{bmatrix}.$$

□

Theorem 5.77 *Assume that $\bar{Q}(t)$ is positive definite for $t \leq t_0$. Then $\mathcal{L}U(t) = 0$ has a recessive solution $U_0(t)$ at $-\infty$ with*

$$U_0^*(t)K(t)U_0(t+1) > 0$$

on $(-\infty, t_0 - 1]$.

Exercise 5.78 *Prove Theorem* 5.77. *Use the ideas of the proof of Theorem* 5.76.

In Example 3.15 on page 91 we saw that if $U(t)$ is a solution of $\mathcal{L}U(t) = 0$ in $[a, b+2]$ and if $P(t)$ is any nonsingular Hermitian matrix, then $U(t)$ is a solution of the generalized self-adjoint equation (3.41) on page 91 where

$$R(t) = K^*(t-1) - P(t) \tag{5.47}$$

for $t \in [a+1, b+2]$ and

$$Q(t) = P(t) + P(t+1) + R(t) + R^*(t) - N(t) \tag{5.48}$$

for $t \in [a+1, b+2]$. Conversely if $U(t)$ is a solution of the generalized self-adjoint equation in $[a, b+2]$ and $P(t)$ is nonsingular, then $U(t)$ is a solution of $\mathcal{L}U(t) = 0$ where

$$K(t) = P(t+1) + R^*(t+1) \tag{5.49}$$

for $t \in [a, b+1]$ and

$$N(t) = P(t) + P(t+1) + R(t) + R^*(t) - Q(t) \tag{5.50}$$

for $t \in [a+1, b+1]$. In Example 3.17 on page 91 we saw that if $U(t)$ is a solution of the generalized self-adjoint difference equation (3.41) on page 91 in $[a, b+2]$ and if we set

$$Y(t) = U(t)$$

for $t \in [a, b+2]$ and

$$Z(t) = P(t+1)\Delta U(t) + R^*(t+1)U(t+1)$$

for $t \in [a, b+1]$, then $Y(t)$, $Z(t)$ is a solution of the Hamiltonian system (3.18), (3.19), on page 82 on $[a, b+1]$ where

$$\begin{aligned} A(t) &= -P^{-1}(t+1)R^*(t+1) && (5.51)\\ B(t) &= P^{-1}(t+1) && (5.52)\\ C(t) &= -Q(t+1) - R(t+1)P^{-1}(t+1)R^*(t+1) && (5.53) \end{aligned}$$

for $t \in [a, b]$. We would now like to find $A(t)$, $B(t)$ and $C(t)$ in terms of $K(t)$, $N(t)$ and $P(t)$. Using (5.51) and (5.47) we get that

$$\begin{aligned} A(t) = -P^{-1}(t+1)[K(t) - P(t+1)]\\ -P^{-1}(t+1)K(t) + I \end{aligned}$$

for $t \in [a, b]$. Formula (5.52) is the desired formula for $B(t)$. Using (5.53) and (5.48) we obtain

$$\begin{aligned} C(t) &= -P(t+1) - P(t+2) - R(t+1) - R^*(t+1) + N(t+1)\\ &\quad -R(t+1)P^{-1}(t+1)R^*(t+1). \end{aligned}$$

With the aid of (5.47) we get that

$$\begin{aligned} C(t) &= -P(t+1) - P(t+2) - K^*(t) + P(t+1)\\ &\quad -K(t) + P(t+1) + N(t+1)\\ &\quad -[K^*(t) - P(t+1)]\, P^{-1}(t+1)\, [K(t) - P(t+1)]\\ &= -P(t+2) - K^*(t) - K(t) + P(t+1) + N(t+1)\\ &\quad -[K^*(t) - P(t+1)][P^{-1}(t+1)K(t) - I]\\ &= -P(t+2) - K^*(t) - K(t) + P(t+1) + N(t+1)\\ &\quad -K^*(t)P^{-1}(t+1)K(t) + K^*(t) + K(t) - P(t+1)\\ &= -P(t+2) + N(t+1) - K^*(t)P^{-1}(t+1)K(t). \end{aligned}$$

Hence if $U(t)$ is a solution of $\mathcal{L}U(t) = 0$ in $[a, b+2]$ and if $P(t)$ is any nonsingular $n \times n$ Hermitian matrix and if we set

$$Y(t) = U(t)$$

for $t \in [a, b+2]$ and

$$Z(t) = P(t+1)\Delta U(t) + R^*(t+1)U(t+1)$$

for $t \in [a, b+1]$, then $Y(t)$, $Z(t)$ is a solution of the Hamiltonian system (3.18), (3.19) on page 82 in $[a, b+1]$ where

$$A(t) = I - P^{-1}(t+1)K(t) \tag{5.54}$$
$$B(t) = P^{-1}(t+1) \tag{5.55}$$
$$C(t) = N(t+1) - P(t+2) - K^*(t)P^{-1}(t+1)K(t) \tag{5.56}$$

for $t \in [a, b]$. In Example 3.17 on page 81 we saw that the Hamiltonian system (3.18), (3.19) on page 82 is equivalent to the symplectic system (3.15) on page 80 with

$$E(t) = [I - A(t)]^{-1} \tag{5.57}$$

$$F(t) = [I - A(t)]^{-1} B(t) = E(t)B(t) \tag{5.58}$$

$$G(t) = C(t)[I - A(t)]^{-1} = C(t)E(t) \tag{5.59}$$

$$H(t) = E^{*-1}(t) + G(t)B(t) \tag{5.60}$$

for $t \in [a, b]$. We would like formulas for $E(t)$, $F(t)$, $G(t)$, and $H(t)$ in terms of $P(t)$, $K(t)$ and $N(t)$. Using (5.57) and (5.54) we get that

$$E(t) = K^{-1}(t)P(t+1) \tag{5.61}$$

for $t \in [a, b]$. Using (5.58), (5.61) and (5.55) we get that

$$F(t) = K^{-1}(t) \tag{5.62}$$

for $t \in [a, b]$. Using (5.59) and (5.56) we obtain

$$G(t) = N(t+1)E(t) - P(t+2)E(t) - K^*(t)P^{-1}(t+1)K(t)E(t).$$

Hence using (5.61)

$$G(t) = -K^*(t) - P(t+2)K^{-1}(t)P(t+1) + N(t+1)K^{-1}(t)P(t+1) \tag{5.63}$$

for $t \in [a, b]$. Using (5.60), (5.55) and (5.59) we get that

$$H(t) = E^{*-1}(t) + G(t)P^{-1}(t+1).$$

With the aid of (5.63) we obtain

$$H(t) = E^{*-1}(t) - K^*(t)P^{-1}(t+1) - P(t+2)K^{-1}(t) + N(t)K^{-1}(t).$$

Using (5.61) we get that

$$H(t) = -P(t+2)K^{-1}(t) + N(t)K^{-1}(t) \tag{5.64}$$

for $t \in [a+1, b+1]$. Hence if $U(t)$ is a solution of $\mathcal{L}U(t) = 0$ and we set

$$Y(t) = U(t)$$

for $t \in [a, b+2]$, and if

$$Z(t) = P(t+1)\Delta U(t) + [K(t) - P(t+1)]\, U(t+1)$$

for $t \in [a, b+1]$ where $P(t)$ is any nonsingular $n \times n$ Hermitian matrix, then $Y(t)$, $Z(t)$ is a solution of the symplectic system (3.15) on page 80 where $E(t)$, $F(t)$, $G(t)$ and $H(t)$ are given by (5.61)–(5.64). From (5.58) and (5.55) we obtain that

$$K(t) = P(t+1)E^{-1}(t) \tag{5.65}$$

for $t \in [a, b+1]$. Using (5.55) and (5.65) we get that

$$N(t) = C(t-1) + P(t+1) + E^{*-1}(t-1)P(t)E^{-1}(t-1) \tag{5.66}$$

for $t \in [a+1, b+1]$. Next we get an expression for $\mathcal{J}[\eta]$ in terms of $P(t)$, $C(t)$ and $E(t)$.

Lemma 5.79 *Assume* (5.65) *and* (5.66) *hold. Then for* $\eta \in \mathcal{F}$,

$$\begin{aligned}\mathcal{J}(\eta) = \sum_{t=a+1}^{b+2} &\{\eta^*(t)C(t-1)\eta(t)\\ &+ \left[\eta^*(t)E^{*-1}(t-1) - \eta^*(t-1)\right] P(t)[E^{-1}(t-1)\eta(t) - \eta(t-1)\}.\end{aligned}$$

Proof: By definition

$$\begin{aligned}\mathcal{J}[\eta] = \sum_{t=a+1}^{b+1} &\{\eta^*(t)N(t)\eta(t) - \eta^*(t-1)K(t-1)\eta(t)\\ &-\eta^*(t)K^*(t-1)\eta(t-1)\}.\end{aligned}$$

Using (5.66) we have

$$\begin{aligned}\mathcal{J}[\eta] &= \sum_{t=a+1}^{b+1} \{\eta^*(t)C(t-1)\eta(t) + \eta^*(t)P(t+1)\eta(t)\\ &\quad + \eta^*(t)E^{*-1}(t-1)P(t)E^{-1}(t-1)\eta(t)\\ &\quad - \eta^*(t-1)K(t-1)\eta(t) - \eta^*(t)K^*(t-1)\eta(t-1)\}.\end{aligned}$$

In the second term we replace t by $t-1$ and in the last two terms we use equation (5.65) to obtain

$$\begin{aligned}\mathcal{J}[\eta] &= \sum_{t=a+2}^{b+2} \eta^*(t-1)P(t)\eta(t-1) \\ &+ \sum_{t=a+1}^{b+1} \{\eta^*(t)C(t-1)\eta(t) + \eta^*(t)E^{*-1}(t-1)P(t)E^{-1}(t-1)\eta(t) \\ &- \eta^*(t-1)P(t)E^{-1}(t-1)\eta(t) - \eta^*(t)E^{*-1}(t-1)P(t)\eta(t-1)\}.\end{aligned}$$

Since $\eta(a) = \eta(b+2) = 0$ we can write that all the sums go from $a+1$ to new upper limit of $b+2$. After some factoring, we have

$$\begin{aligned}\mathcal{J}[\eta] &= \sum_{t=a+1}^{b+2} \{\eta^*(t)C(t-1)\eta(t) \\ &\quad +\eta^*(t)E^{*-1}(t-1)P(t)[E^{-1}(t-1)\eta(t) - \eta(t-1)] \\ &\quad -\eta^*(t-1)P(t)[E^{-1}(t-1)\eta(t) - \eta(t-1)]\}\end{aligned}$$

which groups together as

$$\begin{aligned}\mathcal{J}[\eta] = \sum_{t=a+1}^{b+2} &\{\eta^*(t)C(t-1)\eta(t) \\ &+[\eta^*(t)E^{*-1}(t-1) - \eta^*(t-1)]P(t)[E^{-1}(t-1)\eta(t) - \eta(t-1)]\}\end{aligned}$$

for the desired result. □

Corollary 5.80 *Assume $P(t)$ is a positive definite matrix in $[a+1, b+2]$ such that $C(t)$ given by (5.56) on page 259 is positive semidefinite on $[a, b]$, then $\mathcal{L}u(t) = 0$ is disconjugate on $[a, b+2]$.*

Proof: We see from Lemma 5.79 that $\mathcal{J}[\eta]$ is positive definite in $\mathcal{F}$. It follows that $\mathcal{L}u(t) = 0$ is disconjugate on $[a, b+2]$. □

Exercise 5.81 *Show that Corollary 5.80 implies that if $P(t)$ is positive definite in $[a+1, b+2]$ and $Q(t)$ is negative semidefinite on $[a+1, b+1]$, then the self-adjoint difference equation*

$$\Delta[P(t)\Delta u(t-1)] + Q(t)u(t) = 0$$

is disconjugate on $[a, b+2]$.

Corollary 5.82 *Assume there is an integer* $t_0 \geq a+1$ *such that* $P(t)$ *is a positive definite matrix in* $[t_0, \infty)$ *and such that* $C(t)$ *given by* (5.56) *on page* 259 *is also positive semidefinite in* $[t_0 - 1, \infty)$. *Then* $\mathcal{L}U(t) = 0$ *has a recessive solution* $U_0(t)$ *at* ∞ *with*

$$U_0^*(t)K(t)U_0(t+1) > 0$$

in $[t_0, \infty)$.

Proof: By Corollary 5.80 we get that $\mathcal{L}u(t) = 0$ is disconjugate on $[t_0 - 1, t_1 + 1]$ for all $t_1 \geq t_0$. It follows that $\mathcal{L}u(t) = 0$ is disconjugate on $[t_0 - 1, \infty)$. The conclusion of this theorem follows from Theorem 5.65. □

6

DISCRETE RICCATI EQUATIONS FOR THREE TERM RECURRENCES

6.1 A RICCATI EQUATION FOR $\mathcal{L}U = 0$.

Suppose throughout this chapter that we are given $n \times n$ matrix functions $C(t)$, $P(t)$, and $E(t)$ with the following properties.

(i) $C(t)$ and $P(t)$ are Hermitian for each integer t.

(ii) $E(t)$ and $P(t)$ are nonsingular for each integer t.

The associated three term recurrence of Chapter 5

$$\mathcal{L}u(t) = -K(t)u(t+1) + N(t)u(t) - K^*(t-1)u(t-1) = 0 \tag{6.1}$$

has

$$K(t) = P(t+1)E^{-1}(t) \tag{6.2}$$

and

$$N(t) = C(t-1) + P(t+1) + E^{*-1}(t-1)P(t)E^{-1}(t-1). \tag{6.3}$$

Assume $W(t)$ is an $n \times n$ matrix function defined on $[a+1, b+2]$ such that $W(t) + P(t)$ is nonsingular in $[a+1, b+1]$. Then we define a Riccati operator R by

$$\mathrm{R}W(t) = -W(t+1) + C(t-1) + E^{*-1}(t-1)W(t)[W(t)+P(t)]^{-1}P(t)E^{-1}(t-1).$$

Theorem 6.1 *Assume $U(t)$ is an $n \times n$ nonsingular matrix function defined on the interval $[a, b+2]$. If we make the Riccati substitution*

$$W(t) = K(t-1)U(t)U^{-1}(t-1) - P(t) \tag{6.4}$$

for $t \in [a+1, b+2]$, *then* $W(t) + P(t)$ *is nonsingular and we have the identity*

$$\mathcal{L}U(t) = \mathrm{R}W(t)U(t)$$

for $t \in [a+1, b+1]$.

Proof: Assume $U(t)$ is an $n \times n$ nonsingular matrix function in $[a, b+2]$ and set

$$W(t) = K(t-1)U(t)U^{-1}(t-1) - P(t)$$

for $t \in [a+1, b+2]$. Then

$$W(t) + P(t) = K(t-1)U(t)U^{-1}(t-1) \tag{6.5}$$

in $[a+1, b+2]$. It follows from (6.5) that $W(t)+P(t)$ is nonsingular in $[a+1, b+2]$ and

$$U(t-1)U^{-1}(t) = [W(t) + P(t)]^{-1}K(t-1) \tag{6.6}$$

for $t \in [a+1, b+2]$. For $t \in [a+1, b+1]$ consider

$$\begin{aligned}\mathcal{L}U(t) &= -K(t)U(t+1) + N(t)U(t) - K^*(t-1)U(t-1)\\ &= \left[-K(t)U(t+1)U^{-1}(t) + N(t) - K^*(t-1)U(t-1)U^{-1}(t)\right]U(t).\end{aligned}$$

Using (6.5) and (6.3) we get that

$$\begin{aligned}\mathcal{L}U(t) &= [-W(t+1) + C(t-1) + E^{*-1}(t-1)P(t)E^{-1}(t-1)\\ &\qquad -K^*(t-1)U(t-1)U^{-1}(t)]U(t).\end{aligned}$$

Using (6.2) and (6.6) we conclude that

$$\begin{aligned}\mathcal{L}U(t) &= \{-W(t+1) + C(t-1) + E^{*-1}(t-1)P(t)E^{-1}(t-1)\\ &\qquad -E^{*-1}(t-1)P(t)[W(t) + P(t)]^{-1}P(t)E^{-1}(t-1)\}U(t).\end{aligned}$$

But

$$\begin{aligned}&P(t) - P(t)\left[W(t) + P(t)\right]^{-1}P(t)\\ &= \{I - P(t)\left[W(t) + P(t)\right]^{-1}\}P(t)\\ &= \{[W(t) + P(t)] - P(t)\}\left[W(t) + P(t)\right]^{-1}P(t)\\ &= W(t)\left[W(t) + P(t)\right]^{-1}P(t).\end{aligned}$$

Hence

$$\begin{aligned}\mathcal{L}U(t) &= \{-W(t+1)+C(t-1)+ \\ &\qquad E^{*-1}(t)W(t)[W(t)+P(t)]^{-1}P(t)E^{-1}(t-1)\}U(t) \\ &= \mathrm{R}W(t)U(t)\end{aligned}$$

for $t \in [a+1, b+1]$ which is the desired result. □

The special case where $E(t) \equiv I$ is as follows:

Exercise 6.2 *Show that in the special case when $\mathcal{L}U(t) = 0$ is equivalent to*

$$\Delta[P(t)\Delta U(t-1)] + Q(t)U(t) = 0,$$

then the Riccati substitution is

$$W(t) = P(t)\Delta U(t-1)U^{-1}(t-1).$$

Also show that the Riccati equation $\mathrm{R}W(t) = 0$ in this special case is equivalent to

$$\Delta W(t) = -Q(t) - W(t)[W(t)+P(t)]^{-1}W(t)$$

The following result follows easily from Theorem 6.1.

Corollary 6.3 *Assume $U(t)$ is an $n \times n$ nonsingular matrix function defined on $[a, b+2]$ and $W(t)$ is defined by the Riccati substitution (6.4) for $t \in [a+1, b+2]$. Then $U(t)$ is a solution of the matrix equation $\mathcal{L}U(t) = 0$ on $[a, b+2]$ iff $W(t)$ is a solution of the Riccati equation $\mathrm{R}W(t) = 0$ on $[a+1, b+2]$.*

Theorem 6.4 *Assume $U(t)$ is a nonsingular solution of $\mathcal{L}U(t) = 0$ on $[a, b+2]$ and define $W(t)$ on $[a+1, b+2]$ by the Riccati substitution (6.4). Then $W(t)$ is Hermitian for all t in $[a+1, b+2]$ iff $U(t)$ is a prepared solution of $\mathcal{L}U(t) = 0$. Also if $U(t)$ is a prepared solution, then it has no generalized zeros in $[a, b+2]$ iff*

$$W(t) + P(t) > 0$$

on $[a+1, b+2]$.

Proof: By equation (6.5) we have

$$U^*(t-1)K(t-1)U(t) = U^*(t-1)[W(t)+P(t)]U(t-1). \tag{6.7}$$

It follows that $U(t)$ is a prepared solution iff $U^*(t-1)[W(t)+P(t)]U(t-1)$ is Hermitian for all $t \in [a+1, b+2]$. Since $U(t)$ is nonsingular we get that $U(t)$ is a prepared solution iff $W(t)+P(t)$ is Hermitian for all $t \in [a+1, b+2]$. Next note that a prepared solution $U(t)$ has no generalized zeros on $[a, b+2]$ iff

$$U^*(t-1)K(t-1)U(t) > 0$$

for $t \in [a+1, b+2]$. It then follows from (6.7) that $U(t)$ has no generalized zeros in $[a, b+2]$ iff on $[a+1, b+2]$

$$W(t) + P(t) > 0.$$

□

Lemma 6.5 *Assume $\mathcal{L}u(t) = 0$ is disconjugate on $[a, \infty)$ and $W(t)$ is a Hermitian solution of the Riccati equation $\mathrm{R}W(t) = 0$ on $[a+1, \infty)$. Then there is an integer $t_0 \geq a+1$ such that*

$$W(t) + P(t) > 0$$

on $[t_0, \infty)$.

Proof: Let $U(t)$ be the solution of the initial value problem

$$\begin{aligned} U(t) &= K^{-1}(t-1)[W(t)+P(t)]U(t-1), \quad t \geq a+1 \\ U(a) &= I. \end{aligned}$$

Since $W(t)$ is a solution of $\mathrm{R}W(t) = 0$ on $[a+1, \infty)$, $W(t)+P(t)$ is nonsingular in $[a+1, \infty)$. Hence $U(t)$ is nonsingular on $[a, \infty)$. Solving for $W(t)$ we get the Riccati substitution

$$W(t) = K(t-1)U(t)U^{-1}(t-1) - P(t).$$

From Corollary 6.3, $U(t)$ is a solution of $\mathcal{L}U(t) = 0$ on $[a, \infty)$. By Theorem 6.4 $U(t)$ is a prepared solution of $\mathcal{L}U(t) = 0$. Since $U(a) = I$, $U(t)$ is a prepared basis. Since $\mathcal{L}u(t) = 0$ is disconjugate on $[a, \infty)$ we have by the Sturm Separation Theorem (Theorem 5.30 on page 221) that $U(t)$ has at most $n+1$ generalized zeros on $[a, \infty)$. Hence there is an integer $t_0 \geq a+1$ such that $U(t)$ has no generalized zeros on $[t_0 - 1, \infty)$. By Theorem 6.4 we get that

$$W(t) + P(t) > 0$$

on $[t_0, \infty)$. □

The corresponding left hand result is as follows:

Lemma 6.6 *Assume $\mathcal{L}u(t) = 0$ is disconjugate on $(-\infty, a]$ and $W(t)$ is a Hermitian solution of the Riccati equation* $\mathrm{R}W(t) = 0$ *on $(-\infty, a]$. Then there is an integer $t_0 \le a - 1$ such that*

$$W(t) + P(t) > 0$$

on $(-\infty, t_0]$.

Exercise 6.7 *Assume $Y(t)$ is a nonsingular $n \times n$ matrix solution of $\mathcal{L}U(t) = 0$ on $[t_0 - 1, \infty)$ where $t_0 \ge a + 1$. Show that for any nonsingular matrix C, the solution $W(t)$ of the Riccati equation corresponding to $U(t) = Y(t)C$ under the Riccati transformation* (6.4) *is the same.*

We now prove the following comparison theorem.

Theorem 6.8 *Assume $\mathcal{L}u(t) = 0$ is disconjugate on $[a, \infty)$ and $U_0(t)$ is any recessive solution of $\mathcal{L}U(t) = 0$ at ∞. Let $W_0(t)$ be the solution of the Riccati equation associated with $U_0(t)$. Assume $W(t)$ is a Hermitian solution of the Riccati equation in a neighborhood of ∞. Then*

$$W(t) \ge W_0(t)$$

for all t sufficiently large.

Proof: Since recessive solutions of $\mathcal{L}U(t) = 0$ at ∞ are uniquely determined up to multiplication on the right by a nonsingular $n \times n$ constant matrix we have by Exercise 6.7 that $W_0(t)$ is uniquely determined. By Lemma 6.5 there is an integer $t_0 \ge a + 1$ such that

$$\begin{aligned} W_0(t) + P(t) &> 0 \\ W(t) + P(t) &> 0 \end{aligned}$$

on $[t_0, \infty)$. Because of Exercise 6.7 we can assume $U_0(t_0) = I$. Let $U(t)$ be the solution of the **IVP**

$$\begin{aligned} U(t) &= K^{-1}(t-1)[W(t) + P(t)]U(t-1), \quad t \le t_0 + 1, \qquad (6.8) \\ U(t_0) &= I. \end{aligned}$$

Then $U(t)$ is nonsingular for $t \ge t_0$. Solving for $W(t)$ we get the Riccati transformation (6.4). By Theorem 6.4, $U(t)$ is a prepared solution of $\mathcal{L}U(t) = 0$ on $[t_0, \infty)$. Consider

$$W(t_0 + 1) - W_0(t_0 + 1) = [W(t_0 + 1) + P(t_0 + 1)] - [W_0(t_0 + 1) + P(t_0 + 1)].$$

Using (6.5) we get that

$$\begin{aligned} W(t_0+1) - W_0(t_0+1) = \\ K(t_0)U(t_0+1)U^{-1}(t_0) - K(t_0)U_0(t_0+1)U_0^{-1}(t_0) \\ = K(t_0)[U(t_0+1) - U_0(t_0+1)]. \end{aligned} \tag{6.9}$$

Since $U(t)$ is a nonsingular prepared solution of $\mathcal{L}U(t) = 0$ on $[t_0, \infty)$ we get by the reduction of order theorem that

$$U_0(t) = U(t)[P + S(t)Q]$$

for $t \geq t_0$ where

$$P = U^{-1}(t_0)U_0(t_0) = I$$

and

$$S(t) = \sum_{s=t_0}^{t-1} [U^*(s)K(s)U(s+1)]^{-1}$$

for $t \geq t_0$. Hence

$$U_0(t) = U(t)[I + S(t)Q]$$

and consequently

$$U_0(t) - U(t) = U(t)S(t)Q.$$

Letting $t = t_0 + 1$ we get that

$$\begin{aligned} U_0(t_0+1) - U(t_0+1) &= U(t_0+1)[U^*(t_0)K(t_0)U(t_0+1)]^{-1}Q \\ &= K^{-1}(t_0)Q. \end{aligned} \tag{6.10}$$

Hence, by (6.8)

$$W(t_0+1) - W_0(t_0+1) = -Q. \tag{6.11}$$

Since $\mathcal{L}u(t) = 0$ is disconjugate on $[a, \infty)$ we can let $U(t,s)$ be the unique solution of the **BVP**

$$\begin{aligned} \mathcal{L}U(t) &= 0 \\ U(t_0, s) &= I, \quad U(s,s) = 0 \end{aligned}$$

for $s \geq t_0 + 2$. By Corollary 5.72

$$U_0(t) = \lim_{s\to\infty} U(t,s).$$

By the reduction of order theorem

$$U(t,s) = U(t)[I + S(t)Q_s] \tag{6.12}$$

for $t \geq t_0$. Letting $t = t_0 + 1$ we get that

$$\begin{aligned} U(t_0+1,s) - U(t_0+1) &= U(t_0+1)S(t_0+1)Q_s \\ &= U(t_0+1)[U^*(t_0)K(t_0)U(t_0+1)]^{-1}Q_s \\ &= K^{-1}(t_0)Q_s. \end{aligned}$$

Taking the limit as $s \to \infty$ we get that

$$U_0(t_0+1) - U(t_0+1) = K^{-1}(t_0) \lim_{s\to\infty} Q_s.$$

It follows from (6.10) that

$$\lim_{s\to\infty} Q_s = Q. \tag{6.13}$$

Letting $t = s$ in (6.12) we obtain

$$0 = U(s)[I + S(s)Q_s].$$

Solving for Q_s we have that

$$Q_s = -S^{-1}(s) = -\left[\sum_{\tau=t_0}^{s-1} [U^*(\tau)K(\tau)U(\tau+1)]^{-1}\right]^{-1} < 0$$

by Theorem 6.4. Hence by (6.13) we have $Q \leq 0$. Finally from (6.11) we obtain

$$W(t_0+1) \geq W_0(t_0+1).$$

Since this argument works for any t_0 sufficiently large we get the desired result.

□

6.2 DISTINGUISHED SOLUTIONS OF RICCATI EQUATIONS

Assume $\mathcal{L}u(t) = 0$ has a recessive solution at ∞ which is nonsingular on $[a, \infty)$. The solution $W_0(t)$ of the Riccati equation corresponding to this recessive solution is called the *distinguished solution* at ∞. Note that if $U(t)$ is such a recessive solution and C is a nonsingular constant matrix, then the Riccati solution corresponding to $U(t)C$ is the same as for $U(t)$.

Exercise 6.9 *For each of the following, find the distinguished solution $w_0(t)$ at ∞. Show directly (without using Theorem 6.8) that every real solution $w(t)$ satisfies $w(t) \geq w_0(t)$ for all sufficiently large t.*

(a) $$\Delta w(t) + \frac{w^2(t)}{w(t) + (\frac{1}{4})^{t-1}} = 0,$$

(b) $$\Delta w(t) + \frac{1}{4^t} + \frac{w^2(t)}{w(t) + (\frac{1}{4})^{t-1}} = 0.$$

6.3 PERIODIC COEFFICIENT RICCATI EQUATIONS

Theorem 6.10 *Assume $K(t)$ is a nonsingular, periodic matrix function with integer period $p > 0$ on $(-\infty, \infty)$, $N(t)$ is a Hermitian, periodic matrix function with period $p > 0$ on $(-\infty, \infty)$, and $P(t)$ is a nonsingular Hermitian matrix function which is periodic with period p. If $\mathcal{L}u = 0$ is disconjugate on $(-\infty, \infty)$, then the distinguished solution $W_0(t)$ of the Riccati equation* RW $= 0$ *at ∞ is a Hermitian solution of the Riccati equation on $(-\infty, \infty)$, is periodic with period p and satisfies*

$$W_0(t) + P(t) > 0 \tag{6.14}$$

on $(-\infty, \infty)$. Furthermore, if $K(t)$, $N(t)$, and $P(t)$ have real entries, then $W_0(t)$ is real symmetric for each integer t.

Proof: Since $\mathcal{L}u(t) = 0$ is disconjugate on $(-\infty, \infty)$ it is disconjugate on $[t_0 - 2, \infty)$ for any integer t_0. By Theorem 5.65 on page 251, $\mathcal{L}U = 0$ has a recessive solution $U_0(t)$ at ∞ with

$$U_0^*(t-1)K(t-1)U_0(t) > 0$$

on $[t_0, \infty)$. Let $W_0(t)$ be the corresponding solution of the Riccati equation on $[t_0, \infty)$. Use equation (6.6) on page 264) and the above inequality for

$$U_0^*(t-1)K(t-1)U_0(t) = U_0^*(t-1)\left[W_0(t) + P(t)\right]U_0(t-1).$$

It follows that $W_0(t) + P(t) > 0$ on $[t_0, \infty)$. Since t_0 is arbitrary, inequality (6.14) holds on $(-\infty, \infty)$. Since $C(t)$ and $E(t)$ are given (see equation (5.56) on page 259 and equation (5.61) on page 259) in terms of the periodic functions $K(t)$, $N(t)$, and $P(t)$ we get that $C(t)$ and $E(t)$ are periodic with period p. It follows that

$$\begin{aligned} W_1(t) &\equiv W_0(t+p) \\ W_2(t) &\equiv W_0(t-p) \end{aligned}$$

are solutions of the Riccati equation $RW = 0$ on $(-\infty, \infty)$. Since $W_0(t)$ is Hermitian for all t, so are $W_1(t)$ and $W_2(t)$. Using Theorem 6.8 we get that

$$W_0(t) \leq W_1(t) = W_0(t+p) \tag{6.15}$$
$$W_0(t) \leq W_2(t) = W_0(t-p) \tag{6.16}$$

on $(-\infty, \infty)$. Replacing t by $t+p$ in relation (6.16) we get that

$$W_0(t+p) \leq W_0(t)$$

on $(-\infty, \infty)$. Together with (6.15)we get that

$$W_0(t) \leq W_0(t+p) \leq W_0(t)$$

on $(-\infty, \infty)$. Hence $W_0(t) \equiv W_0(t+p)$ on $(-\infty, \infty)$, so the distinguished solution $W_0(t)$ is periodic with period p.

Since the Hartman type construction employed in the proof of Theorem 5.65 gives a recessive solution with real entries, then the distinguished solution $W_0(t)$ has real entries and is Hermitian. Hence $W_0(t)$ is real symmetric. □

6.4 CONSTANT COEFFICIENT RICCATI EQUATIONS

Theorem 6.11 *Assume $K(t) \equiv K$, $N(t) \equiv N$, $P(t) \equiv P$ on $(-\infty, \infty)$, where K, N, and P are $n \times n$ constant matrices such that N and P are Hermitian and K and P are nonsingular. If $\mathcal{L}u = 0$ is disconjugate on $(-\infty, \infty)$, then the distinguished solution $W_0(t)$ of the Riccati equation $RW = 0$ at ∞ is a constant Hermitian matrix W_0 on $(-\infty, \infty)$, $W_0 + P > 0$ and W_0 is a solution of the "steady state" equation*

$$W = C + E^{*-1}W(W+P)^{-1}PE^{-1} \tag{6.17}$$

where

$$E = K^{-1}P, \qquad C = N - P - E^{*-1}PE^{-1}. \tag{6.18}$$

Also, W_0 is a solution of the "discrete algebraic Riccati equation"

$$WEP^{-1}W + WE - (CEP^{-1} + E^{*-1})W - CE = 0. \tag{6.19}$$

Furthermore, if K, N, and P have real entries, then W_0 is real symmetric.

Proof: In this case $K(t)$, $N(t)$, and $P(t)$ are periodic of period $p = 1$. Hence by Theorem 6.10 the distinguished solution $W_0(t)$ is a constant (Hermitian) matrix W_0 with

$$W_0 + P > 0.$$

Since $W_0(t) \equiv W_0$ is a solution of the Riccati equation $\mathrm{R}W = 0$, we get that W_0 is a matrix solution of the matrix equation (6.17).

To see that (6.17) and (6.19) are equivalent equations, first write (6.17) in the form

$$W - C = E^{*-1}W(W+P)^{-1}PE^{-1}. \tag{6.20}$$

It follows that

$$(W - C)EP^{-1}(W+P) = E^{*-1}W. \tag{6.21}$$

Hence

$$WEP^{-1}W + WE - CEP^{-1}W - CE = E^{*-1}W \tag{6.22}$$

which can be written in the form of equation (6.19). □

From Theorem 6.11 we obtain the following corollary.

Corollary 6.12 *Assume K is a nonsingular constant Hermitian matrix, N is a constant Hermitian matrix, and $C = N - 2K$. Then if either of the equivalent equations*

$$-\Delta\left[K\Delta u(t-1)\right] + Cu(t) = 0 \tag{6.23}$$

or

$$-Ku(t+1) + Nu(t) - Ku(t-1) = 0 \tag{6.24}$$

is disconjugate on $(-\infty, \infty)$, then the distinguished solution at ∞, $W_0(t) \equiv W_0$, of the Riccati equation

$$\Delta W(t) = C - W(t)\left[W(t) + K\right]^{-1} W(t) \tag{6.25}$$

is a constant Hermitian matrix which satisfies $W_0 + K > 0$ and is a solution of the two equivalent equations

$$W = C + W(W+K)^{-1}K \tag{6.26}$$

and

$$WK^{-1}W - CK^{-1}W - C = 0. \tag{6.27}$$

Equation (6.26) is the "steady state equation" and equation (6.27) is the "discrete algebraic Riccati equation" .

Proof: We will only prove here that (6.26) and (6.27) are equivalent equations. The rest of the proof will be left as an exercise.

Assume W is a solution of the steady state equation (6.26). Then $W + K$ is nonsingular and

$$W - C = W(W + K)^{-1}K.$$

It follows that

$$(W - C)K^{-1}(W + K) = W. \tag{6.28}$$

Expanding the product on the left side of (6.28) and simplifying leads easily to equation (6.27).

Conversely, assume W is a solution of equation (6.27). We claim that $W + K$ is nonsingular. Assume this is not the case. Then there is a constant vector α such that

$$(W + K)\alpha = 0. \tag{6.29}$$

Since we can rewrite equation (6.27) in the form of (6.28), we get from the latter form that $W\alpha = 0$. But then by (6.29) we have that $K\alpha = 0$ which contradicts nonsingularity of K. □

Exercise 6.13 *Complete the proof of Corollary* 6.12.

6.5 THE CHARACTERISTIC EQUATION

Exercise 6.14 *The matrix equation*

$$K\Lambda^2 - N\Lambda + K = 0 \tag{6.30}$$

where Λ is an $n \times n$ unknown matrix is called the "characteristic equation" for the constant coefficient matrix difference equation

$$-KU(t+1) + NU(t) - KU(t-1) = 0. \tag{6.31}$$

Show that $U(t) = \Lambda^t$ is a solution of (6.31) *iff Λ is a solution of* (6.30).

Theorem 6.15 *Assume that $n \times n$ constant matrices W and Λ are related by the equation*

$$K\Lambda = W + K, \tag{6.32}$$

where K is a given $n \times n$ nonsingular Hermitian matrix. Assume N is an $n \times n$ Hermitian matrix and $C = N - 2K$. Then W is a solution of the discrete algebraic equation (6.27) iff Λ is a solution of the characteristic equation (6.30). If $K > 0$ and W is a Hermitian solution of (6.26) with $W + K > 0$, then the eigenvalues λ_i, $1 \leq i \leq n$, of Λ are all real and positive. Furthermore, if $W < 0$, then $\lambda_i \in (0,1)$, $1 \leq i \leq n$, while if $W > 0$, then $\lambda_i \in (1,\infty)$ for $1 \leq i \leq n$.

Proof: Assume (6.32) holds and consider

$$\begin{aligned}
&WK^{-1}W - CK^{-1}W - C \\
&\quad = WK^{-1}[W+K] - W - CK^{-1}[W+K] + C - C \\
&\quad = (K\Lambda - K)\Lambda - (K\Lambda - K) - C\Lambda \\
&\quad = K\Lambda^2 - (2K+C)\Lambda + K \\
&\quad = K\Lambda^2 - N\Lambda + K
\end{aligned}$$

if $N = C + 2K$. It follows that W is a matrix solution of (6.27) iff Λ is a solution of (6.30).

Next assume $K > 0$ and W is a solution of (6.26) such that $W + K > 0$. Let λ_i, x_i, $1 \leq i \leq n$ be an eigenvalue–eigenvector pair for Λ. Using (6.32) we get that

$$(W+K)x_i = K\Lambda x_i = \lambda_i K x_i.$$

Hence

$$x_i^*(W+K)x_i = \lambda x_i^* K x_i.$$

Solving for λ_i gives the Rayleigh quotient

$$\lambda_i = \frac{x_i^*(W+K)x_i}{x_i^* K x_i}. \tag{6.33}$$

Taking the complex conjugate of both sides gives

$$\bar{\lambda}_i = \frac{x_i^*(W^* + K^*)x_i}{x_i^* K^* x_i} = \frac{x_i^*(W+K)x_i}{x_i^* K x_i} = \lambda_i.$$

Thus each eigenvalue of Λ is real. Since $K > 0$, $W + K > 0$ and $x_i \neq 0$ it follows from (6.33) that $\lambda_i > 0$, for $1 \leq i \leq n$. If $W > 0$, then $\lambda_i > 1$; while if $W < 0$, then $0 < \lambda_i < 1$, $1 \leq i \leq n$. □

Exercise 6.16 *Verify directly that the conclusions of Theorem 6.14 are true in the scalar case when*

(a) $K = 1, \quad N = 2$

(b) $K = 2, \quad N = 5.$

In each case find the Riccati equation and show directly that the solutions of the discrete algebraic Riccati equation are solutions of the Riccati equation.

6.6 MINIMALITY OF THE DISTINGUISHED SOLUTION AT ∞

Theorem 6.17 *Assume $K > 0$ and $C = N - 2K > 0$. Then the distinguished solution W_0 at ∞ for the Riccati equation for* (6.24) *is the unique negative definite solution W of the discrete algebraic Riccati equation* (6.27) *with $W + K > 0$. Also $-K < W_0 < 0$ and if W is a Hermitian solution of* (6.26) *with $W + K > 0$, then*

$$W_0 \leq W.$$

Proof: Let $W(t)$ be the solution of the initial value problem (see equation (6.25))

$$\Delta W(t) = C - W(t)\,[W(t) + K]^{-1}\,W(t), \qquad W(0) = 0. \tag{6.34}$$

We claim that $W(t)$ is a Hermitian solution of this Riccati equation on the whole interval $[0, \infty)$ and $W(t) > 0$ on $[1, \infty)$. We prove this by induction on t. Letting $t = 0$ in the Riccati equation (6.34) we see that $W(1)$ is well defined and

$$W(1) = C > 0.$$

Now assume $W(t)$ is well defined for an integer value of t, $t \geq 1$, and $W(t) > 0$. Then from (6.34)

$$W(t+1) = W(t) + C - W(t)\,[W(t) + K]^{-1}\,W(t)$$

$$= C + W(t)\,[W(t) + K]^{-1}\,[W(t) + K] \\ -W(t)\,[W(t) + K]^{-1}\,W(t) \tag{6.35}$$

$$= C + W(t)\,[W(t) + K]^{-1}\,K \tag{6.36}$$

$$= C + \left\{K^{-1}\left[W(t)+K\right]W^{-1}(t)\right\}^{-1}$$
$$= C + \left[K^{-1}+W^{-1}(t)\right]^{-1} > 0$$

and the induction is complete. Thus $W(t)$ is a positive definite solution of (6.34) on $[1,\infty)$.

Since $K>0$ and $C>0$ we have from Corollary 5.80 on page 261 that equation (6.24) is disconjugate on $(-\infty,\infty)$. From Corollary 6.12, the distinguished solution of the Riccati equation $\mathrm{R}W=0$ at ∞ is a constant Hermitian matrix W_0 with $W_0+K>0$. Also, by Corollary 6.12, W_0 is a matrix solution of the discrete algebraic Riccati equation (6.27) on page 272. From Theorem 6.8 on page 267 we obtain

$$W(t) \geq W_0$$

for $t \geq 0$. Setting $t=0$ we get $0 = W(0) \geq W_0$ so W_0 is negative semidefinite. But from (6.36) with $W(t) = W_0$, we have

$$W_0 - C = W_0\left[W_0+K\right]^{-1}K.$$

This can be written in the form

$$(W_0-C)K^{-1}(W_0+K) = W_0.$$

This last equation implies that W_0 is nonsingular. But W_0 negative semidefinite and W_0 nonsingular implies that W_0 has all negative eigenvalues and is negative definite.

The uniqueness part of this theorem remains to be proved. Assume W is also a negative definite solution of the discrete algebraic Riccati equation (6.27) on page 272. Define Λ and Λ_0 by

$$K\Lambda = W + K \tag{6.37}$$
$$K\Lambda_0 = W_0 + K. \tag{6.38}$$

By Theorem 6.15, Λ and Λ_0 are solutions of the characteristic equation (6.30). Hence $U(t) \equiv \Lambda^t$ and $U_0(t) \equiv \Lambda_0^t$ are solutions of (6.30). From equations (6.37) and (6.38) we obtain

$$W - W_0 = K(\Lambda - \Lambda_0). \tag{6.39}$$

Since $\mathcal{L}u(t) = 0$ is disconjugate on $(-\infty,\infty)$ there is a unique solution $U(t,s)$ of the boundary value problem

$$-KU(t+1) + NU(t) - KU(t-1) = 0$$
$$U(0,s) = I, \qquad U(s,s) = 0$$

for $s \geq 2$. By the reduction of order theorem, Theorem 5.35 on page 228,

$$U(t,s) = U(t)\left[I + S(t)Q_s\right] \tag{6.40}$$

for $t \geq 0$ where

$$Q_s = \{U(t); U(t,s)\}$$

and

$$\begin{aligned} S(t) &= \sum_{\tau=0}^{t-1} \left[U^*(\tau)KU(\tau+1)\right]^{-1} \\ &= \sum_{\tau=0}^{t-1} \left[(\Lambda^*)^\tau K\Lambda^{\tau+1}\right]^{-1} \\ &= \sum_{\tau=0}^{t-1} \left[(\Lambda^*)^\tau (K\Lambda)\Lambda^\tau\right]^{-1} \\ &= \sum_{\tau=0}^{t-1} \left\{(\Lambda^*)^\tau \left[W+K\right]\Lambda^\tau\right\}^{-1} \end{aligned} \tag{6.41}$$

by equation (6.37). Note that

$$\begin{aligned} W_0 - W = K(\Lambda_0 - \Lambda) &= \{U; U_0\}(1) \\ &= \{U; U_0\} \\ &= \left\{U(t); \lim_{s\to\infty} U(t,s)\right\} \\ &= \lim_{s\to\infty} Q_s \end{aligned} \tag{6.42}$$

by Corollary 5.72. Letting $t = s$ in (6.40) gives

$$0 = U(s)\left[I + S(s)Q_s\right]$$

which implies

$$Q_s = -S^{-1}(s).$$

From equations (6.41) and (6.42) we have

$$W_0 - W = -\lim_{s\to\infty}\left\{\sum_{\tau=0}^{s-1}\left\{(\Lambda^*)^\tau\left[W+K\right]\Lambda^\tau\right\}^{-1}\right\}^{-1}.$$

Now $W_0 \leq W$ and if $W \neq W_0$, then it follows that an eigenvalue of

$$\left\{\sum_{\tau=0}^{s-1}\left\{(\Lambda^*)^\tau\left[W+K\right]\Lambda^\tau\right\}^{-1}\right\}^{-1}$$

does not go to zero as $s \to \infty$. It follows that an eigenvalue of

$$M(s) \equiv \sum_{\tau=0}^{s-1} \{(\Lambda^*)^\tau [W+K] \Lambda^\tau\}^{-1}$$

converges to a finite number. Hence there are eigenvectors $\alpha(s)$, with $||\alpha(s)||_2 = 1$ such that $\alpha^*(s)M(s)\alpha(s)$ converges to a finite number. It follows that there is a unit vector α such that

$$\sum_{\tau=0}^{\infty} \alpha^* \{(\Lambda^*)^\tau [W+K] \Lambda^\tau\}^{-1} \alpha$$

converges. It follows that

$$\lim_{\tau\to\infty} \alpha^* \{(\Lambda^*)^\tau [W+K] \Lambda^\tau\}^{-1} \alpha = 0.$$

Since $W + K > 0$ we get that

$$\lim_{\tau\to\infty} \alpha^* (\Lambda^*)^{-\tau} (\Lambda)^{-\tau} \alpha = 0$$

and consequently

$$\lim_{\tau\to\infty} ||\Lambda^{-\tau}\alpha||_2 = 0.$$

But, since $W < 0$ by Theorem 6.15 on page 273 the eigenvalues of Λ satisfy $0 < \lambda_i < 1$ which gives a contradiction. □

6.7 THE REVERSE RICCATI EQUATION

If one has proven theorems concerning the behavior of solutions of $\mathcal{L}U(t) = 0$ in a neighborhood of ∞ and one wants to prove the corresponding theorems concerning the behavior of solutions of $\mathcal{L}U(t) = 0$ in a neighborhood of $-\infty$, then one might be interested in the reverse Riccati equation which we introduce in this section . This reverse Riccati equation was studied by Ahlbrandt and Hooker in [18].

First we motivate how we are led to the reverse Riccati transformation that leads to the reverse Riccati equation. To this end assume t_0 is a negative integer and $U(t)$ is a square matrix solution of

$$\mathcal{L}U(t) = -K(t)U(t+1) + N(t)U(t) - K^*(t-1)U(t-1) = 0$$

on $(-\infty, t_0]$. Letting $t = -\tau$ for $\tau \geq -t_0 = |t_0|$ we get the equation

$$-K(-\tau)U(-\tau+1) + N(-\tau)U(-\tau) - K^*(-\tau-1)U(-\tau-1) = 0.$$

Defining

$$\begin{aligned}\bar{K}(\tau) &= K^*(-\tau-1)\\ \bar{N}(\tau) &= N(-\tau)\\ \bar{U}(\tau) &= U(-\tau)\end{aligned}$$

for $\tau \geq |t_0|$ we observe that $\bar{U}(\tau)$ is a solution of the three term difference equation

$$-\bar{K}(\tau)\bar{U}(\tau+1) + \bar{N}(\tau)\bar{U}(\tau) - \bar{K}^*(\tau-1)\bar{U}(\tau-1) = 0 \tag{6.43}$$

on $[|t_0|, \infty)$. Now we make the standard Riccati transformation for (6.43), namely,

$$\begin{aligned}\bar{W}(\tau) &= \bar{K}(\tau-1)\Delta\bar{U}(\tau-1)\bar{U}^{-1}(\tau-1)\\ &= \bar{K}(\tau-1)\left[\bar{U}(\tau) - \bar{U}(\tau-1)\right]\bar{U}^{-1}(\tau-1).\end{aligned}$$

In terms of t we obtain

$$\bar{W}(-t) = K^*(t)\left[U(t) - U(t+1)\right]U^{-1}(t+1).$$

Hence

$$\bar{W}(-t) = -K^*(t)\left[\Delta U(t)\right]U^{-1}(t+1)$$

and consequently

$$\bar{W}(-t+1) = -K^*(t-1)\left[\Delta U(t-1)\right]U^{-1}(t).$$

This motivates the reverse Riccati transformation which is

$$Z(t) = -\bar{W}(-t+1) = K(t-1)\left[\Delta U(t-1)\right]U^{-1}(t). \tag{6.44}$$

We now show how the reverse Riccati transformation leads to the reverse Riccati equation. First note that the three term equation

$$-K(t)U(t+1) + N(t)U(t) - K(t-1)U(t-1) = 0$$

where we assume that $K(t)$ is nonsingular and Hermitian on $(-\infty, b+1]$ and $N(t)$ is Hermitian on $(-\infty, b+1]$ is equivalent to the self–adjoint second order difference equation

$$-\Delta\left[K(t-1)\Delta U(t-1)\right] + C(t)U(t) = 0 \tag{6.45}$$

where

$$C(t) = N(t) - K(t) - K(t-1)$$

for $t \in (-\infty, b+1]$. Assume $U(t)$ is a nonsingular solution of (6.45) on $(-\infty, b+2]$ and make the reverse Riccati substitution

$$\begin{aligned} Z(t) &= K(t-1)\,[\Delta U(t-1)]\,U^{-1}(t) \\ &= K(t-1) - K(t-1)U(t-1)U^{-1}(t). \end{aligned}$$

Hence

$$K(t-1) - Z(t) = K(t-1)U(t-1)U^{-1}(t) \tag{6.46}$$

for $t \le b+2$. It follows that $K(t-1) - Z(t)$ is invertible on $(-\infty, b+2]$ and

$$U(t)U^{-1}(t-1)K^{-1}(t-1) = [K(t-1) - Z(t)]^{-1}. \tag{6.47}$$

From (6.44) we obtain

$$\Delta Z(t) = \Delta\,[K(t-1)\Delta U(t-1)]\,U^{-1}(t) + K(t)\,[\Delta U(t)]\,\Delta U^{-1}(t).$$

We now want to use the following exercise.

Exercise 6.18 *Show that if $U(t)$ is nonsingular at t and $t+1$, then*

$$\Delta U^{-1}(t) = -U^{-1}(t)\,[\Delta U(t)]\,U^{-1}(t+1)$$

and

$$\Delta U^{-1}(t) = -U^{-1}(t+1)\,[\Delta U(t)]\,U^{-1}(t).$$

This exercise gives

$$\begin{aligned} \Delta Z(t) &= \Delta\,[K(t-1)\Delta U(t-1)]\,U^{-1}(t) \\ &\quad -K(t)\Delta U(t)\left[U^{-1}(t)\,(\Delta U(t))\,U^{-1}(t+1)\right]. \end{aligned}$$

Since $U(t)$ is a solution of (6.45) and using (6.44) we obtain

$$\Delta Z(t) = C(t) - Z(t+1)U(t+1)U^{-1}(t)\,(\Delta U(t))\,U^{-1}(t+1).$$

Using (6.46) we have

$$\Delta Z(t) = C(t) - Z(t+1)\,[K(t) - Z(t+1)]^{-1}\,K(t)\,(\Delta U(t))\,U^{-1}(t+1)$$

and by (6.44) we get the *Reverse Riccati Equation*

$$\Delta Z(t) = C(t) - Z(t+1)\,[K(t) - Z(t+1)]^{-1}\,Z(t+1). \tag{6.48}$$

We say that $Z(t)$ is a *solution* of the Reverse Riccati Equation (6.48) on the interval $(-\infty, b+2]$ provided $K(t) - Z(t+1)$ is invertible on $(-\infty, b+1]$ and (6.48) holds for $t \in (-\infty, b+1]$.

We now write (6.48) in another form. Adding and subtracting a term in (6.48) yields

$$\begin{aligned}\Delta Z(t) = C(t) - Z(t+1)\left[K(t) - Z(t+1)\right]^{-1}\left[K(t) - Z(t+1)\right] \\ +Z(t+1) - Z(t+1)\left[K(t) - Z(t+1)\right]^{-1} Z(t+1).\end{aligned}$$

This simplifies to the desired form

$$Z(t) = -C(t) + Z(t+1)\left[K(t) - Z(t+1)\right]^{-1} K(t). \tag{6.49}$$

We have proven part of the following theorem.

Theorem 6.19 *Assume $K(t)$ is nonsingular and Hermitian on $(-\infty, b+1]$ and $N(t)$ is Hermitian on $(-\infty, b+1]$. If $U(t)$ is a nonsingular solution of $\mathcal{L}U(t) = 0$ or equivalently (6.45) on $(-\infty, b+2]$ and $Z(t)$ is defined by the reverse Riccati substitution (6.44) on $(-\infty, b+2]$, then $Z(t)$ is a solution of the equivalent Riccati equations (6.48), (6.49) on $(-\infty, b+2]$. Conversely, if $Z(t)$ is a solution of the reverse Riccati equation (6.48), or equivalently, (6.49) on $(-\infty, b+2]$ and $U(t)$ is defined by the reverse Riccati substitution (6.44) on $(-\infty, b+2]$, where $U(b+2)$ is some nonsingular matrix, then $U(t)$ is a nonsingular matrix solution of $\mathcal{L}U(t) = 0$ and the difference equation (6.45) on $(-\infty, b+2]$. In addition, $U(t)$ is prepared iff $Z(t)$ is Hermitian. Also, $U(t)$ has no generalized zeros in $(-\infty, b+2]$ iff*

$$K(t) - Z(t+1) > 0$$

on $(-\infty, b+1]$.

Proof: We already proved the second sentence of this theorem. Now assume $Z(t)$ is a solution of the reverse Riccati equation on $(-\infty, b+2]$. Solving the reverse Riccati transformation (6.44) for $U(t-1)$ yields

$$U(t-1) = K^{-1}(t-1)\left[K(t-1) - Z(t)\right] U(t). \tag{6.50}$$

Let $U(t)$ be the solution of (6.50) such that $U(b+2)$ is some given nonsingular matrix. It follows from (6.50) that $U(t)$ is a nonsingular solution of (6.50) on $(-\infty, b+2]$. From (6.44) we get that for $t \in (-\infty, b+2]$

$$K(t-1)\Delta U(t-1) = Z(t)U(t).$$

Hence

$$\Delta\left[K(t-1)\Delta U(t-1)\right] = \left(\Delta Z(t)\right)U(t) + Z(t+1)\Delta U(t)$$

for $t \in (-\infty, b+1]$. From (6.48)

$$\begin{aligned}\Delta\left[K(t-1)\Delta U(t-1)\right] &= C(t)U(t)\\ &-Z(t+1)\left\{\left[K(t)-Z(t+1)\right]^{-1}Z(t+1)U(t) - \Delta U(t)\right\}.\end{aligned}$$

To get the desired result we will show that

$$\left[K(t)-Z(t+1)\right]^{-1}Z(t+1)U(t) = \Delta U(t).$$

To see this note by (6.47) and (6.44) that

$$\begin{aligned}&\left[K(t)-Z(t+1)\right]^{-1}Z(t+1)U(t)\\ &= U(t+1)U^{-1}(t)K^{-1}(t)K(t)\left(\Delta U(t)\right)U^{-1}(t+1)U(t)\\ &= U(t+1)U^{-1}(t)\left[U(t+1)-U(t)\right]U^{-1}(t+1)U(t)\\ &= U(t+1)-U(t) = \Delta U(t).\end{aligned}$$

Hence $U(t)$ is a nonsingular solution of (6.45) which is equivalent to $\mathcal{L}U(t) = 0$ on $(-\infty, b+2]$. From (6.50)

$$K(t-1)U(t-1) = \left[K(t-1) - Z(t)\right]U(t)$$

for $t \in (-\infty, b+2]$. Multiplying both sides by $U^*(t)$ gives

$$U^*(t)K(t-1)U(t-1) = U^*(t)\left[K(t-1)-Z(t)\right]U(t)$$

for $t \in (-\infty, b+2]$. Taking the conjugate transpose of both sides and using the fact that $K(t)$ is Hermitian we obtain

$$U^*(t-1)K(t-1)U(t) = U^*(t)\left[K(t-1)-Z^*(t)\right]U(t) \tag{6.51}$$

for $t \in (-\infty, b+2]$. It follows from (6.51) that $U(t)$ is prepared iff $Z(t)$ is Hermitian on $(-\infty, b+2]$. It also follows from (6.51) that if $U(t)$ and $Z(t)$ are related by the reverse Riccati transformation (6.44) on $(-\infty, b+2]$ and either $U(t)$ is a prepared solution of $\mathcal{L}U(t) = 0$ or $Z(t)$ is a Hermitian solution of the reverse Riccati equation, then $U(t)$ has no generalized zeros in $(-\infty, b+2]$ iff

$$K(t) - Z(t+1) > 0$$

on $(-\infty, b+1]$. □

Assume $Z(t)$ is an $n \times n$ matrix function defined on $[a+1, b+2]$ such that $K(t) - Z(t+1)$ is nonsingular on $[a, b+1]$, then we define the reverse Riccati operator $\tilde{\mathrm{R}}$ by

$$\tilde{\mathrm{R}}Z(t) = Z(t) + C(t) - Z(t+1)\left[K(t) - Z(t+1)\right]^{-1} K(t)$$

for $t \in [a+1, b+1]$, where $K(t)$ is Hermitian and nonsingular on $[a, b+1]$, and $N(t)$ is Hermitian on $[a+1, b+1]$.

Theorem 6.20 *Assume either $U(t)$ is a given nonsingular $n \times n$ matrix function on $[a, b+2]$ and $Z(t)$ is defined on $[a+1, b+2]$ by the reverse Riccati substitution* (6.44) *or $Z(t)$ is a given $n\times n$ matrix function with $K(t) - Z(t+1)$ nonsingular on $[a, b+1]$ and $U(t)$ is defined by* (6.50) *where $U(b+2)$ is some given nonsingular matrix. Then $U(t)$ is nonsingular on $[a, b+2]$, $Z(t)$ is defined on $[a+1, b+2]$ with $K(t) - Z(t+1)$ nonsingular on $[a, b+1]$,* (6.50) *holds for $t \in [a+1, b+2]$. Furthermore, we get the factorization*

$$\mathcal{L}U(t) = \{\tilde{\mathrm{R}}Z(t)\}U(t) \tag{6.52}$$

for $t \in [a+1, b+1]$.

Proof: We will leave the details of this proof to the reader except for (6.52) which we now prove. For $t \in [a+1, b+1]$ consider

$$\begin{aligned}\mathcal{L}U(t) &= -K(t-1)U(t-1) + N(t)U(t) - K(t)U(t+1)\\ &= \left[-K(t-1)U(t-1)U^{-1}(t) + N(t) - K(t)U(t+1)U^{-1}(t)\right] U(t).\end{aligned}$$

Use of equation (6.46) yields

$$\mathcal{L}U(t) = [Z(t) - K(t-1) + N(t) - K(t)U(t+1)U^{-1}(t)]U(t).$$

Since $C(t) = N(t) - K(t) - K(t-1)$ and

$$U(t+1)U^{-1}(t) = \left[K(t) - Z(t+1)\right]^{-1} K(t),$$

which follows from (6.50), we get that

$$\begin{aligned}\mathcal{L}U(t) &= \left\{Z(t) + C(t) + K(t) - K(t)\left[K(t) - Z(t+1)\right]^{-1} K(t)\right\} U(t)\\ &= \left\{Z(t) + C(t) + \left[K(t) - Z(t+1)\right]\left[K(t) - Z(t+1)\right]^{-1} K(t)\right.\\ &\qquad \left. - K(t)\left[K(t) - Z(t+1)\right]^{-1} K(t)\right\}U(t)\\ &= \left\{Z(t) + C(t) - Z(t+1)\left[K(t) - Z(t+1)\right]^{-1} K(t)\right\} U(t)\\ &= \{\tilde{\mathrm{R}}Z(t)\}U(t).\end{aligned}$$

□

Exercise 6.21 *Show that*

$$\tilde{\mathrm{R}}Z(t) = Z(t) + C(t) - K(t)\,[K(t) - Z(t+1)]^{-1}\, Z(t+1)$$

and

$$\tilde{\mathrm{R}}Z(t) = -\Delta Z(t) + C(t) - Z(t+1)\,[K(t) - Z(t+1)]^{-1}\, Z(t+1).$$

Theorem 6.22 *If $\mathcal{L}u(t) = 0$ is disconjugate on $(-\infty, b+2]$, then the reverse Riccati equation $\tilde{\mathrm{R}}Z(t) = 0$ has a Hermitian solution $Z(t)$ on $(-\infty, b+1]$ with $K(t) - Z(t+1) > 0$ on $(-\infty, b]$. A partial converse is as follows: If $\tilde{\mathrm{R}}Z(t) = 0$ has a Hermitian solution $Z(t)$ on $(-\infty, b+2]$ with $K(t) - Z(t+1) > 0$ on $(-\infty, b+1]$, then $\mathcal{L}u(t) = 0$ is disconjugate on $(-\infty, b+2]$.*

Proof: Assume $\mathcal{L}u(t) = 0$ is disconjugate on $(-\infty, b+2]$. Let $U(t)$ be the solution of the initial value problem

$$\mathcal{L}U(t) = 0, \qquad U(b+1) = I, \qquad U(b+2) = 0.$$

By the Reid Roundabout Theorem 5.20 on page 215

$$U^*(t-1)K(t-1)U(t) > 0$$

on $(-\infty, b+1]$. Define $Z(t)$ on $(-\infty, b+1]$ by the reverse Riccati substitution (6.44) on page 279, that is,

$$Z(t) = K(t-1)\,[\Delta U(t-1)]\, U^{-1}(t)$$

for $t \in (-\infty, b+1]$. By Theorem 6.19, we know that $Z(t)$ is a Hermitian solution of $\tilde{\mathrm{R}}Z(t) = 0$ on $(-\infty, b+1]$ and satisfies

$$K(t) - Z(t+1) > 0$$

on $(-\infty, b]$.

For the partial converse, assume $Z(t)$ is a Hermitian solution of $\tilde{\mathrm{R}}Z(t) = 0$ on $(-\infty, b+2]$ and

$$K(t) - Z(t+1) > 0$$

on $(-\infty, b+1]$. Let $U(t)$ be the solution of the **IVP**

$$U(t-1) = K^{-1}(t-1)\,[K(t-1) - Z(t)]\, U(t), \qquad U(b+2) = I. \tag{6.53}$$

(Refer to equation (6.50) on page 281.) By Theorem 6.19 $U(t)$ is a prepared solution of $\mathcal{L}U(t) = 0$ with no generalized zeros in $(-\infty, b+2]$. Hence by Theorem 5.20, $\mathcal{L}u(t) = 0$ is disconjugate on $(-\infty, b+2]$. □

Theorem 6.23 *Assume $\mathcal{L}u(t) = 0$ is disconjugate on $(-\infty, b+2]$ and $Z(t)$ is a solution of the reverse Riccati equation $\tilde{\text{R}}Z(t) = 0$ on $(-\infty, b+2]$. Then there is an integer $t_0 \leq b+1$ such that*

$$K(t) - Z(t+1) > 0$$

on $(-\infty, t_0]$.

Proof: Let $U(t)$ be the solution of the initial value problem (6.53). Since $Z(t)$ is a solution of the reverse Riccati equation $\tilde{\text{R}}Z(t) = 0$ on $(-\infty, b+2]$, we know that $K(t) - Z(t+1)$ is nonsingular on $(-\infty, b+2]$. Then by (6.53) the solution $U(t)$ is nonsingular on $(-\infty, b+2]$. By the Sturm Separation Theorem 5.30 on page 221, there is a $t_0 \leq b+1$ such that

$$U^*(t)K(t)U(t+1) > 0$$

on $(-\infty, t_0]$. So $U(t)$ has no generalized zeros in $(-\infty, t_0+1]$ and by Theorem 6.19 on page 281

$$K(t) - Z(t+1) > 0$$

on $(-\infty, t_0]$. □

6.8 UPPER SOLUTIONS OF THE REVERSE RICCATI EQUATION

Whenever $\mathcal{L}U(t) = 0$ has a recessive solution $U_0(t)$ at $-\infty$ which is nonsingular in some neighborhood of $-\infty$ then we define the *distinguished solution at $-\infty$* of the reverse Riccati equation $\tilde{\text{R}}Z(t) = 0$ to be the corresponding solution $Z_0(t)$ defined by the reverse Riccati substitution

$$Z_0(t) = K(t-1)\left[\Delta U_0(t-1)\right] U_0^{-1}(t).$$

Note that by Theorem 5.66 on page 253 disconjugacy of $\mathcal{L}u(t) = 0$ in some neighborhood of $-\infty$ implies the existence of such a recessive solution $U_0(t)$ at $-\infty$. Also, any recessive solution at $-\infty$ which is nonsingular in a neighborhood of $-\infty$ must be a nonsingular constant multiple of $U_0(t)$.

Theorem 6.24 *Assume $\mathcal{L}u(t) = 0$ is disconjugate on $(-\infty, b+2]$. Let $Z_0(t)$ be the distinguished solution of the reverse Riccati equation at $-\infty$. If $Z(t)$ is* *a Hermitian solution of the reverse Riccati equation in a neighborhood of* $-\infty$, *then on any discrete interval* $(-\infty, t_0]$, $t_0 \leq b+1$, *such that* $K(t) - Z(t+1) > 0$ *on* $(-\infty, t_0]$, *(there is such an interval by Theorem* 6.23*) we have*

$$Z_0(t) \geq Z(t)$$

on $(-\infty, t_0]$.

Proof: By Theorem 5.66 on page 253 and Theorem 6.22 on page 284, $Z_0(t)$ is a solution of the reverse Riccati equation $\tilde{R}Z(t) = 0$ on $(-\infty, b+1]$ and

$$K(t) - Z_0(t+1) > 0$$

on $(-\infty, b+1]$. Let $Z(t)$ be a Hermitian solution of the reverse Riccati equation in a neighborhood of $-\infty$. (See equation (6.48) on page 281.) Assume $t_0 \leq b+1$ and

$$K(t) - Z(t+1) > 0$$

on $(-\infty, t_0]$. Fix $t_1 \leq t_0$. Let $U_0(t)$ be the recessive soltuion of $\mathcal{L}U(t) = 0$ at $-\infty$ with $U_0(t_1) = I$. Since $Z_0(t)$ is the distinguished solution at $-\infty$ we have

$$Z_0(t) = K(t-1)[\Delta\left(U_0(t-1)\right)]U_0^{-1}(t).$$

Let $U(t)$ be the solution of the **IVP** (6.53) on page 284. By Theorem 6.22, $U(t)$ is a prepared solution of $\mathcal{L}U(t) = 0$ with

$$U^*(t-1)K(t-1)U(t) > 0$$

on $(-\infty, t_1]$. Consider the difference of these two solutions at t_1

$$\begin{aligned} &Z_0(t_1) - Z(t_1) = \\ &\qquad K(t_1-1)(\Delta U_0(t_1-1))U_0^{-1}(t_1) - K(t_1-1)(\Delta U(t_1-1))U^{-1}(t_1) \end{aligned}$$

which simplifies to

$$Z_0(t_1) - Z(t_1) = K(t_1-1)\left[U(t_1-1) - U_0(t_1-1)\right]. \tag{6.54}$$

By the backwards reduction of order theorem, Theorem 5.38 on page 232,

$$U_0(t) = U(t)\left[I - \bar{S}(t)Q\right] \tag{6.55}$$

where

$$Q = \{U; U_0\}$$

and

$$\bar{\mathcal{S}}(t) = \sum_{s=t}^{t_1-1} [U^*(s)K(s)U(s+1)]^{-1}$$

for $t \leq t_1$. Note that

$$\begin{aligned}\bar{\mathcal{S}}(t_1-1) &= [U^*(t_1-1)K(t_1-1)U(t_1)]^{-1}\\ &= [U^*(t_1)K(t_1-1)U(t_1-1)]^{-1}\end{aligned}$$

since $U(t)$ is a prepared solution. Simplifying we get that

$$U(t_1-1) - U_0(t_1-1) = U(t_1-1)\bar{\mathcal{S}}(t_1-1)Q = K^{-1}(t_1-1)Q.$$

Hence from the difference of solutions (6.54) we have

$$Z_0(t_1) - Z(t_1) = Q. \tag{6.56}$$

Let $U(t,s)$ be the solution of the **BVP**

$$\mathcal{L}U(t) = 0, \qquad U(t_1) = I, \quad U(s) = 0.$$

By the backwards reduction of order theorem, Theorem 5.38 on page 232, we have

$$U(t,s) = U(t)\left[I - \bar{\mathcal{S}}(t)Q_s\right], \tag{6.57}$$

where

$$Q_s = \{U(t); U(t,s)\}.$$

Letting $t = s$ in (6.57) gives

$$0 = U(s)\left[I - \bar{\mathcal{S}}(s)Q_s\right].$$

Hence

$$Q_s = \bar{\mathcal{S}}^{-1}(s) = \left\{\sum_{\tau=s}^{t_1-1} [U^*(\tau)K(\tau)U(\tau+1)]^{-1}\right\}^{-1} > 0.$$

But

$$\begin{aligned}Q = \{U(t); U_0(t)\} &=\\ \left\{U(t); \lim_{s\to-\infty} U(t,s)\right\} &=\\ \lim_{s\to-\infty}\{U(t); U(t,s)\} &= \lim_{s\to-\infty} Q_s \geq 0.\end{aligned}$$

Therefore, from (6.56) we get that $Z_0(t_1) \geq Z(t_1)$. Since t_1 is an arbitrary integer in $(-\infty, t_0]$, we have

$$Z_0(t) \geq Z(t) \qquad \text{for } t \leq t_0.$$

□

Theorem 6.25 *Assume $K(t)$ is a nonsingular, Hermitian $n \times n$ matrix function on $(-\infty, \infty)$ which is periodic with period a positive integer p. If $\mathcal{L}u(t) = 0$ is disconjugate on $(-\infty, \infty)$, then the distinguished solution $Z_0(t)$ of the reverse Riccati equation at $-\infty$ is a solution on $(-\infty, \infty)$, has period p and satisfies*

$$K(t) - Z_0(t+1) > 0 \tag{6.58}$$

on $(-\infty, \infty)$.

Proof: Since $\mathcal{L}u(t) = 0$ is disconjugate on $(-\infty, \infty)$, it is disconjugate on $(-\infty, t_0]$ for any integer t_0. Hence by Theorem 5.66 on page 253, $\mathcal{L}U(t) = 0$ has a recessive solution $U_0(t)$ at $t = -\infty$ with

$$U_0^*(t)K(t)U_0(t+1) > 0$$

on $(-\infty, t_0 - 1]$. It follows that the corresponding solution $Z_0(t)$ of the reverse Riccati equation exists on $(-\infty, t_0]$ and consequently satisfies inequality (6.58) on $(-\infty, t_0 - 1]$. Since t_0 is arbitrary $Z_0(t)$ is a solution of $\tilde{\mathrm{R}}Z(t) = 0$ on $(-\infty, \infty)$ which satisfies inequality (6.58) on $(-\infty, \infty)$. Since

$$C(t) = N(t) - K(t) - K(t-1)$$

we get that $C(t)$ is a Hermitian matrix function on $(-\infty, \infty)$ which is periodic of period p. Because the coefficients in $\tilde{\mathrm{R}}Z(t) = 0$ are periodic with period p it follows that

$$Z_1(t) \equiv Z_0(t+p)$$
$$Z_2(t) \equiv Z_0(t-p)$$

are also Hermitian solutions of the reverse Riccati equation on $(-\infty, \infty)$. Also,

$$K(t) - Z_i(t+1) > 0$$

on $(-\infty, \infty)$ for $i = 1, 2$. By Theorem 6.24 on page 285 the bounds

$$Z_1(t) = Z_0(t+p) \leq Z_0(t) \tag{6.59}$$
$$Z_2(t) = Z_0(t-p) \leq Z_0(t) \tag{6.60}$$

are satisfied on $(-\infty, \infty)$. Replacing t in (6.60) by $t + p$ implies

$$Z_0(t) \leq Z_0(t+p).$$

Combining this with (6.59) gives the bounds

$$Z_0(t) \leq Z_0(t+p) \leq Z_0(t)$$

on $(-\infty, \infty)$. Hence $Z_0(t+p) = Z_0(t)$ for all t and $Z_0(t)$ is periodic with period p. □

Theorem 6.26 (Constant Coefficients) *Assume $K(t)$ and $N(t)$ are constant $n \times n$ Hermitian matrices with $K(t) \equiv K$ nonsingular, $N(t) \equiv N$, and set $C \equiv N - 2K$. If $\mathcal{L}u(t) = 0$ is disconjugate on $(-\infty, \infty)$, then the distinguished solution $Z_0(t)$ of the reverse Riccati equation at $-\infty$ is a constant Hermitian matrix $Z_0(t) \equiv Z_0$ satisfying the inequality*

$$K - Z_0 > 0. \tag{6.61}$$

Furthermore Z_0 is a matrix solution of the "reverse steady state equation"

$$Z(K - Z)^{-1}K = Z + C \tag{6.62}$$

and the "reverse discrete algebraic Riccati equation"

$$ZK^{-1}Z + CK^{-1}Z - C = 0. \tag{6.63}$$

Proof: From Theorem 6.25 we get that $Z_0(t)$ is a Hermitian solution of $\tilde{R}Z(t) = 0$ on $(-\infty, \infty)$ (see Exercies 6.21 on page 284) which is periodic with period $p = 1$ satisfying $K - Z_0(t) > 0$ on $(-\infty, \infty)$. Hence $Z_0(t) \equiv Z_0$, a constant Hermitian matrix satisfying (6.61). The fact that $Z_0(t) \equiv Z_0$ is a solution of $\tilde{R}Z(t) = 0$ readily implies that Z_0 is a solution of the reverse steady state equation (6.62) and it follows that

$$(Z_0 + C)K^{-1} = Z_0(K - Z_0)^{-1},$$

i.e.,

$$(Z_0 + C)K^{-1}(K - Z_0) = Z_0$$

which multiplies out and simplifies to

$$Z_0K^{-1}Z_0 + CK^{-1}Z_0 - C = 0 \tag{6.64}$$

which is the desired result.

Exercise 6.27 *Show that if $Z = -W$, then W is a solution of the discrete algebraic Riccati equation* (6.27) *on page 272 iff Z is a solution of the reverse discrete algebraic Riccati equation* (6.63).

Exercise 6.28 *Show that the reverse steady state equation* (6.62) *and the reverse algebraic Riccati equation* (6.63) *are equivalent. (In this verification you have to show that if Z is a solution of* (6.63)*, then $K - Z$ is invertible.)*

Theorem 6.29 (Characteristic Equation) *Assume K is a constant Hermitian nonsingular $n \times n$ matrix, N is an $n \times n$ constant Hermitian matrix and $C = N - 2K$. If Z and Λ are $n \times n$ matrices related by the equation*

$$(K - Z)\Lambda = K \tag{6.65}$$

then Z is a solution of the reverse discrete algebraic Riccati equation (6.63) *with $K - Z$ nonsingular iff Λ is a nonsingular matrix solution of the characteristic equation* (6.30). *If $K > 0$, and Z is a Hermitian solution of* (6.63) *with $K - Z > 0$, then the eigenvalues λ_i, $1 \leq i \leq n$, of Λ are real and positive. If $Z > 0$, then $\lambda_i > 1$, $1 \leq i \leq n$, while if $Z < 0$, then $0 < \lambda_i < 1$ for $1 \leq i \leq n$.*

Proof: Assume Z and Λ are $n \times n$ matrices related by (6.65). Note that Λ is nonsingular iff $K - Z$ is nonsingular. In this case we can solve (6.65) for Z to get that

$$Z = K\left(I - \Lambda^{-1}\right).$$

Hence,

$$\begin{aligned} ZK^{-1}Z &+ CK^{-1}Z - C \\ &= K\left(I - \Lambda^{-1}\right)^2 + C\left(I - \Lambda^{-1}\right) - C \\ &= K - 2K\Lambda^{-1} + K\Lambda^{-2} + C - C\Lambda^{-1} - C \\ &= \left[K\Lambda^2 - (2K + C)\Lambda + K\right]\Lambda^{-2} \\ &= \left(K\Lambda^2 - N\Lambda + K\right)\Lambda^{-2}. \end{aligned}$$

It follows that Z is a matrix solution of (6.63) with $K - Z$ nonsingular iff Λ is a nonsingular matrix solution of (6.30).

For the rest of the proof assume $K > 0$, Z is a solution of (6.63) with $K - Z > 0$. Let λ_i, x_i be eigenpairs for Λ, $1 \leq i \leq n$. From (6.65) we get that

$$(K - Z)\Lambda x_i = Kx_i.$$

Hence

$$\lambda_i(K - Z)x_i = Kx_i.$$

Premultiplying both sides by x_i^* and solving for λ_i gives the Rayleigh quotient representation of λ_i of

$$\lambda_i = \frac{x_i^* K x_i}{x_i^*(K - Z)x_i}$$

which must be real and positive. Rewrite this as

$$\lambda_i = \frac{x_i^* K x_i}{x_i^* K x_i - x_i^* Z x_i}$$

in order to conclude that if $Z > 0$, then each eigenvalue of Λ satisfies $\lambda_i > 1$, while $Z < 0$ implies $0 < \lambda_i < 1$, for $1 \leq i \leq n$. □

Theorem 6.30 *Assume $K > 0$ and $C = N - 2K > 0$. Then the distinguished solution Z_0 at $-\infty$ of the reverse Riccati equation is the unique positive definite solution of the reverse discrete algebraic Riccati equation* (6.63) *satisfying $K - Z > 0$. Consequently, $0 < Z_0 < K$ and if Z is a Hermitian solution of* (6.63) *with $K - Z > 0$, then*

$$Z \leq Z_0.$$

Proof: Let $Z(t)$ be the solution of the **IVP**

$$Z(t) = -C + Z(t+1)[K - Z(t+1)]^{-1}K, \quad t \leq -1, \quad \text{with} \quad Z(0) = 0.$$

We claim that $Z(t)$ is a solution of the above reverse Riccati equation on the whole interval $(-\infty, 0]$ and $Z(t) < 0$ on $(-\infty, -1]$. We prove this claim by induction to the left on t. Let $t = -1$ in the above reverse Riccati equation to get

$$Z(-1) = -C < 0.$$

Now assume $t < -1$, $Z(t)$ is well defined and satisfies $Z(t) < 0$. Since $K - Z(t)$ is positive definite it has an inverse and the equation

$$Z(t-1) = -C + Z(t)\,[K - Z(t)]^{-1}\,K$$

defines $Z(t-1)$. Also,

$$\begin{aligned} Z(t-1) &= -C + \left\{K^{-1}\,[K - Z(t)]\,Z^{-1}(t)\right\}^{-1} \\ &= -C + \left\{Z^{-1}(t) - K^{-1}\right\}^{-1} < 0 \end{aligned}$$

and the induction is complete.

Since $K > 0$ and $C > 0$ we have from Corollary 5.80 on page 261 that (6.24) on page 272 is disconjugate on $(-\infty, \infty)$. Therefore by Theorem 6.24 on page 285 we get that

$$Z_0 \geq Z(t)$$

on $(-\infty, 0]$. Setting $t = 0$ we get that

$$Z_0 \geq Z(0) = 0.$$

Hence Z_0 is positive semidefinite. But it follows from (6.64) on page 289 that Z_0 is nonsingular. Thus Z_0 has positive eigenvalues and must be positive definite.

The uniqueness part of this theorem remains to be established. Assume Z is also a reverse discrete algebraic Riccati equation (6.63) on page 289 satisfying $K - Z > 0$. Let Λ and Λ_0 be matrices defined by

$$(K - Z)\Lambda = K \tag{6.66}$$
$$(K - Z_0)\Lambda_0 = K. \tag{6.67}$$

By Theorem 6.29 and Exercise 6.14 on page 273 we get that

$$U(t) = \Lambda^t$$
$$U_0(t) = \Lambda_0^t$$

are solutions of (6.31) on page 273 corresponding to $Z(t)$ and $Z_0(t)$, resp., via the reverse Riccati transformation (6.44). Solving (6.66) and (6.67) for Z and Z_0 we obtain

$$Z = K - K\Lambda^{-1}$$
$$Z_0 = K - K\Lambda_0^{-1}.$$

It follows that

$$Z_0 - Z = K\left(\Lambda^{-1} - \Lambda_0^{-1}\right). \tag{6.68}$$

Since $\mathcal{L}u(t) = 0$ is disconjugate on $(-\infty, \infty)$ there is a unique solution $U(t,s)$ of the boundary value problem

$$\mathcal{L}U(t) = 0, \qquad U(t,s) = I, \qquad U(s,s) = 0$$

for $s \leq -2$. By the "backwards reduction of order" Theorem 5.38 (on page 232) for $t \leq 0$

$$U(t,s) = U(t)[I - \bar{\mathcal{S}}(t)Q_s] \tag{6.69}$$

where

$$Q_s = \{U(t); U(t,s)\}$$

and

$$\begin{aligned}
\bar{\mathcal{S}}(t) &= \sum_{\tau=t}^{1} [U^*(\tau)KU(\tau+1)]^{-1} \\
&= \sum_{\tau=t}^{1} [U^*(\tau+1)KU(\tau)]^{-1} \\
&= \sum_{\tau=t}^{1} \left[(\Lambda^*)^{\tau+1} K\Lambda^{\tau}\right]^{-1} \\
&= \sum_{\tau=t}^{1} \left[(\Lambda^*)^{\tau+1} \left(K\Lambda^{-1}\right) \Lambda^{\tau+1}\right]^{-1}.
\end{aligned}$$

Use (6.66) to substitute for $K\Lambda^{-1}$ in the above to obtain

$$\bar{S}(t) = \sum_{\tau=t}^{1} \left[(\Lambda^*)^{\tau+1} (K - Z) \Lambda^{\tau+1} \right]^{-1}. \tag{6.70}$$

Letting $t = s$ in (6.69) and solving for Q_s gives

$$Q_s = \bar{S}^{-1}(s). \tag{6.71}$$

Hence

$$\begin{aligned}\lim_{s\to-\infty} \bar{S}^{-1}(s) &= \lim_{s\to-\infty} Q_s \\ &= \lim_{s\to-\infty} \{U(t); U(t,s)\} \\ &= \{U(t); U_0(t)\} = \{U; U_0\}(0) \\ &= \Lambda^{*-1}K - K\Lambda_0^{-1} = K\Lambda^{-1} - K\Lambda_0^{-1} = Z_0 - Z\end{aligned}$$

since $K\Lambda^{-1} = K - Z$ is Hermitian and equation (6.68) holds. Use (6.70) with $\tau = s$ to get that

$$Z - Z_0 = \lim_{s\to-\infty} \left\{ \sum_{\tau=s}^{1} \left[(\Lambda^*)^{\tau+1} (K - Z) \Lambda^{\tau+1} \right]^{-1} \right\}^{-1}.$$

This implies $Z_0 \geq Z$ and if $Z_0 \neq Z$, then it follows that an eigenvalue of

$$\left\{ \sum_{\tau=s}^{1} \left[(\Lambda^*)^{\tau+1} (K - Z) \Lambda^{\tau+1} \right]^{-1} \right\}^{-1}$$

does not go to zero as $s \to -\infty$. It follows that an eigenvalue of

$$M(s) \equiv \sum_{\tau=s}^{1} \left[(\Lambda^*)^{\tau+1} (K - Z) \Lambda^{\tau+1} \right]^{-1}$$

converges to a finite number as s goes to $-\infty$. Hence there are eigenvectors $\alpha(s)$, $||\alpha(s)||_2 = 1$ such that $\alpha^*(s)M(s)\alpha(s)$ converges to a finite number as s goes to $-\infty$. It follows that there is a vector α with $||\alpha||_2 = 1$ such that

$$\sum_{\tau=s}^{1} \alpha^* \left\{ (\Lambda^*)^{\tau+1} (K - Z) \Lambda^{\tau+1} \right\}^{-1} \alpha$$

converges. It follows that

$$\lim_{\tau\to-\infty} \alpha^* \left\{ (\Lambda^*)^{\tau+1} (K - Z) \Lambda^{\tau+1} \right\}^{-1} \alpha = 0.$$

Since $K - Z > 0$ we get that

$$\lim_{\tau \to -\infty} ||\Lambda^{-(\tau+1)}|| = 0.$$

But since $Z > 0$, Theorem 6.29 implies that all the eigenvalues of Λ are larger than 1, which gives a contradiction. □

Theorem 6.31 *Assume $K > 0$ and $C = N - 2K > 0$. Then if W_0 is the distinguished solution of the Riccati equation $\mathrm{R}W = 0$ at ∞ and if Z_0 is the distinguished solution of the reverse Riccati equation $\tilde{\mathrm{R}}Z = 0$, then*

$$W_0 = -Z_0.$$

Proof: By Exercise 6.27 on page 289 $W = -Z_0$ is a solution of the discrete algebraic Riccati equation (6.27) on page 272. By Theorem 6.17 on page 275 and Theorem 6.30, W_0 and $W = -Z_0$ are Hermitian negative definite solutions of equation (6.26) satisfying $W + K > 0$ and $W_0 + K > 0$. By the uniqueness in Theorem 6.17 on page 275 we get that $W_0 = -Z_0$. □

7

GREEN'S FUNCTIONS FOR NONHOMOGENEOUS SECOND ORDER DIFFERENCE EQUATIONS

7.1 INTRODUCTION

In this chapter we will be concerned with second order nonhomogeneous difference equations. We will be studying equations of the form

$$\mathcal{L}y(t) = \Delta\left[P(t)\Delta y(t-1)\right] + Q(t)y(t) = h(t)$$

where $P(t)$ and $Q(t)$ are given $n \times n$ Hermitian matrix functions on the discrete intervals $[a+1, b+2]$ and $[a+1, b+1]$, respectively. We also assume $P(t)$ is nonsingular for $t \in [a+1, b+2]$. In Section 2 of this chapter we will derive a variation of constants formula for the above nonhomogeneous problem. In Sections 3 and 4 we will be concerned with Green's matrix functions for the conjugate boundary value problem and for the right focal boundary value problem. In Section 5 we will consider a Green's matrix function for a more general two point boundary value problem.

7.2 VARIATION OF CONSTANTS FORMULA

In this section we define the Cauchy matrix function for $\mathcal{L}y(t) = 0$ and we use this function to derive a variation of constants formula for $\mathcal{L}y(t) = h(t)$. Expanding out the difference operators in $\mathcal{L}y(t)$ we obtain a symmetric three term vector recurrence relation

$$\mathcal{L}y(t) = P(t+1)y(t+1) + B(t)y(t) + P(t)y(t-1) \tag{7.1}$$

where

$$B(t) = Q(t) - P(t) - P(t-1)$$

for $t \in [a+1, b+1]$. Note that it follows from (7.1) that the symmetric vector equation $\mathcal{L}y(t) = 0$ can be written in the form of (5.1) with $E(t) \equiv I$ and

$$K(t) = K^*(t) = P(t+1) \tag{7.2}$$

$$N(t) = -B(t) = P(t) + P(t-1) - Q(t). \tag{7.3}$$

Hence many of the definitions and results of Chapter 5 hold for the self-adjoint case $\mathcal{L}y(t) = 0$ where we assume equations (7.2) and (7.3) hold. In particular, note that a generalized zero t_0 of a prepared solution of $\mathcal{L}y(t) = 0$ is defined in terms of the expression $y^*(t_0 - 1)P(t_0)y(t_0)$.

First we define the *Cauchy matrix function* for $\mathcal{L}y(t) = 0$. The matrix operator $\mathcal{L}Y(t)$ is defined by (7.1) if the vector function $y(t)$ is replaced by the $n \times n$ matrix function $Y(t)$. For each fixed $s \in [a+1, b+1]$ and $t \in [s, b+2]$ define the Cauchy matrix function $Y(t, s)$ as the solution of the **IVP**

$$\mathcal{L}Y(t,s) = 0, \qquad Y(s,s) = 0, \quad Y(s+1,s) = P^{-1}(s+1).$$

Throughout the subsequent results we will use the *summation convention* that the sum is zero if the upper index is less than the lower index.

Exercise 7.1 *Show that*

$$Y(t,s) = \sum_{\tau=s+1}^{t} P^{-1}(\tau)$$

for $a \le t \le b+2, \quad a+1 \le s \le b+1,\ t \ge s$ *is the Cauchy matrix function for*

$$\mathcal{L}y(t) = \Delta\left[P(t)\Delta y(t-1)\right] = 0.$$

We now wish to derive the variation of constants formula for our nonhomogeneous problem. The Cauchy matrix function will be used in this derivation.

Theorem 7.2 *Assume* $h(t)$ *is an* $n \times 1$ *vector function defined on* $[a+1, b+1]$. *Then the solution of the* **IVP**

$$\mathcal{L}y(t) = h(t), \qquad t \in [a+1, b+1],$$

$$y(a) = 0, \qquad y(a+1) = 0$$

is given by

$$y(t) = \sum_{s=a+1}^{t-1} Y(t,s)h(s)$$

for $t \in [a, b+2]$, *where* $Y(t,s)$ *is the Cauchy matrix function for* $\mathcal{L}y(t) = 0$.

Proof: By our present convention on sums that the sum is zero if the upper index is less than the lower index, we have

$$y(a) = y(a+1) = 0.$$

It is easy to check that $\mathcal{L}y(t) = h(t)$ is valid for $t = a+1$. Next consider the case where $a+2 \le t \le b+1$. Then from (7.1)

$$\begin{aligned}
\mathcal{L}y(t) &= P(t+1)\left[\sum_{s=a+1}^{t-1} Y(t+1,s)h(s) + Y(t+1,t)h(t)\right] \\
&\quad + B(t)\sum_{s=a+1}^{t-1} Y(t,s)h(s) + P(t)\sum_{s=a+1}^{t-2} Y(t-1,s)h(s) \\
&= \sum_{s=a+1}^{t-1} \mathcal{L}Y(t,s)h(s) + P(t+1)P^{-1}(t+1)h(t) \\
&= h(t).
\end{aligned}$$

□

It is easy to generalize the last result to get what also is called a variation of constants formula for $\mathcal{L}y(t) = h(t)$.

Corollary 7.3 (Variation of Constants Formula) *Assume that* α, β *are given* $n \times 1$ *constant vectors and* $h(t)$ *is an* $n \times 1$ *vector function on* $[a+1, b+1]$. *Then the solution of the* **IVP**

$$\mathcal{L}y(t) = h(t), \quad t \in [a+1, b+1], \tag{7.4}$$

$$y(a) = \alpha, \tag{7.5}$$

$$y(a+1) = \beta \tag{7.6}$$

is given by

$$y(t) = u(t) + \sum_{s=a+1}^{t-1} Y(t,s)h(s) \tag{7.7}$$

for $t \in [a, b+2]$ *where* $u(t)$ *is the solution of the* **IVP**

$$\mathcal{L}u(t) = 0, \qquad u(a) = \alpha, \quad u(a+1) = \beta.$$

Exercise 7.4 *Prove Corollary* 7.3.

7.3 THE GREEN'S MATRIX FUNCTION FOR THE CONJUGATE PROBLEM

Our main interest in this section is what we will call the Green's matrix function for the conjugate **BVP**

$$\mathcal{L}y(t) = 0, \quad t \in [a+1, b+1] \tag{7.8}$$
$$y(a) = 0 = y(b+2). \tag{7.9}$$

Most of the results in this section can be found in [117].

Lemma 7.5 *Assume* $G(t, s)$ *is an* $n \times n$ *matrix function satisfying the following properties*

a) $G(t, s)$ *is defined for* $a \le t \le b+2$, $a+1 \le s \le b+1$.

b) $\mathcal{L}G(t, s) = \delta_{ts} I$, $a+1 \le t$, $s \le b+1$, *where* δ_{ts} *is the Kronecker delta function (defined by* $\delta_{ts} = 0$ *if* $t \ne s$ *and* $\delta_{ss} = 1$*) and* I *is the* $n \times n$ *identity matrix.*

c) $G(a, s) = G(b+2, s) = 0$ *for* $a+1 \le s \le b+1$.

If $h(t)$ *is a given* $n \times 1$ *vector function defined on* $[a+1, b+1]$, *then*

$$y(t) = \sum_{s=a+1}^{b+1} G(t, s) h(s)$$

is a solution of the conjugate **BVP**

$$\mathcal{L}y(t) = h(t), \quad t \in [a+1, b+1] \tag{7.10}$$
$$y(a) = 0 = y(b+2). \tag{7.11}$$

Proof: Assume $G(t,s)$ is an $n \times n$ matrix function satisfying properties **(a)**–**(c)** and define

$$y(t) = \sum_{s=a+1}^{b+1} G(t,s)h(s)$$

for $t \in [a, b+2]$. First note that

$$y(a) = \sum_{s=a+1}^{b+1} G(a,s)h(s) = 0$$

by property **(c)**. Similarly, $G(b+2,s) = 0$, $a+1 \le s \le b+1$ gives $y(b+2) = 0$. Finally, using **(b)** we get that for $t \in [a+1, b+1]$

$$\mathcal{L}y(t) = \sum_{s=a+1}^{b+1} \mathcal{L}G(t,s)h(s) = \sum_{s=a+1}^{b+1} \delta_{ts}h(s) = h(t).$$

□

Lemma 7.6 *Let $Y(t,s)$ be the Cauchy matrix function for $\mathcal{L}y(t) = 0$. If the* **BVP** *$\mathcal{L}y(t) = 0$, $y(a) = 0 = y(b+2)$ has only the trivial solution, then we can define an $n \times n$ matrix function $G(t,s)$ on the set of points (t,s) in $[a, b+2] \times [a+1, b+1]$ by*

$$G(t,s) = \begin{cases} -Y(t,a)Y^{-1}(b+2,a)Y(b+2,s), & t \le s, \\ Y(t,s) - Y(t,a)Y^{-1}(b+2,a)Y(b+2,s), & s < t. \end{cases}$$

This function $G(t,x)$ has the properties of Lemma 7.5 *and is the unique matrix function satisfying conditions* **(a)**–**(c)** *of Lemma* 7.6

Proof: If $Y(b+2,a)$ is singular, then there is a vector $\xi \neq 0$ such that

$$Y(b+2,a)\xi = 0.$$

Set

$$y(t) = Y(t,a)\xi.$$

Then $y(t)$ is a nontrivial solution of $\mathcal{L}y(t) = 0$ with

$$y(a) = 0 = y(b+2)$$

which is a contradiction. Hence $Y(b+2,a)$ is nonsingular and $G(t,s)$ is well defined. Now we show that $G(t,s)$ satisfies properties **(a)**–**(c)** in Lemma 7.5. Clearly property **(a)** is satisfied. Also for $a+1 \leq s \leq b+1$

$$G(a,s) = -Y(a,a)Y^{-1}(b+2,a)Y(b+2,s) = 0$$

and

$$G(b+2,s) = Y(b+2,s) - Y(b+2,a)Y^{-1}(b+2,a)Y(b+2,s) = 0.$$

Thus property **(c)** holds.

We now show that $G(t,s)$ satisfies property **(b)**. If $a+1 \leq t < s \leq b+1$, then

$$\mathcal{L}G(t,s) = -\mathcal{L}Y(t,a)Y^{-1}(b+2,a)Y(b+2,s) = 0 = \delta_{ts}I.$$

Next assume $a+1 \leq s < t \leq b+1$. Then

$$\mathcal{L}G(t,s) = \mathcal{L}Y(t,s) - \mathcal{L}Y(t,a)Y^{-1}(b+2,a)Y(b+2,s) = 0 = \delta_{ts}I.$$

Finally assume $a+1 \leq s = t \leq b+1$. Then by (7.1)

$$\begin{aligned} \mathcal{L}G(t,s)\big|_{s=t} &= P(t+1)G(t+1,t) + B(t)G(t,t) + P(t)G(t-1,t) \\ &= P(t+1)Y(t+1,t) - \mathcal{L}Y(t,a)Y^{-1}(b+2,a)Y(b+2,s) \\ &= I = \delta_{ts}I. \end{aligned}$$

Hence $G(t,s)$ satisfies **(c)**.

In order to establish uniqueness, let $H(t,s)$ be an $n \times n$ matrix function which satisfies properties **(a)**–**(c)**. Fix $s \in [a+1, b+1]$ and define

$$Y(t) = G(t,s) - H(t,s)$$

for $t \in [a, b+2]$. It follows from property **(b)** that $Y(t)$ is a solution of $\mathcal{L}Y(t) = 0$. But from **(c)**

$$Y(a) = Y(b+2) = 0.$$

It follows that each of the columns of $Y(t)$ is the zero vector. Hence

$$Y(t) = G(t,s) - H(t,s) = 0$$

for $t \in [a, b+2]$. Since $s \in [a+1, b+1]$ is arbitrary

$$G(t,s) = H(t,s)$$

on $[a, b+2] \times [a+1, b+1]$ and uniqueness is established. □

When the **BVP** $\mathcal{L}y(t) = 0$, $y(a) = 0 = y(b+2)$ has only the trivial solution, then the unique matrix function $G(t, s)$ satisfing properties **(a)**–**(c)** in Lemma 7.5 is called the *Green's matrix function* for the conjugate **BVP**

$$\begin{aligned} \mathcal{L}y(t) &= 0, \quad t \in [a+1, b+1], \\ y(a) &= 0 = y(b+2). \end{aligned}$$

Exercise 7.7 *Let $Y(t, s)$ be the Cauchy matrix function for $\mathcal{L}y(t) = 0$. Show that the* **BVP** *$\mathcal{L}y(t) = 0$, $y(a) = 0 = y(b+2)$ has only the trivial solution iff $Y(b+2, a)$ is nonsingular.*

We have proven most of the results in the following theorem.

Theorem 7.8 (Green's matrix function for the conjugate problem) *If the* **BVP** *$\mathcal{L}y(t) = 0$, $y(a) = 0 = y(b+2)$ has only the trivial solution then the Green's matrix function $G(t, s)$ for this* **BVP** *exists. It is the unique matrix function satisfying properties* **(a)**–**(c)** *in Lemma 7.5 and is given by the formula in Lemma 7.6. If α and β are given $n \times 1$ constant vectors and $h(t)$ is a given $n \times 1$ vector function on $[a+1, b+1]$, then the unique solution of the nonhomogeneous conjugate* **BVP** *on $[a+1, b+1]$*

$$\begin{aligned} \mathcal{L}y(t) &= h(t), \quad t \in [a+1, b+1], \\ y(a) &= \alpha, \qquad y(b+2) = \beta \end{aligned}$$

is given by

$$y(t) = u(t) + \sum_{s=a+1}^{b+1} G(t, s)h(s)$$

where $u(t)$ is the unique solutions of the conjugate **BVP**

$$\begin{aligned} \mathcal{L}u(t) &= 0, \quad t \in [a+1, b+1], \\ u(a) &= \alpha, \quad u(b+2) = \beta. \end{aligned}$$

Exercise 7.9 *Prove Theorem 7.8.*

Corollary 7.10 *If $\sum_{t=a+1}^{b+2} P^{-1}(t)$ is nonsingular, then the Green's matrix function $G(t,s)$ for the conjugate* **BVP**

$$\Delta[P(t)\Delta y(t-1)] = 0 \tag{7.12}$$
$$y(a) = 0 = y(b+2) \tag{7.13}$$

exists and is given by

$$G(t,s) = \begin{cases} -\sum_{\tau=a+1}^{t} P^{-1}(\tau) A \sum_{\tau=s+1}^{b+2} P^{-1}(\tau), & t \leq s, \\ -\sum_{\tau=t+1}^{b=2} P^{-1}(\tau) A \sum_{\tau=a+1}^{s} P^{-1}(\tau), & s < t, \end{cases}$$

where

$$A = \left\{ \sum_{t=a+1}^{b+2} P^{-1}(t) \right\}^{-1}. \tag{7.14}$$

Proof: By Exercise 7.1 the Cauchy matrix function for (7.12) is given by

$$Y(t,s) = \sum_{\tau=s+1}^{t} P^{-1}(\tau).$$

Hence

$$Y(b+2,a) = \sum_{\tau=a+1}^{b+2} P^{-1}(\tau)$$

is nonsingular by assumption. Consequently, by Exercise 7.7 the **BVP** (7.12), (7.13) has only the trivial solution. Then by Theorem 7.8 the Green's matrix function for (7.12), (7.13) exists and is given by the formula in Lemma 7.6. In particular for $t \leq s$,

$$\begin{aligned} G(t,s) &= -Y(t,a)Y^{-1}(b+2,a)Y(b+2,s) \\ &= -\sum_{\tau=a+1}^{t} P^{-1}(\tau) A \sum_{\tau=s+1}^{b+2} P^{-1}(\tau), \end{aligned}$$

which is the desired result. By the formula for $G(t,s)$ in Lemma 7.6 for $t > s$

$$\begin{aligned} G(t,s) &= Y(t,s) - Y(t,a)Y^{-1}(b+2,a)Y(b+2,s) \\ &= \sum_{\tau=s+1}^{t} P^{-1}(\tau) - \sum_{\tau=a+1}^{t} P^{-1}(\tau) A \sum_{\tau=s+1}^{b+2} P^{-1}(\tau) \end{aligned}$$

$$= \left[\sum_{\tau=s+1}^{b+2} P^{-1}(\tau) - \sum_{\tau=t+1}^{b+2} P^{-1}(\tau)\right] A \left[\sum_{\tau=a+1}^{s} P^{-1}(\tau) + \sum_{\tau=s+1}^{b+2} P^{-1}(\tau)\right]$$
$$- \sum_{\tau=a+1}^{t} P^{-1}(\tau) A \sum_{\tau=s+1}^{b+2} P^{-1}(\tau)$$
$$= \sum_{\tau=s+1}^{b+2} P^{-1}(\tau) A \sum_{\tau=a+1}^{b+2} P^{-1}(\tau) - \sum_{\tau=t+1}^{b+2} P^{-1}(\tau) A \sum_{\tau=a+1}^{s} P^{-1}(\tau)$$
$$- \sum_{\tau=t+1}^{b+2} P^{-1}(\tau) A \sum_{\tau=s+1}^{b+2} P^{-1}(\tau) - \sum_{\tau=a+1}^{t} P^{-1}(\tau) A \sum_{\tau=s+1}^{b+2} P^{-1}(\tau)$$
$$= - \sum_{\tau=t+1}^{b+2} P^{-1}(\tau) A \sum_{\tau=a+1}^{s} P^{-1}(\tau)$$

which is the desired result. □

Exercise 7.11 *Show that for $s \in [a+1, b+1]$*

$$\sum_{\tau=a+1}^{s} P^{-1}(\tau) A \sum_{\tau=s+1}^{b+2} P^{-1}(\tau) = \sum_{\tau=s+1}^{b+2} P^{-1}(\tau) A \sum_{\tau=a+1}^{s} P^{-1}(\tau)$$

where A is given by (7.14).

7.4 GREEN'S MATRIX FUNCTION FOR THE RIGHT FOCAL BVP

The primary objective of this section is to construct the Green's matrix function for the *right focal* **BVP**

$$\mathcal{L}y(t) = 0, \quad t \in [a+1, b+1], \tag{7.15}$$
$$y(a) = 0 = \Delta y(b+1). \tag{7.16}$$

Lemma 7.12 *Assume $H(t,s)$ is an $n \times n$ matrix function satisfying*

(a') *$H(t,s)$ is defined for $a \le t \le b+2$, $a+1 \le s \le b+1$.*

(b') *$\mathcal{L}H(t,s) = \delta_{ts}I$ for $a+1 \le t, s \le b+1$, where δ_{ts} is the Kronecker delta function and I is the identity matrix.*

(c') *$H(a,s) = \Delta H(b+1,s) = 0$ for $a+1 \le s \le b+1$.*

If $h(t)$ is a given $n \times 1$ vector function defined on $[a+1, b+1]$ then

$$y(t) = \sum_{s=a+1}^{b+1} H(t,s)h(s)$$

is a solution of the right focal **BVP**

$$\begin{aligned} \mathcal{L}y(t) &= h(t), \quad t \in [a+1, b+1] \\ y(a) &= 0 = \Delta y(b+1). \end{aligned}$$

Proof: Define

$$y(t) = \sum_{s=a+1}^{b+1} H(t,s)h(s)$$

where $H(t,s)$ and $h(t)$ are as in the statement of this lemma. By **(c')**

$$y(a) = \sum_{s=a+1}^{b+1} H(a,s)h(s) = 0$$

and

$$\Delta y(b+1) = \sum_{s=a+1}^{b+1} \Delta H(b+1,s)h(s) = 0.$$

Finally using **(b')** we get that

$$\begin{aligned} \mathcal{L}y(t) &= \sum_{s=a+1}^{b+1} \mathcal{L}H(t,s)h(s) \\ &= \sum_{s=a+1}^{b+1} \delta_{ts}h(s) = h(t) \end{aligned}$$

for $t \in [a+1, b+1]$. □

Theorem 7.13 *Let $Y(t,s)$ be the Cauchy matrix function (defined on page 296) for $\mathcal{L}y(t) = 0$. Then the* **BVP** *(7.15), (7.16) has only the trivial solution if and only if*

$$\Delta Y(b+1, a)$$

is nonsingular.

Proof: Let $Z(t,a)$ be the solution of the matrix **IVP**

$$\mathcal{L}Z(t) = 0$$
$$Z(a) = I, \quad Z(a+1) = 0.$$

Then a general solution of $\mathcal{L}y(t) = 0$ is

$$y(t) = Y(t,a)u + Z(t,a)v$$

where u and v are arbitrary $n \times 1$ constant vectors. Note that $y(a) = 0$ if and only if $v = 0$. Hence $y(t)$ satisfies the boundary conditions (7.16) if and only if

$$\Delta Y(b+1,a)u = 0.$$

If follows that the **BVP** (7.15), (7.16) has only the trivial solution iff $\Delta Y(b+1,a)$ is nonsingular. □

Lemma 7.14 *Let $Y(t,s)$ be the Cauchy matrix function for $\mathcal{L}y(t) = 0$. If $\Delta Y(b+1,a)$ is nonsingular and $H(t,s)$ is defined on $[a,b+2] \times [a+1,b+1]$ by*

$$H(t,s) = \begin{cases} -Y(t,a)[\Delta Y(b+1,a)]^{-1}\Delta Y(b+1,s), & t \le s, \\ Y(t,s) - Y(t,a)[\Delta Y(b+1,a)]^{-1}\Delta Y(b+1,s), & s < t, \end{cases}$$

then $H(t,s)$ is the unique matrix function satisfying properties **(a')**–**(c')** *in Lemma* 7.12.

Proof: Obviously $H(t,s)$ satisfies property **(a')**. Also

$$H(a,s) = -Y(a,a)[\Delta Y(b+1,a)]^{-1}\Delta Y(b+1,s) = 0$$

and

$$\Delta H(b+1,s) = \Delta Y(b+1,s) - \Delta Y(b+1,a)[\Delta Y(b+1,a)]^{-1}\Delta Y(b+1,s) = 0$$

imply that **(c')** is satisfied. Next we show that $H(t,s)$ satisfies **(b')**. First assume $a+1 \le t < s \le b+1$. Then

$$\mathcal{L}H(t,s) = -\mathcal{L}Y(t,a)[\Delta Y(b+1,a)]^{-1}\Delta Y(b+1,s) = 0 = \delta_{ts}I.$$

Next assume $a+1 \le s < t \le b+1$. Then

$$\mathcal{L}H(t,s) = \mathcal{L}Y(t,s) - \mathcal{L}Y(t,a)[\Delta Y(b+1,a)]^{-1}\Delta Y(b+1,s) = 0 = \delta_{ts}I.$$

Finally assume $a+1 \leq s = t \leq b+1$. Then by (7.1)

$$\begin{aligned}
\mathcal{L}H(t,s)\big|_{s=t} &= P(t+1)H(t+1,t) + B(t)H(t,t) + P(t)H(t+1,t) \\
&= P(t+1)Y(t+1,t) - \mathcal{L}Y(t,a)[\Delta Y(b+1,a)]^{-1}\Delta Y((t+1,s) \\
&= I = \delta_{ts}I
\end{aligned}$$

and $H(t,s)$ satisfies **(b')**. To establish uniqueness assume $K(t,s)$ is an $n \times n$ matrix function that satisfies properties **(a')**–**(c')**. Fix $s \in [a+1, b+1]$ and set

$$Y(t) = H(t,s) - K(t,s).$$

Then

$$\begin{aligned}
\mathcal{L}Y(t) &= \mathcal{L}H(t,s) - \mathcal{L}K(t,s) \\
&= \delta_{ts}I - \delta_{ts}I = 0
\end{aligned}$$

for $t \in [a+1, b+1]$. By **(c')**

$$Y(a) = H(a,s) - K(a,s) = 0$$

and

$$\Delta Y(b+1) = \Delta H(b+1,s) - \Delta K(b+1,s) = 0.$$

Thus the columns of $Y(t)$ are solutions of the **BVP** (7.15), (7.16) on page 303. Since $\Delta Y(b+1,a)$ is nonsingular we have by Theorem 7.13 that the columns of $Y(t)$ are the zero vector function on $[a, b+2]$. Hence

$$Y(t) = H(t,s) - K(t,s) = 0$$

for $t \in [a, b+2]$. Since $s \in [a+1, b+1]$ is arbitrary

$$H(t,s) = K(t,s)$$

on $[a, b+2] \times [a+1, b+1]$ and the uniqueness is proved. □

If the **BVP** (7.15), (7.16) has only the trivial solution then the unique function $H(t,s)$ satisfying properties **(a')**–**(c')** in Lemma 7.12 is called the *Green's matrix function* for the *right focal* **BVP** (7.15), (7.16). We have proved most of the statements in the following theorem.

Theorem 7.15 (Green's matrix for right focal BVP) *If $\Delta Y(b+1,a)$ is invertible then the Green's matrix function $H(t,s)$ for the right focal* **BVP** *(7.15), (7.16) exists. It is the unique matrix function satisfying properties* **(a')**–**(c')** *of Lemma 7.12 and is given by the formula in Lemma 7.14. If α and β*

are given $n \times 1$ constant vectors and $h(t)$ is a given $n \times 1$ vector function on $[a+1, b+1]$, then the right focal **BVP**

$$\begin{aligned} \mathcal{L}y(t) &= h(t), \quad t \in [a+1, b+1] \\ y(a) &= \alpha, \quad \Delta y(b+1) = \beta \end{aligned}$$

has a unique solution $y(t)$ given by

$$y(t) = u(t) + \sum_{s=a+1}^{b+1} H(t,s)h(s),$$

where $u(t)$ is the unique solution of the right focal **BVP**

$$\begin{aligned} Lu(t) &= 0 \\ u(a) &= \alpha, \quad \Delta u(b+1) = \beta. \end{aligned}$$

Exercise 7.16 *Prove Theorem* 7.15.

Corollary 7.17 *The Green's matrix function $H(t,s)$ for the right focal* **BVP**

$$\Delta[P(t)\Delta y(t-1)] = 0, \tag{7.17}$$

$$y(a) = 0 = \Delta y(b+1) \tag{7.18}$$

exists and is given by

$$H(t,s) = \begin{cases} -\sum_{\tau=a+1}^{t} P^{-1}(\tau), & t \le s, \\ -\sum_{\tau=a+1}^{s} P^{-1}(\tau), & s < t, \end{cases}$$

where $a \le t \le b+2$, $a+1 \le s \le b+1$.

Proof: By Exercise 7.1

$$Y(t,s) = \sum_{\tau=s+1}^{t} P^{-1}(\tau)$$

is the Cauchy matrix function for (7.17). Since

$$\Delta Y(b+1, a) = P^{-1}(b+2)$$

is nonsingular the **BVP** (7.17), (7.18) has a Green's matrix function $H(t,s)$ which is given by the formula in Lemma 7.14. In particular for $t \leq s$,

$$\begin{aligned} H(t,s) &= -Y(t,a)[\Delta Y(b+1,a)]^{-1}\Delta Y(b+1,s) \\ &= -\sum_{\tau=a+1}^{t} P^{-1}(\tau)P(b+2)P^{-1}(b+2) \\ &= -\sum_{\tau=a+1}^{t} P^{-1}(\tau). \end{aligned}$$

For $s < t$,

$$\begin{aligned} H(t,s) &= Y(t,s) - Y(t,a)[\Delta Y(b+1,a)]^{-1}\Delta Y(b+1,s) \\ &= \sum_{\tau=s+1}^{t} P^{-1}(\tau) - \sum_{\tau=a+1}^{t} P^{-1}(\tau) \\ &= -\sum_{\tau=a+1}^{s} P^{-1}(\tau), \end{aligned}$$

which is the desired result. □

Exercise 7.18 *Find the Green's matrix function for the vector* **BVP**

$$\Delta^2 y(t-1) = 0$$
$$y(a) = 0 = \Delta y(b+1).$$

Use your answer and Theorem 7.15 *to solve the scalar* **BVP**

$$\Delta^2 y(t-1) = t, \quad y(0) = 0, \quad \Delta y(b+1) = 1.$$

7.5 A GREEN'S MATRIX FUNCTION FOR A GENERAL TWO-POINT BVP

In this section we will be concerned with what we will call the Green's matrix function for the two point **BVP**

$$\mathcal{L}y(t) = 0 \tag{7.19}$$

$$\alpha y(a) - \beta \Delta y(a) = 0 \tag{7.20}$$

$$\gamma y(b+1) + \delta \Delta y(b+1) = 0. \qquad (7.21)$$

We assume $\alpha \neq -\beta$ so that the boundary condition (7.20) involves $y(a)$ and we assume $\delta \neq 0$ so that the boundary conditon (7.21) involves $y(b+2)$. Most of the results in this section are due to Ferhan Merdivenci Atici [100]. First we state the following lemma.

Lemma 7.19 *Assume $K(t,s)$ is an $n \times n$ matrix function which satisfies the following properties*

(a") *$K(t,s)$ is defined for $a \leq t \leq b+2$, $a+1 \leq s \leq b+1$.*

(b") *$\mathcal{L}K(t,s) = \delta_{ts} I$ for $a+1 \leq t, s \leq b+1$.*

(c") *$\alpha K(a,s) - \beta \Delta K(a,s) = 0, \qquad \gamma K(b+1,s) + \delta \Delta K(b+1,s) = 0$ for $a+1 \leq s \leq b+1$.*

If $h(t)$ is a given $n \times 1$ vector function on $[a+1, b+1]$, then

$$y(t) = \sum_{s=a+1}^{b+1} K(t,s)h(s)$$

is a solution of the BVP

$$\mathcal{L}y(t) = h(t)$$

$$\alpha y(a) - \beta \Delta y(a) = 0$$

$$\gamma y(b+1) + \delta \Delta y(b+1) = 0.$$

Exercise 7.20 *Prove Theorem 7.19.*

Exercise 7.21 *Prove that if the* **BVP** (7.19), (7.20), (7.21) *has only the trivial solution, then there is at most one matrix function $K(t,s)$ satisfying properties* **(a")–(c")** *in Lemma 7.19.*

If the **BVP** (7.19), (7.20), (7.21) has only the trivial solution and $K(t,s)$ is a function satisfying the properties **(a")–(c")**, then $K(t,s)$ is called the Green's matrix function for the **BVP** (7.19), (7.20), (7.21).

Theorem 7.22 *The matrix*

$$D \equiv \beta\gamma P^{-1}(a+1) + \alpha\delta P^{-1}(b+2) + \alpha\gamma \sum_{\tau=a+1}^{b+1} P^{-1}(\tau)$$

is nonsingular iff the **BVP** (7.17), (7.20), (7.21) *has only the trivial solution.*

Proof: Assume $y(t)$ is a solution of (7.17) on $[a, b+2]$. Then

$$\Delta[P(t)\Delta y(t-1)] = 0$$

for $t \in [a+1, b+1]$. Summing from $a+1$ to $t-1$ we obtain

$$P(t)\Delta y(t-1) = u$$

where u is a constant vector. It follows that

$$\Delta y(t-1) = P^{-1}(t)u$$

for $a+1 \leq t \leq b+2$. Summing from $a+1$ to t yields

$$y(t) = \sum_{\tau=a+1}^{t} P^{-1}(\tau)u + v$$

where v is a constant vector. The boundary conditions (7.20), (7.21) lead to the equations

$$\alpha v - \beta P^{-1}(a+1)u = 0, \tag{7.22}$$

$$\gamma v + [\delta p^{-1}(b+2) + \gamma \sum_{\gamma=a+1}^{b+1} P^{-1}(\tau)]u = 0. \tag{7.23}$$

We consider two cases.

Case 1. $\alpha \neq 0$. In this case we get from (7.22)

$$v = \frac{\beta}{\alpha} P^{-1}(a+1)u. \tag{7.24}$$

Using (7.24) and (7.23) we obtain

$$\left[\frac{\beta}{\alpha} P^{-1}(a+1) + \delta P^{-1}(b+2) + \gamma \sum_{\tau=a+1}^{b+1} P^{-1}(\tau)\right] u = 0.$$

Multiplying by α and using the definition of D we have that

$$Du = 0. \tag{7.25}$$

It follows from (7.25), (7.24) that D is nonsingular iff $u = 0 = v$ iff (7.17), (7.20), (7.21) has only the trivial solution.

Case 2. $\alpha = 0$. In this case we have $\beta u = 0$ from (7.22). Since $\alpha \neq -\beta$, β is nonzero and hence $u = 0$. But then from (7.23)

$$\gamma v = 0.$$

If $\gamma \neq 0$, then we get that $u = v = 0$ and hence (7.17), (7.20), (7.21) has only the trivial solution. But in this case

$$D = \beta\gamma P^{-1}(a+1)$$

is nonsingular. Finally if $\gamma = 0$, then $D = 0$ is singular. But the **BVP** (7.17), (7.20), (7.21) becomes $\Delta[P(t)\Delta y(t-1)] = 0$, $\Delta y(a) = 0$, $\Delta y(b+1) = 0$ which has the nontrivial solution $y(t) \equiv 1$. □

Theorem 7.23 *If the matrix D in Theorem 7.22 is nonsingular, then the Green's matrix function $K(t,s)$ for the* **BVP** *(7.19), (7.20), (7.21) where $\alpha \neq -\beta$, $\gamma \neq 0$ is given by*

$$K(t,s) = -U(t)D^{-1}V(s), \qquad \textit{for} \quad t \leq s$$

and for $s < t$,

$$K(t,s) = -\left[P^{-1}(a+1)D^{-1}V(a)P(a+1)D - \gamma \sum_{\tau=a+1}^{t} P^{-1}(\tau)\right] D^{-1}U(s),$$

where

$$U(t) = \alpha \sum_{\tau=a+1}^{t} P^{-1}(\tau) + \beta P^{-1}(a+1)$$

and

$$V(t) = \gamma \sum_{\tau=t+1}^{b+1} P^{-1}(\tau) + \delta P^{-1}(b+2).$$

Proof: We want to find matrix functions $M(s)$, $N(s)$, $R(s)$ and $Q(s)$ so that

$$K(t,s) = \begin{cases} M(s) + \sum_{\tau=a+1}^{t} P^{-1}(\tau)N(s), & t \leq s, \\ R(s) + \sum_{\tau=a+1}^{t} P^{-1}(\tau)Q(s), & s < t, \end{cases} \tag{7.26}$$

where $a \le t \le b+2$, $a+1 \le s \le b+1$, satisfies properties **(a")–(c")** in Lemma 7.19. Note for each fixed $s \in [a+1, b+1]$, $K(t,s)$ is a solution of the matrix analogue of (7.17) on $[a,s]$ and $[s,b+2]$. From property **(b")** we want

$$\left[R(s) + \sum_{\tau=a+1}^{t} P^{-1}(\tau)Q(s)\right] - \left[M(s) + \sum_{\tau=a+1}^{t} P^{-1}(\tau)N(s)\right] = Y(t,s) \quad (7.27)$$

where $Y(t,s) = \sum_{\tau=s+1}^{t} P^{-1}(\tau)$ is the Cauchy matrix function for (7.17). Letting $t = s$ in (7.27) we obtain

$$[R(s) - M(s)] + \sum_{\tau=a+1}^{s} P^{-1}(\tau)\,[Q(s) - N(s)] = 0. \quad (7.28)$$

Subtracting both sides of (7.28) from both sides of (7.27) respectively we get

$$\sum_{\tau=s+1}^{t} P^{-1}(\tau)\,[Q(s) - N(s)] = \sum_{\tau=s+1}^{t} P^{-1}(\tau).$$

Let $t = s+1$ in this last equation to obtain

$$Q(s) - N(s) = I. \quad (7.29)$$

Equations (7.28) and (7.29) give

$$R(s) - M(s) = -\sum_{\tau=a+1}^{s} P^{-1}(\tau). \quad (7.30)$$

We want $K(t,s)$ to satisfy **(c")** (the boundary conditions). This leads to the equations

$$\alpha M(s) - \beta P^{-1}(a+1)N(s) = 0 \quad (7.31)$$

and

$$\gamma\left[R(s) + \sum_{\tau=a+1}^{b+1} P^{-1}(\tau)Q(s)\right] + \delta P^{-1}(b+2)Q(s) = 0. \quad (7.32)$$

We now solve (7.29)–(7.32) for $Q(s)$, $N(s)$, $M(s)$ and $R(s)$, respectively. From (7.30) and (7.31)

$$\alpha R(s) = \beta P^{-1}(a+1)N(s) - \alpha \sum_{\tau=a+1}^{s} P^{-1}(\tau). \quad (7.33)$$

Multiplying both sides of (7.32) by α we get

$$\alpha PR(s) + \alpha\gamma \sum_{\tau=a+1}^{b+1} P^{-1}(\tau)Q(s) + \alpha\delta P^{-1}(b+2)Q(s) = 0.$$

Using (7.33) and this last equation yields

$$\gamma\beta P^{-1}(a+1)N(s) - \alpha\gamma \sum_{\tau=a+1}^{s} P^{-1}(\tau)$$
$$+\alpha\gamma \sum_{\tau=a+1}^{b+1} P^{-1}(\tau)Q(s) + \alpha\delta P^{-1}(b+2)Q(s) = 0.$$

From (7.29) and this last equation

$$\left[\gamma\beta P^{-1}(a+1) + \alpha\gamma \sum_{\tau=a+1}^{b+1} P^{-1}(\tau) + \alpha\delta P^{-1}(b+2)\right] N(s)$$
$$= \alpha\gamma \sum_{\tau=a+1}^{s} P^{-1}(\tau) - \alpha\gamma \sum_{\tau=a+1}^{b+1} P^{-1}(\tau) - \alpha\delta P^{-1}(b+2).$$

Hence

$$\begin{aligned} DN(s) &= -\alpha\gamma \sum_{\tau=s+1}^{b+1} P^{-1}(\tau) - \alpha\delta P^{-1}(b+2) \\ &= -\alpha V(s). \end{aligned}$$

Therefore

$$N(s) = -\alpha D^{-1}V(s) \tag{7.34}$$

which is the desired expression for $N(s)$. Next we find $Q(s)$. From (7.29) and (7.34) we get

$$\begin{aligned} Q(s) &= I - \alpha D^{-1}V(s) \\ &= D^{-1}D - D^{-1}\left[\alpha\gamma \sum_{\tau=s+1}^{b+1} P^{-1}(\tau) + \alpha\delta P^{-1}(b+2)\right] \\ &= D^{-1}\left[D - \alpha\gamma \sum_{\tau=s+1}^{b+1} P^{-1}(\tau) - \alpha\delta P^{-1}(b+2)\right]. \end{aligned}$$

Hence by the definition of D

$$Q(s) = D^{-1}[\alpha\gamma \sum_{\tau=a+1}^{s} P^{-1}(\tau) + \beta\gamma P^{-1}(a+1)].$$

From the definition of $U(t)$

$$Q(s) = \gamma D^{-1} U(s) \tag{7.35}$$

which is the desired expression for $Q(s)$. It remains to solve for $M(s)$ and $R(s)$. From (7.31)

$$\alpha M(s) = \beta P^{-1}(a+1) N(s)$$

and thus by (7.34)

$$\alpha M(s) = -\alpha\beta P^{-1}(a+1) D^{-1} V(s).$$

If $\alpha \neq 0$ we get the desired expression for $M(s)$

$$M(s) = -\beta P^{-1}(a+1) D^{-1} V(s). \tag{7.36}$$

We now show that even when $\alpha = 0$, (7.36) gives the correct expression for $M(s)$. Assume $\alpha = 0$. Then by (7.34)

$$N(s) = 0$$

and therefore by (7.29)

$$Q(s) = I.$$

Letting $Q(s) = I$ in (7.32) we obtain

$$\gamma R(s) = -\gamma \sum_{\tau=a+1}^{b+1} P^{-1}(\tau) - \delta P^{-1}(b+2) = -V(a). \tag{7.37}$$

Since $\alpha = 0$, D has the value $D = \beta\gamma P^{-1}(a+1)$. But D nonsingular implies $\gamma \neq 0$ and (7.37) gives

$$R(s) = -\frac{1}{\gamma} V(a). \tag{7.38}$$

From (7.30) and (7.38)

$$\begin{aligned} M(s) &= R(s) + \sum_{\tau=a+1}^{s} P^{-1}(\tau) \\ &= -\frac{1}{\gamma}\left[V(a) - \gamma \sum_{\tau=a+1}^{s} P^{-1}(\tau)\right] \\ &= -\frac{1}{\gamma}\left[\gamma \sum_{\tau=s+1}^{b+1} P^{-1}(\tau) + \delta P^{-1}(b+2)\right]. \end{aligned}$$

Hence by the definition of $V(t)$

$$M(s) = -\frac{1}{\gamma}V(s).$$

But if we let $\alpha = 0$ in (7.36), then this last expression for $M(s)$ is exactly what we obtain. Hence (7.36) holds in all cases ($\alpha = 0$ or $\alpha \neq 0$). Finally we find $R(s)$. From (7.30)

$$R(s) = M(s) - \sum_{\tau=a+1}^{s} P^{-1}(\tau)$$

and by (7.36)

$$R(s) = -\beta P^{-1}(a+1)D^{-1}V(s) - \sum_{\tau=a+1}^{s} P^{-1}(\tau).$$

From the definition of $V(t)$, we have

$$\begin{aligned} R(s) = & -\beta P^{-1}(a+1)D^{-1}[\gamma \sum_{\tau=s+1}^{b+1} P^{-1}(\tau) + \delta P^{-1}(b+2)] \\ & -P^{-1}(a+1)D^{-1}DP(a+1) \sum_{\tau=a+1}^{s} P^{-1}(\tau). \end{aligned}$$

Use the definition of D for the result

$$\begin{aligned} R(s) &= -P^{-1}(a+1)D^{-1}\{[\beta\gamma \sum_{\tau=s+1}^{b+1} P^{-1}(\tau) + \beta\gamma P^{-1}(b+2)] \\ &\quad + [\alpha\gamma \sum_{\tau=a+1}^{b+1} P^{-1}(\tau) + \alpha\delta P^{-1}(b+2) \\ &\quad + \beta\gamma P^{-1}(a+1)]P(a+1) \sum_{\tau=a+1}^{s} P^{-1}(\tau)\} \\ &= -P^{-1}(a+1)D^{-1}\Big\{\beta\Big[\gamma \sum_{\tau=a+1}^{b+1} P^{-1}(\tau) + \delta P^{-1}(b+2)\Big] \\ &\quad + \alpha\Big[\gamma \sum_{\tau=a+1}^{b+1} P^{-1}(\tau) + \delta P^{-1}(b+2)\Big] P(a+1) \sum_{\tau=a+1}^{s} P^{-1}(\tau)\Big\} \\ &= -P^{-1}(a+1)D^{-1}\Big\{\beta V(a) + \alpha V(a)P(a+1) \sum_{\tau=a+1}^{s} P^{-1}(\tau)\Big\} \end{aligned}$$

by the definition of $V(t)$. But then by the definition of $V(t)$

$$R(s) = -P^{-1}(a+1)D^{-1}V(a)P(a+1)U(s) \tag{7.39}$$

which is the desired formula for $R(s)$. We now show that the formulas (7.34), (7.35), (7.36), (7.39) for $N(s)$, $Q(s)$, $M(s)$, and $R(s)$ when used in (7.26) on page 312 give the desired formula for $K(t,s)$. First by (7.26) for $t \leq s$

$$\begin{aligned} K(t,s) &= M(s) + \sum_{\tau=a+1}^{t} P^{-1}(\tau)N(s) \\ &= -\beta P^{-1}(a+1)D^{-1}V(s) - \alpha \sum_{\tau=a+1}^{t} P^{-1}(\tau)D^{-1}V(s) \\ &= -[\beta P^{-1}(a+1) + \alpha \sum_{\tau=a+1}^{t} P^{-1}(\tau)]D^{-1}V(s). \end{aligned}$$

Hence by the definition of $U(t)$

$$K(t,s) = -U(t)D^{-1}V(s)$$

which is the desired expression for $K(t,s)$ when $t \leq s$. Finally use (7.26) for $s < t$, namely,

$$K(t,s) = R(s) + \sum_{\tau=a+1}^{t} P^{-1}(\tau)Q(s).$$

Using (7.39) and (7.35)

$$\begin{aligned} K(t,s) &= -\Big[P^{-1}(a+1)D^{-1}V(a)P(a+1) - \gamma \sum_{\tau=a+1}^{t} P^{-1}(\tau)D^{-1}\Big]U(s) \\ &= -\Big[P^{-1}(a+1)D^{-1}V(a)P(a+1)D - \gamma \sum_{\tau=a+1}^{t} P^{-1}(\tau)\Big]D^{-1}U(s) \end{aligned}$$

which is the desired result for $t > s$. □

Exercise 7.24 *Show that Corollaries* 7.10 *and* 7.17 *follow from Theorem* 7.23.

Corollary 7.25 *If D is nonsingular and $P(t)$ is a diagonal matrix function, then the Green's matrix function for the* **BVP** (7.19)–(7.21), *where $\alpha \neq -\beta$ and $\delta \neq 0$, exists and is given by*

$$K(t,s) = \begin{cases} -U(t)D^{-1}V(s), & t \leq s \\ -V(t)D^{-1}U(s), & s < t. \end{cases}$$

Exercise 7.26 *Prove Corollary* 7.25.

7.6 NOTES

Recessive and dominant solutions of nonhomogeneous difference equations have been studied by Olver [108], Patula [111], and by Ahlbrandt and Patula [19]. Since those papers are reasonably accessible, we will not include that work here.

8

DISCONJUGACY CRITERIA

8.1 INTRODUCTION

In this chapter we will study second order difference equations of the form

$$\mathcal{L}y(t) = \Delta\left[P(t)\Delta y(t-1)\right] + Q(t)y(t) = 0, \tag{8.1}$$

where $P(t)$ and $Q(t)$ are given $n \times n$ Hermitian matrix functions on the discrete intervals $[a+1, b+2]$ and $[a+1, b+1]$, respectively. As in Chapter 7, we also assume $P(t)$ is nonsingular for $t \in [a+1, b+2]$.

8.2 A SUFFICIENT CONDITION FOR DISCONJUGACY

In this section we will prove a sufficient condition for $\mathcal{L}y(t) = 0$ to be disconjugate on $[a, b+2]$. The proof of the corresponding result for differential equations is due to W.T. Reid [133]. The proof of the discrete analogue given here is due to A. Peterson and J. Ridenhour [123]. First we establish the matrix analogue of Lemma 1.27 on page 21. [See Lemma 11, [44]].

Lemma 8.1 *If M and N are $n \times n$ matrices such that $M > 0$ and $N > 0$ (i.e. M and N are positive definite) then*

$$M^{-1} + N^{-1} \geq 4(M+N)^{-1}.$$

Proof. Assume $C > 0$ and I is the identity matrix. Then C is Hermitian and

$$(C - I)^2 \geq 0.$$

Thus

$$C^2 - 2C + I \geq 0$$

and by adding $4C$ to both sides, we have

$$C^2 + 2C + I \geq 4C,$$

i.e.,

$$(C + I)^2 \geq 4C. \tag{8.2}$$

Since C is Hermitian, there exist a unitary matrix U and a diagonal matrix D such that

$$C = U^* DU.$$

Substitution in (8.2) gives

$$(U^* DU + I)^2 \geq 4U^* DU.$$

Hence

$$(U^*[D + I]U)^2 \geq 4U^* DU$$

and it follows that

$$U^*(D + I)^2 U \geq U^*(4D)U$$

and hence

$$(D + I)^2 \geq 4D.$$

Since D is diagonal and positive definite, (since C is Hermitian and positive definite), it follows that $D + I > I$ and $D + I$ is a diagonal matrix with positive diagonal entries and

$$(D + I) \geq 4D(D + I)^{-1}.$$

Therefore

$$U^*(D + I)U \geq 4U^* D(D + I)^{-1} U.$$

It follows that

$$\begin{aligned} C + I &\geq 4CU^*(D + I)^{-1} U \\ &= 4C(C + I)^{-1} \\ &= 4[(C + I)C^{-1}]^{-1} \\ &= 4(I + C^{-1})^{-1}. \end{aligned}$$

Apply this to the choice of

$$C = N^{1/2}M^{-1}N^{1/2}.$$

for the inequalities

$$\begin{aligned} N^{1/2}M^{-1}N^{1/2} + I &\leq 4(I + N^{-1/2}MN^{-1/2})^{-1} \\ &= 4[N^{-1/2}(N+M)N^{-1/2}]^{-1} \\ &= 4N^{1/2}(M+N)^{-1}N^{1/2}. \end{aligned}$$

It follows that

$$M^{-1} + N^{-1} \geq 4(M+N)^{-1}.$$

□

We will generalize Theorem 1.29 to the vector case. First we prove three useful Theorems.

Theorem 8.2 *The quadratic form* $\mathcal{J}$ *associated with equation* (8.1) *can be written in the form*

$$\mathcal{J}[\eta] = \sum_{t=a+1}^{b+2} \Delta\eta^*(t-1)P(t)\Delta\eta(t-1) - \sum_{t=a+1}^{b+1} \eta^*(t)Q(t)\eta(t)$$

for η *in*

$$\mathcal{F} \equiv \{\eta : [a, b+2] \to \mathrm{C}^n \text{ such that } \eta(a) = 0 = \eta(b+2)\}.$$

Proof. In Section 5.1 we defined the quadratic form $\mathcal{J}$ associated with the three term equation (5.1) by

$$\mathcal{J}[\eta] = \sum_{t=a+1}^{b+1} \{\eta^*(t)N(t)\eta(t) - \eta^*(t-1)K(t-1)\eta(t) - \eta^*(t)K^*(t-1)\eta(t-1)\}.$$

As pointed out in Section 7.1 the self-adjoint equation $\mathcal{L}y(t) = 0$ is a special case of $\mathcal{L}y(t) = 0$ where for $E(t) \equiv I$, $K(t)$ and $N(t)$ are given by (7.2) and (7.3), (on page 296), respectively. Use those equations to write the above form as

$$\begin{aligned} \mathcal{J}[\eta] = \sum_{t=a+1}^{b+1} &\{\eta^*(t)[P(t) + P(t+1) - Q(t)]\eta(t) \\ &-\eta^*(t-1)P(t)\eta(t) - \eta^*(t)P(t)\eta(t-1)\}. \end{aligned}$$

But the term involving $P(t+1)$ may be written as

$$\sum_{t=a+1}^{b+1} \eta^*(t)P(t+1)\eta(t) = \sum_{t=a+2}^{b+2} \eta^*(t-1)P(t)\eta(t-1)$$

$$= \eta^*(b+1)P(b+2)\eta(b+1) + \sum_{t=a+1}^{b+1} \eta^*(t-1)P(t)\eta(t-1)$$

since $\eta(a) = 0$. Use $\eta(b+2) = 0$ to conclude that

$$\mathcal{J}[\eta] = \sum_{t=a+1}^{b+2} \Delta\eta^*(t-1)P(t)\Delta\eta(t-1) - \sum_{t=a+1}^{b+1} \eta^*(t)Q(t)\eta(t),$$

which is the desired result. □

Theorem 8.3 *If $P(t)$ is positive definite and $Q(t)$ is negative semidefinite on $[a+1, b+1]]$, then $\mathcal{J}$ is positive definitne on $\mathcal{F}$ and $\mathcal{L}u = 0$ is disconjugate on $[a, b+2]$.*

The strong sign conditions make $\mathcal{J}[\eta] \geq 0$ on $\mathcal{F}$ with equality iff $\eta = 0$. Disconjugacy on $[a, b+2]$ is a consequence of the Reid Roundabout Theorem, Theorem 5.13 on page 208. □

The analogue of Lemma 1.37 on page 29 for the self-adjoint operator $\mathcal{L}$ is as follows:

Theorem 8.4 *If $\eta \in \mathcal{F}$, then*

$$\mathcal{J}[\eta] = -\sum_{t=a+1}^{b+1} \eta^*(t)\mathcal{L}\eta(t).$$

The analogue of Corollary 5.10 for $\mathcal{L}y(t) = 0$ is obtained by replacing $K(t)$ by $P(t+1)$. We will use this notational change in the proof of the next theorem. This result generalizes Theorem 1.29 on page 22 to the vector case. Note that this result allows Q to have some positivity which depends upon the size of P.

Theorem 8.5 *Assume $P(t) > 0$ on $[a+1, b+2]$ and $q(t)$ is real-valued with $q(t) \geq 0$ and $Q(t) \leq q(t)I$ on $[a+1, b+1]$. If*

$$\sum_{t=a+1}^{b+1} q(t)I < 4\left\{\sum_{t=a+1}^{b+2} P^{-1}(t)\right\}^{-1}, \tag{8.3}$$

then $\mathcal{L}y(t) = 0$ is disconjugate on $[a, b+2]$.

Proof: Assume $\mathcal{L}y(t) = 0$ is not disconjugate on $[a, b+2]$. Then there is a nontrivial prepared solution $y(t)$ of $\mathcal{L}y(t) = 0$ such that $y(t)$ has two generalized zeros in $[a, b+2]$. It follows that there are integers c, d with $a+1 \leq c < d \leq b+2$ such that $y(c) \neq 0$, $y(d-1) \neq 0$,

$$y^*(c-1)P(c)y(c) \leq 0, \tag{8.4}$$

and

$$(8.5)$$

$$y^*(d-1)P(d)y(d) \leq 0. \tag{8.6}$$

Define η on $[a, b+2]$ by

$$\eta(t) = \begin{cases} y(t), & c \leq t \leq d-1, \\ 0, & \text{otherwise.} \end{cases}$$

By Corollary 5.11 with $K(t) = P(t+1)$ we have from (8.4), (8.6) that

$$\mathcal{J}[\eta] = y^*(c-1)P(c)y(c) + y^*(d-1)P(d)y(d) \leq 0.$$

In the remainder of the proof we will use the inequality (8.3) to show that $\mathcal{J}\eta > 0$ which will give us the desired contradiction. Pick $t_0 \in [c, d-1]$ such that

$$|y(t_0)|_2 = \max_{t \in [c,d-1]} |y(t)|_2 = \max_{t \in [c,d-1]} |\eta(t)|_2$$

where $|\cdot|_2$ is the Euclidean norm. Since $y(c) \neq 0$, we have $y(t_0) \neq 0$. Define the matrix function $U(t)$ and the vector function $u(t)$ by

$$U(t) = \sum_{\tau=a+1}^{t} P^{-1}(\tau)$$

and

$$u(t) = U(t)U^{-1}(t_0)y(t_0),$$

respectively. It follows that

$$P(t)\Delta u(t-1) = U^{-1}(t_0)y(t_0), \tag{8.7}$$

$$u(a) = 0, \qquad u(t_0) = y(t_0). \tag{8.8}$$

Since $P(t) > 0$ on $[a+1, b+2]$

$$\begin{aligned} 0 &\leq \sum_{t=a+1}^{t_0} [\Delta[\eta^*(t-1) - u^*(t-1)]P(t)\Delta[\eta(t-1) - u(t-1)] \\ &= \sum_{t=a+1}^{t_0} \Delta\eta^*(t-1)P(t)\Delta\eta(t-1) - \sum_{t=a+1}^{t_0} \Delta\eta^*(t-1)P(t)\Delta u(t-1) \\ &\quad - \sum_{t=a+1}^{t_0} \Delta u^*(t-1)P(t)\Delta\eta(t-1) + \sum_{t=a+1}^{t_0} \Delta u^*(t-1)P(t)\Delta u(t-1). \end{aligned}$$

Using (8.7) we get that

$$\begin{aligned} 0 \leq & \sum_{t=a+1}^{t_0} \Delta\eta^*(t-1)P(t)\Delta\eta(t-1) - \sum_{t=a+1}^{t_0} \Delta\eta^*(t-1)U^{-1}(t_0)y(t_0) \\ & - \sum_{t=a+1}^{t_0} y^*(t_0)U^{-1}(t_0)\Delta\eta(t-1) + \sum_{t=a+1}^{t_0} \Delta u^*(t-1)U^{-1}(t_0)y(t_0). \end{aligned}$$

It follows that

$$\begin{aligned} 0 \leq & \sum_{t=a+1}^{t_0} \Delta\eta^*(t-1)P(t)\Delta\eta(t-1) - \eta^*(t_0)U^{-1}(t_0)y(t_0) \\ & - y^*(t_0)U^{-1}(t_0)\eta(t_0) + u^*(t_0)U^{-1}(t_0)y(t_0). \end{aligned}$$

Since $y(t_0) = \eta(t_0) = u(t_0)$ we get that

$$\sum_{t=a+1}^{t_0} \Delta\eta^*(t-1)P(t)\Delta y(t-1) \geq y^*(t_0)U^{-1}(t_0)y(t_0). \tag{8.9}$$

Next define the matrix function $V(t)$ and the vector function $v(t)$ by

$$V(t) = -\sum_{\tau=t+1}^{b+2} P^{-1}(\tau)$$

and

$$v(t) \quad = \quad V(t)V^{-1}(t_0)y(t_0),$$

respectively. It follows that

$$P(t)\Delta v(t-1) = V^{-1}(t_0)y(t_0) \tag{8.10}$$

and

$$v(t_0) = y(t_0) = \eta(t_0), \quad v(b+2) = 0. \tag{8.11}$$

Since $P(t) > 0$ on $[a+1, b+2]$,

$$\begin{aligned}
0 \leq & \sum_{t=t_0+1}^{b+2} \Delta[\eta^*(t-1) - v^*(t-1)]P(t)\Delta[\eta(t-1) - v(t-1)] \\
& = \sum_{t=t_0+1}^{b+2} \Delta\eta^*(t-1)P(t)\Delta\eta(t-1) - \sum_{t=t_0+1}^{b+2} \Delta\eta^*(t-1)P(t)\Delta v(t-1) \\
& \quad - \sum_{t=t_0+1}^{b+2} \Delta v^*(t-1)P(t)\Delta\eta(t-1) + \sum_{t=t_0+1}^{b+2} \Delta v^*(t-1)P(t)\Delta v(t-1).
\end{aligned}$$

Using (8.10) we obtain

$$\begin{aligned}
0 \leq & \sum_{t=t_0+1}^{b+2} \Delta\eta^*(t-1)P(t)\Delta\eta(t-1) - \sum_{t=t_0+1}^{b+2} \Delta\eta^*(t-1)V^{-1}(t_0)y(t_0) \\
& - \sum_{t=t_0+1}^{b+2} y^*(t_0)V^{-1}(t_0)\Delta\eta(t-1) + \sum_{t=t_0+1}^{b+2} \Delta v^*(t-1)V^{-1}(t_0)y(t_0) \\
& = \sum_{t=t_0+1}^{b+2} \Delta y^*(t-1)P(t)\Delta\eta(t-1) + \eta^*(t_0)V^{-1}(t_0)y(t_0) \\
& \quad + y^*(t_0)V^{-1}(t_0)\eta(t_0) - v^*(t_0)V^{-1}(t_0)y(t_0) \\
& = \sum_{t=t_0+1}^{b+2} \Delta\eta^*(t-1)P(t)\Delta\eta(t-1) + y^*(t_0)V^{-1}(t_0)y(t_0).
\end{aligned}$$

Hence

$$\sum_{t=t_0+1}^{b+2} \Delta\eta^*(t-1)P(t)\Delta\eta(t-1) \geq -y^*(t_0)V^{-1}(t_0)y(t_0).$$

Combining this last inequality with (8.9) we have that

$$\sum_{t=a+1}^{b+2} \Delta\eta^*(t-1)P(t)\Delta\eta(t-1) \geq y^*(t_0)[U^{-1}(t_0) - V^{-1}(t_0)]y(t_0).$$

Using Lemma 8.1 we get that

$$\sum_{t=a+1}^{b+2} \Delta\eta^*(t-1)P(t)\Delta\eta(t-1) \geq y^*(t_0)4[U(t_0)-V(t_0)]^{-1}y(t_0).$$

Hence from the definitions of $U(t)$ and $V(t)$

$$\sum_{t=a+1}^{b+2} \Delta\eta^*(t-1)P(t)\Delta\eta(t-1) \geq y^*(t_0)4\left\{\sum_{t=a+1}^{b+2} P^{-1}(t)\right\}^{-1} y(t_0). \quad (8.12)$$

Also,

$$\begin{aligned}\sum_{t=a+1}^{b+1} \eta^*(t)Q(t)\eta(t) &\leq \sum_{t=a+1}^{b+1} \eta^*(t)q(t)I\eta(t)\\ &\leq y^*(t_0)\sum_{t=a+1}^{b+1} q(t)Iy(t_0).\end{aligned}$$

This last inequality together with (8.12) gives us that

$$\mathcal{J}\eta \geq y^*(t_0)\left[4\left\{\sum_{t=1+1}^{b+2} P^{-1}(t)\right\}^{-1} - \sum_{t=a+1}^{b+1} q(t)\right] y(t_0).$$

So by (8.3)

$$\mathcal{J}\eta > 0$$

which is our desired contradiction. □

8.3 A SUFFICIENT CONDITION FOR RIGHT DISFOCALITY

In this section we will prove a sufficient condition for $\mathcal{L}y(t) = 0$ to be right disfocal on $[a, b+2]$. This result was proved by A. Peterson and J. Ridenhour in [123].

We say the $\mathcal{L}y(t) = 0$ is *right disfocal* on $[a, b+2]$ provided there is no nontrivial prepared solution $y(t)$ of $\mathcal{L}y(t) = 0$ and an integer $d \in [a+1, b+2]$ such that $\Delta y(d) = 0$ and y has a generalized zero in $[a, d]$. Our proof of the next theorem uses a quadratic functional J_1 which was studied by Peil and Peterson [113]. For square symmetric matrices A and B we will use the notation $A \leq B$ to denote that $B - A$ is positive semi-definite.

Theorem 8.6 *Assume $P(t)$ is positive definite on $[a+1, b+2]$ and there is an associated function $q(t) \geq 0$ such that*

$$Q(t) \leq q(t)I \tag{8.13}$$

on $[a+1, b+1]$, i.e, $q(t)$ is an upper bound for the eigenvalues of $Q(t)$. If

$$\sum_{t=a+1}^{b+1} q(t)I \leq \left\{ \sum_{t=a+1}^{b+1} P^{-1}(t) \right\}^{-1}, \tag{8.14}$$

then $\mathcal{L}y(t) = 0$ is right disfocal on $[a, b+2]$.

Proof: Assume $\mathcal{L}y(t) = 0$ is not right disfocal on $[a, b+2]$. Then there is a nontrivial prepared solution $y(t)$ and an integer $d \in [a+1, b+1]$ such that $\Delta y(d) = 0$ and $y(t)$ has a generalized zero in $[a, d]$. Let t_0 be the first generalized zero of $y(t)$ in $[a, d]$. If $y(t_0) = 0$, set $c = t_0 + 1$, while if $y(t_0) \neq 0$, set $c = t_0$. It follows that $a + 1 \leq c \leq d$,

$$y^*(c-1)P(c)y(c) \leq 0 \tag{8.15}$$

and $y(c) \neq 0$. Define $\eta(t)$ on $[a, d+1]$by

$$\eta(t) = \begin{cases} 0, & a \leq t \leq c-1 \\ y(t), & c \leq t \leq d+1 \end{cases}$$

and define $\mathcal{J}_1[\eta]$ by

$$\mathcal{J}_1[\eta] = \sum_{t=a+1}^{d} \Delta\eta^*(t-1)P(t)\Delta\eta(t-1) - \sum_{t=a+1}^{d} \eta^*(t)Q(t)\eta(t).$$

From the definition of $y(t)$

$$\begin{aligned} \mathcal{J}_1[\eta] = {} & \Delta\eta^*(c-1)P(c)\Delta\eta(c-1) \\ & + \sum_{t=c+1}^{d} \Delta y^*(t-1)P(t)\Delta y(t-1) - \sum_{t=c}^{d} y^*(t)Q(t)y(t). \end{aligned}$$

Using a summation by parts formula on the second term we obtain

$$\begin{aligned} \mathcal{J}_1[\eta] = {} & y^*(c)P(c)y(c) + \left\{ y^*(t-1)P(t)\Delta y(t-1) \right\}_{c+1}^{d+1} \\ & - \sum_{t=c+1}^{d} \Delta y^*(t)\Delta\left[P(t)\Delta y(t-1)\right] \\ & - \sum_{t=c}^{d} y^*(t)Q(t)y(t). \end{aligned}$$

Since $\Delta y(d) = 0$ and $y(t)$ is a solution of $\mathcal{L}y(t) = 0$,

$$\begin{aligned}\mathcal{J}_1[\eta] &= y^*(c)P(c)y(c) - y^*(c)P(c+1)\Delta y(c) \\ &\quad - y^*(c)Q(c)y(c) \\ &= -y^*(c)\{[P(c+1)\Delta y(c) - P(c)\Delta y(c-1)] + Q(c)y(c)\} \\ &\quad + y^*(c)P(c)y(c-1) \\ &= -y^*(c)\mathcal{L}y(c) + y^*(c)P(c)y(c-1).\end{aligned}$$

Since $y(t)$ is a prepared solution

$$\mathcal{J}_1[\eta] = y^*(c-1)P(c)y(c) \le 0 \tag{8.16}$$

by (8.15). In the rest of the proof we will use (8.13) and (8.14) to show that $\mathcal{J}_1[\eta] > 0$ which would contradict (8.16). Choose $t_1 \in [c,d]$ such that

$$|y(t_1)|_2 = \max_{t\in[c,d]} |y(t)|_2 = \max_{t\in[a,b+2]} |\eta(t)|_2,$$

where $|\ |_2$ denotes the Euclidean norm. Set

$$u(t) = U(t)U^{-1}(t_1)y(t_1)$$

for $a \le t \le t_1$, where

$$U(t) \equiv \sum_{s=a+1}^{t} P^{-1}(s).$$

Note that $u(t)$ satisfies

$$P(t)\Delta u(t-1) = D^{-1}(t_1)y(t_1), \qquad u(a) = 0, \quad u(t_1) = y(t_1). \tag{8.17}$$

Since $P(t) > 0$ on $[a+1, b+2]$ we have

$$\begin{aligned}0 &\le \sum_{t=a+1}^{t_1} \Delta[\eta^*(t-1) - u^*(t-1)]P(t)\Delta[\eta(t-1) - u(t-1)] \\ &= \sum_{t=a+1}^{t_1} \Delta\eta^*(t-1)P(t)\Delta\eta(t-1) - \sum_{t=a+1}^{t_1} \Delta\eta^*(t-1)P(t)\Delta u(t-1) \\ &\quad - \sum_{t=a+1}^{t_1} \Delta u^*(t-1)P(t)\Delta\eta(t-1) + \sum_{t=a+1}^{t_1} \Delta u^*(t-1)P(t)\Delta u(t-1).\end{aligned}$$

Using (8.17)

$$\begin{aligned}0 \le \sum_{t=a+1}^{t_1} \Delta\eta^*(t-1)P(t)\Delta\eta(t-1) - \Big[\sum_{t=a+1}^{t_1} \Delta\eta^*(t-1)\Big]U^{-1}(t_1)y(t_1) \\ -y^*(t_1)U^{-1}(t_1)\sum_{t=a+1}^{t_1} \Delta\eta(t-1) + \Big[\sum_{t=a+1}^{t_1} \Delta u^*(t-1)\Big]U^{-1}(t_1)y(t_1).\end{aligned}$$

It follows that

$$0 \le \sum_{t=a+1}^{t_1} \Delta\eta^*(t-1)P(t)\Delta\eta(t-1) - y^*(t_1)U^{-1}(t_1)y(t_1).$$

Hence,

$$\sum_{t=a+1}^{t_1} \Delta\eta^*(t-1)P(t)\Delta\eta(t-1) \ge y^*(t_1)U^{-1}(t_1)y(t_1). \tag{8.18}$$

Now consider

$$\begin{aligned}\mathcal{J}_1[\eta] &= \sum_{t=a+1}^{d} \Delta\eta^*(t-1)P(t)\Delta\eta(t-1) - \sum_{t=a+1}^{d} \eta^*(t)Q(t)\eta(t)\\ &\ge \sum_{t=a+1}^{t_1} \Delta\eta^*(t-1)P(t)\Delta\eta(t-1) - \sum_{t=a+1}^{d} \eta^*(t)q(t)\eta(t)\end{aligned}$$

using (8.13). Hence by (8.18)

$$\begin{aligned}\mathcal{J}_1[\eta] &\ge y^*(t_1)U^{-1}(t_1)y(t_1) - \sum_{t=a+1}^{d} y^*(t_1)q(t)y(t_1)\\ &= y^*(t_1)\{\Big[\sum_{s=a+1}^{t_1} P^{-1}(s)\Big]^{-1}\} - \sum_{t=a+1}^{d} y^*(t_1)q(t)y(t_1)\\ &\ge y^*(t_1)\{\Big[\sum_{s=a+1}^{b+1} P^{-1}(s)\Big]^{-1} - \sum_{t=a+1}^{b+1} q(t)I\}y(t_1) \;> 0\end{aligned}$$

by (8.14). This is a contradiction. □

Exercise 8.7 *We say that $\mathcal{L}y(t) = 0$ is left disfocal on $[a, b+2]$ provided there is no nontrivial prepared solution $y(t)$ and an integer $c \in [a, b+1]$ such that $\Delta y(c) = 0$ and $y(t)$ has a generalized zero in $[c, b+2]$. State and prove a theorem like Theorem 8.6 on page 327 which contains a sufficient condition for $\mathcal{L}y(t) = 0$ to be left disfocal on $[a, b+2]$.*

8.4 NOTES

The disconjugacy and disfocality sufficiency criteria of this chapter are qualitative in the sense that they are testable pointwise or by summations analogous to integral type tests for differential equations. They do not assume any knowledge about the solution of any equation, but depend wholly and explicitly on the coeffiecients. These criteria are distinct from many results for differential equations of the type that are concerned with eventual disconjugacy. We saw in Chapter 2 that convergence of continued fractions can be assured if there is a recessive solution which is nonzero (or nonsingular in the matrix case). However, for systems equivalent to a symmetric three term recurrence, disconjugacy on a slightly larger interval implies that there is a recessive solution which is nonsingular on the required interval.

The history of focal points and disfocality suffers from a confusion introduced into the differential equation literature which is contrary to the geometrical meaning of a focal point. The variational origins stem from a family of solutions of the Jacobi equation having a common zero which is called the *focal point* of the family (as rays of light focus at a point). Somehow this terminology became reversed in the differential equations literature about 30–40 years ago. The left and right disfocal terminology used here is consistent with recent differential equations literature. However, the correct terminology for focal points is still used in variational theory and differential geometry.

9

DISCRETE LINEAR HAMILTONIAN SYSTEMS

9.1 PRELIMINARIES

This chapter is an introduction to Martin Bohner's approach to the discrete linear Hamiltonian system

$$\begin{aligned} \Delta y(t) &= A(t)y(t+1) + B(t)z(t) \\ \Delta z(t) &= C(t)y(t+1) - A^*(t)z(t). \end{aligned} \tag{9.1}$$

The first equation, which Bohner calls the *equation of motion*, holds for t on $[a+1, b+2]$ and the second vector equation, which Bohner calls the *Euler equation*, holds for t on $[a+1, b+1]$. We assume that $A(t)$ is an $n \times n$ matrix function such that $I - A(t)$ is nonsingular on $[a+1, b+2]$, $B(t)$ is an $n \times n$ Hermitian matrix function on $[a+1, b+2]$, and $C(t)$ is an $n \times n$ Hermitian matrix function on $[a+1, b+1]$. Bohner's methods were developed for his program of solving the problem [6] of a formulation of a Reid Roundabout Theorem for the case of singular B. Bohner's papers [27]–[35] give more results than we present here.

We repeat some of our results of Chapter 3 to minimize looking back and to put those results in Bohner's terminology.

Exercise 9.1 *Show that if $y(t)$, $z(t)$ is a solution of* (9.1), *then $y(t)$ is defined on all of $[a+1, b+3]$ and $z(t)$ is defined on all of $[a+1, b+2]$.*

Exercise 9.2 *Show that for Hermitian $P(t)$ and $Q(t)$ with $P(t)$ nonsingular, the self-adjoint vector difference equation*

$$\Delta[P(t)\Delta u(t-1)] + Q(t)u(t) = 0,$$

for $t \in [a+1, b+1]$, with the change of variables

$$\begin{aligned} y(t) &= u(t-1), \quad t \in [a+1, b+3], \\ z(t) &= P(t)\Delta u(t-1), \quad t \in [a+1, b+2], \end{aligned}$$

can be written as an equivalent Hamiltonian system (9.1) *with*

$$A(t) = 0, \qquad B(t) = P^{-1}(t)$$

for $t \in [a+1, b+2]$ and $C(t) = -Q(t)$ for $t \in [a+1, b+1]$.

The matrix analogue of (9.1) is

$$\begin{aligned} \Delta Y(t) &= A(t)Y(t+1) + B(t)Z(t), \\ \Delta Z(t) &= C(t)Y(t+1) - A^*(t)Z(t), \end{aligned} \tag{9.2}$$

where the first equation holds for t in $[a+1, b+2]$ and the second equation holds for t in $[a+1, b+1]$. In Example 3.10 on page 82 we saw that if $Y(t)$, $Z(t)$ is a solution of (9.2) and if we set

$$X(t) = \begin{bmatrix} Y(t) \\ Z(t) \end{bmatrix}$$

then $X(t)$ is a solution of the symplectic system

$$X(t+1) = M(t)X(t) \tag{9.3}$$

on $[a+1, b+2]$. Here

$$M(t) = \begin{bmatrix} E(t) & F(t) \\ G(t) & H(t) \end{bmatrix},$$

with $E(t)$ nonsingular on $[a+1, b+2]$,

$$E(t) = [I - A(t)]^{-1} \tag{9.4}$$

$$F(t) = E(t)B(t), \tag{9.5}$$

for $t \in [a+1, b+2]$, and

$$G(t) = C(t)E(t) \tag{9.6}$$

$$H(t) = C(t)F(t) + E^{*^{-1}}(t) \tag{9.7}$$

for $t \in [a+1, b+1]$. Of course we could write (9.3) in the form

$$\begin{aligned} Y(t+1) &= E(t)Y(t) + F(t)Z(t) \\ Z(t+1) &= G(t)Y(t) + H(t)Z(t) \end{aligned} \tag{9.8}$$

where we want the first equation to hold for $t \in [a+1, b+2]$ and the second equation holds for $t \in [a+1, b+1]$. Conversely, if $Y(t)$, $Z(t)$ is a solution of (9.8) and if $E(t)$ is nonsingular on $[a+1, b+2]$, then (see Theorem 3.15 on page 91) $Y(t)$, $Z(t)$ is a solution of the linear Hamiltonian system (9.2) with

$$\begin{aligned} A(t) &= I - E^{-1}(t) \\ B(t) &= E^{-1}(t)F(t) \end{aligned}$$

for $t \in [a+1, b+2]$ and

$$C(t) = G(t)E^{-1}(t)$$

for $t \in [a+1, b+1]$. Note that $B(t)$ is Hermitian on $[a+1, b+2]$, $C(t)$ is Hermitian on $[a+1, b+1]$ and $I - A(t)$ is invertible on $[a+1, b+2]$ as desired when we study a linear Hamiltonian system. It is easy to see that any **IVP** consisting of system (9.2) (or of system (9.8)) with initial conditions

$$Y(t_0) = Y_0, \quad Z(t_0) = Z_0,$$

where Y_0, Z_0 are given $n \times m$ constant matrices, has a unique solution $Y(t)$, $Z(t)$ where $Y(t)$ is defined on $[a+1, b+3]$ and $Z(t)$ is defined on $[a+1, b+2]$.

Exercise 9.3 *Show that the* $2n$-th *order scalar difference equation* (3.21) *on page* 85 *can be written as an equivalent Hamiltonian system* (9.1) *where* $y(t)$ *and* $z(t)$ *are given in Example* 3.14 *on page* 85, A *is the superdiagonal* $n \times n$ *matrix with zeros everywhere except for ones on the superdiagonal,* $B(t)$ *is an* $n \times n$ *matrix function with zeros everywhere except in the* n, n *position where that entry is* $1/r_n(t)$, *and* $C(t)$ *is the* $n \times n$ *diagonal matrix with* j-th *diagonal entry of* $(-1)^{n-j+1} r_{j-1}(t)$.

In Theorem 3.19 on page 94 we showed that if $Y_1(t)$, $Z_1(t)$ and $Y_2(t)$, $Z_2(t)$ are $n \times m$ and $n \times p$ matrix solution pairs of (9.2) (or (9.8)), then the *Wronskian* of these two pairs of solutions defined by

$$\left\{ \begin{array}{cc} Y_1(t) & Y_2(t) \\ Z_1(t) & Z_2(t) \end{array} \right\} = Y_1^*(t)Z_2(t) - Z_1^*(t)Y_2(t) \tag{9.9}$$

is a constant $m \times p$ matrix on $[a+1, b+2]$.

Exercise 9.4 *Prove the last statement by taking the difference of* (9.9) *and using* (9.2).

If $Y(t)$, $Z(t)$ is a solution of (9.2) (or (9.8)) such that

$$\left\{ \begin{array}{cc} Y(t) & Y(t) \\ Z(t) & Z(t) \end{array} \right\} = 0$$

for $t \in [a+1, b+2]$, then we say $Y(t)$, $Z(t)$ is a *prepared solution* of (9.2) (of (9.8)). If, in addition,

$$\text{rank} \left[\begin{array}{c} Y(t) \\ Z(t) \end{array} \right] = n,$$

for $t \in [a+1, b+2]$, then we say $Y(t)$, $Z(t)$ is a *prepared basis* of (9.2) (of (9.8)). If $Y_i(t)$, $Z_i(t)$, $i = 1, 2$, are prepared bases of (9.2) such that

$$\left\{ \begin{array}{cc} Y_1(t) & Y_2(t) \\ Z_1(t) & Z_2(t) \end{array} \right\} = I,$$

for $t \in [a+1, b+2]$, then we say that this pair of solutions is a *normalized prepared* basis of (9.2). Let $Y_i(t, a+1)$, $Z_i(t, a+1)$, $i = 1, 2$, be the solutions of the **IVP**'s defined by system (9.2) which respectively satisfy the initial conditions

$$\begin{array}{ll} Y_1(a+1, a+1) = 0, & Z_1(a+1, a+1) = I, \\ Y_2(a+1, a+1) = -I, & Z_2(a+1, a+1) = 0. \end{array}$$

It is easy to see that this pair of solutions of (9.2) is a normalized pair of solutions of (9.2). We call $Y_1(t, a+1)$, $Z_1(t, a+1)$ the principal solution at $a+1$ and call $Y_2(t, a+1)$, $Z_2(t, a+1)$ the coprincipal solution at $a+1$. This language is consistent with previous definitions [3] of principal and coprincipal solutions at ∞ used for continuous systems.

The following theorem gives us some important properties of normalized prepared bases of (9.2).

Theorem 9.5 *If* $Y_1(t)$, $Z_1(t)$ *and* $Y_2(t)$, $Z_2(t)$ *is a normalized prepared basis for* (9.2), *then the matrix function*

$$X(t) = \left[\begin{array}{cc} Y_1(t) & Y_2(t) \\ Z_1(t) & Z_2(t) \end{array} \right]$$

is symplectic for $t \in [a+1, b+2]$. *Furthermore*

$$\begin{aligned} Y_1^*(t)Z_1(t) &= Z_1^*(t)Y_1(t) && (9.10)\\ Y_2^*(t)Z_2(t) &= Z_2^*(t)Y_2(t) && (9.11)\\ Y_1^*(t)Z_2(t) &- Z_1^*(t)Y_2(t) = I && (9.12)\\ Y_1(t)Y_2^*(t) &= Y_2(t)Y_1^*(t) && (9.13)\\ Z_1(t)Z_2^*(t) &= Z_2(t)Z_1^*(t) && (9.14)\\ Y_1(t)Z_1^*(t) &- Y_2(t)Z_1^*(t) = I && (9.15) \end{aligned}$$

for $t \in [a+1, b+2]$ *and*

$$\begin{aligned} Y_1(t+1)Z_2^*(t) - Y_2(t+1)Z_1^*(t) &= E(t) && (9.16)\\ Y_2(t+1)Y_1^*(t) - Y_1(t+1)Y_2^*(t) &= F(t) && (9.17)\\ Z_1(t+1)Z_2^*(t) - Z_2(t+1)Z_1^*(t) &= G(t) && (9.18)\\ Z_2(t+1)Y_1^*(t) - Z_1(t+1)Y_2^*(t) &= H(t) && (9.19) \end{aligned}$$

for $t \in [a+1, b+1]$.

Proof: Equations (9.10) and (9.11) hold because $Y_1(t)$, $Z_1(t)$ and $Y_2(t)$, $Z_2(t)$ are prepared solutions of (9.2). Since $Y_1(t)$, $Z_1(t)$ and $Y_2(t)$, $Z_2(t)$ is a normalized prepared basis for (9.2)

$$\left\{ \begin{matrix} Y_1(t) & Y_2(t) \\ Z_1(t) & Z_2(t) \end{matrix} \right\} = I$$

for $t \in [a+1, b+2]$. This implies (9.12) holds for $t \in [a+1, b+2]$. But then (9.10)–(9.12) imply that the conditions (3.6) in Theorem 3.4 on page 74 hold for

$$X(t) = \begin{bmatrix} Y_1(t) & Y_2(t) \\ Z_1(t) & Z_2(t) \end{bmatrix}$$

for $t \in [a+1, b+2]$. Hence by Theorem 3.4, (see page 74), $X(t)$ is symplectic for $t \in [a+1, b+2]$. Equations (9.13)–(9.15) follow from equations (3.7) applied to $X(t)$. It remains to prove equations (9.16)–(9.19). But

$$X(t+1) = M(t)X(t),$$

for $t \in [a+1, b+1]$, implies

$$X(t+1)X^{-1}(t) = M(t).$$

Using equation (3.5) on page 74 applied to $X(t)$ we get that

$$\begin{bmatrix} Y_1(t+1) & Y_2(t+1) \\ Z_1(t+1) & Z_2(t+1) \end{bmatrix} \begin{bmatrix} Z_2^*(t) & -Y_2^*(t) \\ -Z_1^*(t) & Y_1^*(t) \end{bmatrix} = \begin{bmatrix} E(t) & F(t) \\ G(t) & H(t) \end{bmatrix}$$

for $t \in [a+1, b+1]$. Equations (9.16)–(9.19) follow easily from this last equation. □

Exercise 9.6 *Establish equations* (9.16) *and* (9.17) *directly by using the first equation in* (9.8). *Note equations* (9.16) *and* (9.17) *actually hold on* $[a+1, b+2]$.

Exercise 9.7 *Establish equations* (9.18) *and* (9.19) *directly by using the second equation in* (9.8).

Exercise 9.8 *Use Remark* 3.38 *on page* 115 *to show that if* $Y(t)$, $Z(t)$ *is a prepared basis of* (9.2), *then there is a prepared basis* $\hat{Y}(t)$, $\hat{Z}(t)$ *such that* $Y(t)$, $Z(t)$ *and* $\hat{Y}(t)$, $\hat{Z}(t)$ *is a normalized pair of solutions of* (9.2).

Theorem 9.9 *If* $Y(t)$, $Z(t)$ *is a prepared basis for* (9.2) *such that for some* $t_0 \in [a+1, b+2]$,

$$\ker Y(t_0+1) \quad \subset \quad \ker Y(t_0),$$

then

$$\ker Y^*(t_0+1) \quad \subset \quad \ker F^*(t_0).$$

Proof: By Exercise 9.8 there is a prepared basis $\hat{Y}(t)$, $\hat{Z}(t)$ such that $Y(t)$, $Z(t)$ and $\hat{Y}(t)$, $\hat{Z}(t)$ is a normalized prepared basis for (9.2). Fix

$$\xi \in \ker Y^*(t_0+1).$$

Then

$$Y^*(t_0+1)\xi = 0$$

and consequently

$$\hat{Y}(t_0+1)Y^*(t_0+1)\xi = 0.$$

Hence, by (9.13)

$$Y(t_0+1)\hat{Y}^*(t_0+1)\xi = 0.$$

Since $\ker Y(t_0+1) \subset \ker Y(t_0)$, we have

$$Y(t_0)\hat{Y}^*(t_0+1)\xi = 0.$$

Using (9.17) we get that

$$Y(t_0)\hat{Y}^*(t_0+1) = \hat{Y}(t_0)Y^*(t_0+1) + F^*(t_0).$$

Hence

$$\hat{Y}(t_0)Y^*(t_0+1)\xi + F^*(t_0)\xi = 0.$$

But $\xi \in \ker Y^*(t_0+1)$ gives

$$F^*(t_0)\xi = 0.$$

Therefore we have proven that

$$\ker Y^*(t_0+1) \subset \ker F^*(t_0).$$

□

Exercise 9.10 *Find* $\ker F^*(t)$ *in Example* 3.16.

9.2 THE MOORE–PENROSE INVERSE

In this section we give a brief introduction to the Moore–Penrose Pseudo–Inverse of a matrix. In the next theorem we define the Moore–Penrose Pseudo–Inverse and state some of its properties.

Theorem 9.11 (The Pseudo–Inverse) *Let A be an $n \times m$ matrix. Then there is a unique $m \times n$ matrix $A^\dagger$, (read as A–dagger), called the Moore–Penrose Pseudo–Inverse of A, (or the pseudo–inverse) satisfying the following properties:*

$$A^\dagger A A^\dagger = A^\dagger \tag{9.20}$$

$$A A^\dagger A = A \tag{9.21}$$

$$(AA^\dagger)^* = (AA^\dagger) \quad \text{and} \quad (A^\dagger A)^* = A^\dagger A. \tag{9.22}$$

Furthermore, if A has real entries, then $A^\dagger$ has real entries.

Proof: Existence may be quickly established from the *singular value decomposition* (**SVD**). See Stewart [143, pg. 325] or Golub and Van Loan [68, pp. 242-243] for that construction. Furthermore, since the construction in terms of the **SVD** is in real arithmetic if A is real, the Pseudo–Inverse of a real matrix is real.

We now establish uniqueness of the matrix $A^\dagger$. The proof is patterned after the proof that in a ring with identity e, an element y which has both a left inverse

x and a right inverse z must have $x = z$. Indeed, in that case, we have from $yz = e$ and $xy = e$ that

$$x = xe = x(yz) = (xy)z = ez = z.$$

Here that proof will is modified to show that $yx = yz$ and $xy = zy$ give

$$x = xyx = x(yx) = x(yz) = (xy)z = (zy)z = zyz = z.$$

Assume $A^\dagger$ and $\tilde{A}$ are Moore–Penrose inverses of A. First we show that

$$AA^\dagger = A\tilde{A}. \tag{9.23}$$

From (9.22)

$$\begin{aligned} AA^\dagger &= [AA^\dagger]^* = [(A\tilde{A}A)A^\dagger]^* \\ &= [(A\tilde{A}(AA^\dagger)]^* = (AA^\dagger)^*(A\tilde{A})^* \\ &= AA^\dagger A\tilde{A} = A\tilde{A}. \end{aligned}$$

Hence (9.23) holds. Next we show that

$$A^\dagger A = \tilde{A}A. \tag{9.24}$$

From (9.22)

$$\begin{aligned} A^\dagger A &= [A^\dagger A]^* = [A^\dagger(A\tilde{A}A)]^* \\ &= [(A^\dagger A)(\tilde{A}A)]^* = (\tilde{A}A)^*(A^\dagger A)^* \\ &= (\tilde{A}A)(A^\dagger A) = \tilde{A}(AA^\dagger A) = \tilde{A}A. \end{aligned}$$

Hence (9.24) holds. Finally using (9.23) and (9.24) we have that

$$A^\dagger = A^\dagger AA^\dagger = A^\dagger A\tilde{A} = \tilde{A}A\tilde{A} = \tilde{A}.$$

Hence the proof of the uniqueness is complete. □

Exercise 9.12 *By using the defining properties* (9.20)–(9.22) *of the Moore–Penrose inverse find* $A^\dagger$ *if*

$$\text{(a)}\quad A = \begin{bmatrix} 0 & 0 \\ 0 & 1 \end{bmatrix} \qquad \text{(b)}\quad A = \begin{bmatrix} 1 & 1 \end{bmatrix}$$

$$\text{(c)}\quad A = \begin{bmatrix} 1 & 0 & 1 \\ 0 & 0 & -1 \end{bmatrix} \quad \text{(d)}\quad A = \begin{bmatrix} 1 & 0 \\ 0 & i \\ 0 & 1 \end{bmatrix}.$$

We give several important properties of the Moore–Penrose inverse in the following theorem. Here we use the notation $\operatorname{Im} A$ for the *image* set

$$\operatorname{Im} A = \{y \mid y = Ax, \text{ for some } x\}.$$

This set is also called the *range* of A.

Theorem 9.13 *Let A be an $n \times m$ matrix.*

(a) *If A is invertible, then $A^\dagger = A^{-1}$*

(b) $(A^\dagger)^\dagger = A$

(c) *For $k \neq 0$, $(kA)^\dagger = \frac{1}{k}A^\dagger$*

(d) *If 0 is the $m \times n$ zero matrix, then $0^\dagger = 0$*

(e) $(A^*)^\dagger = (A^\dagger)^*$

(f) *rank $A^\dagger$ = rank A*

(g) $\ker A = \ker[(A^\dagger)^*]$

(h) *If $\alpha \in \operatorname{Im} A$, then $AA^\dagger\alpha = \alpha$.*

Proof: We will only establish that **(g)** holds. The proof of the rest of the theorem is left as the next exercise. Note that by (9.20)

$$(A^\dagger)^* = (A^\dagger A A^\dagger)^* = (A^\dagger)^*(A^\dagger A)^*.$$

Hence by (9.22)

$$(A^\dagger)^* = (A^\dagger)^* A^\dagger A. \tag{9.25}$$

Furthermore by (9.21) and (9.22)

$$A = AA^\dagger A = A(A^\dagger A)^*.$$

Hence

$$A = AA^*(A^\dagger)^*. \tag{9.26}$$

It follows from (9.25) and (9.26) that

$$\ker A = \ker[(A^\dagger)^*].$$

□

Exercise 9.14 *Complete the proof of Theorem* 9.13.

The **SVD** construction of the Pseudo–Inverse gives a basis for numerical computation. The numerical package "Matlab" has a built–in estimate of the pseudo–inverse called "pinv". However, numerical computation of Pseudo-Inverses of a rank defective matrix A is ill–defined. Indeed, for such a matrix A and a positive ϵ smaller than the smallest positive singular value of A, there exists a matrix B of higher rank with $||B - A||_2 = \epsilon$ and $||B^\dagger - A^\dagger||_2 = \frac{1}{\epsilon}$. This is accomplished by changing the first zero singular value of A in the **SVD** of A to ϵ to define B. (Consider, for example, $B = [\,\epsilon\,]$ as compared to $A = [0]$.) In particular, the pseudo–inverse function "$\dagger$", which maps the vector space of $n \times m$ matrices onto the vector space of $m \times n$ matrices, is not continuous at any rank defective A.

Our experience with matrix inverses suggests that the Moore–Penrose inverse of a product AB might be $B^\dagger A^\dagger$. However, the following example refutes that conjecture.

Exercise 9.15 *Show that for*

$$A = \begin{bmatrix} 1 & 1 & 3 \\ 1 & 1 & 3 \end{bmatrix}, \quad B = \begin{bmatrix} 1 & 1 \\ 2 & 0 \\ 0 & 1 \end{bmatrix},$$

we have the unexpected result

$$(AB)^\dagger \neq B^\dagger A^\dagger.$$

Exercise 9.16 *Show that if A is an $n \times n$ unitary matrix (so $AA^* = I$) and B is an $n \times m$ matrix, then*

$$(AB)^\dagger = B^\dagger A^{-1}.$$

Theorem 9.17 *Assume V is an $m \times n$ matrix and W is a $p \times n$ matrix. Then the following are equivalent*

(a) $\ker V \subset \ker W$

(b) $W = WV^\dagger V$

(c) $W^\dagger = V^\dagger V W^\dagger$.

Proof: First we show that **(a)** implies **(b)**. Assume **(a)** holds and $\alpha \in \mathrm{C}^n$. Since

$$V[\alpha - V^\dagger V\alpha] = V\alpha - VV^\dagger V\alpha = V\alpha - V\alpha = 0$$

we have $\alpha - V^\dagger V\alpha \in \ker V$. It follows from **(a)** that

$$\alpha - V^\dagger V\alpha \in \ker W.$$

Hence

$$0 = W[\alpha - V^\dagger V\alpha] = W\alpha - WV^\dagger V\alpha.$$

Therefore for all $\alpha \in \mathrm{C}^n$

$$W\alpha = WV^\dagger V\alpha.$$

This implies

$$W = WV^\dagger V.$$

The fact that **(b)** implies **(a)** is trivial. Hence **(a)** and **(b)** are equivalent.

We now show that **(a)** implies **(c)**. By part **(g)** of Theorem 9.13

$$\ker V = \ker[(V^\dagger)^*] \qquad \text{and} \qquad \ker W = \ker[(W^\dagger)^*].$$

Since we are assuming **(a)** holds we must have that

$$\ker[(V^\dagger)^*] \subset \ker[(W^\dagger)^*]. \tag{9.27}$$

Using the fact that **(a)** and **(b)**, with V replaced by $(V^\dagger)^*$ and W replaced by $(W^\dagger)^*$, are equivalent, we arrive at the conclusion

$$(W^\dagger)^* = (W^\dagger)^*[(V^\dagger)^*]^\dagger (V^\dagger)^*. \tag{9.28}$$

Taking the conjugate transpose of both sides of (9.28) and using the facts that $V^{\dagger\dagger} = V$ and that the operators * and $\dagger$ commute (parts **(b)** and **(e)** of Theorem 9.13) we get the desired result

$$W^\dagger = V^\dagger V W^\dagger.$$

Conversely assume **(c)** holds. But this implies (9.28) holds which implies (9.27) which implies

$$\ker V = \ker[(V^\dagger)^*] \subset \ker[(W^\dagger)^*] = \ker W$$

and **(a)** holds. □

It is now easy to prove the following linear algebra result by using the Moore–Penrose inverse. Again, we use the notation Im A for the *image* set.

Corollary 9.18 *If V is an $m \times n$ matrix and W is a $p \times n$ matrix such that*

$$\ker V \subset \ker W$$

then

$$\text{Im } V^* \supset \text{Im } W^*.$$

Proof: Assume

$$\alpha \in \text{Im } W^*.$$

Then there is a $\xi \in C^p$ such that

$$W^*\xi = \alpha.$$

Since $\ker V \subset \ker W$ we have by the previous theorem that

$$W = WV^\dagger V.$$

Hence

$$\begin{aligned} \alpha &= W^*\xi \\ &= (WV^\dagger V)^*\xi \\ &= V^*(V^\dagger)^*W^*\xi. \end{aligned}$$

It follows that

$$\alpha \in \text{Im } V^*$$

and we have shown

$$\text{Im } W^* \subset \text{Im } V^*$$

□

Corollary 9.19 *If $\ker Y(t_0+1) \subset \ker Y(t_0)$, then*

$$Y(t_0) = Y(t_0)Y^\dagger(t_0+1)Y(t_0+1) \tag{9.29}$$

and

$$Y^\dagger(t_0) = Y^\dagger(t_0+1)Y(t_0+1)Y^\dagger(t_0). \tag{9.30}$$

If, in addition, $Y(t)$, $Z(t)$ is a prepared basis of (9.2), *then*

$$F(t_0) = Y(t_0+1)Y^\dagger(t_0+1)F(t_0). \tag{9.31}$$

Proof: Assume

$$\ker Y(t_0 + 1) \subset \ker Y(t_0).$$

Using **(a)** implies **(b)** in Theorem 9.17 we get that (9.29) holds. Using **(a)** implies **(c)** in Theorem 9.17 we get that (9.30) holds.

Next assume $Y(t)$, $Z(t)$ is a prepared basis of (9.2) and $\ker Y(t_0+1) \subset \ker Y(t_0)$ for some $t_0 \in [a+1, b+2]$. By Theorem 9.9

$$\ker Y^*(t_0 + 1) \subset \ker F^*(t_0).$$

Using **(a)** implies **(b)** in Theorem 9.17 we obtain

$$F^*(t_0) = F^*(t_0)\,[Y^*(t_0 + 1)]^\dagger\, Y^*(t_0 + 1).$$

Taking the conjugate transpose of both sides we get that (9.31) holds. □

9.3 THE QUADRATIC FORM

We will say that the pair of vector functions $y(t)$, $z(t)$ is an *admissible pair* provided they are related by the equation of motion on $[a + 1, b + 2]$, that is

$$\Delta y(t) = A(t)y(t + 1) + B(t)z(t)$$

for $t \in [a + 1, b + 2]$. Thus $y(t)$, $z(t)$ is an admissible pair iff

$$y(t + 1) = E(t)y(t) + F(t)z(t)$$

for $t \in [a+1, b+2]$ where $E(t)$ and $F(t)$ are defined by equations (9.4) and (9.5), respectively. Note that if $B(t)$ and $F(t)$ are singular as they are in Exercise 9.3 on page 333, (or see Example 3.14 on page 85), we can not solve for $z(t)$ in terms of $y(t)$ from the equation of motion. Because of this we define our quadratic form for the Hamiltonian system (*symplectic system*) in terms of admissible pairs. On the set of all admissible pairs $y(t)$, $z(t)$ we define a quadratic form J associated with the Hamiltonian system (9.1) by

$$J[y, z] = \sum_{t=a+1}^{b+2} \{y^*(t + 1)C(t)y(t + 1) + z^*(t)B(t)z(t)\}.$$

We define

$$\mathcal{A} = \{\text{admissible pairs } y(t),\ z(t) \text{ such that } y(a + 1) = 0 = y(b + 3)\}.$$

We say J is *positive definite* on $\mathcal{A}$ provided $J[y,z] \geq 0$ for all y, z in $\mathcal{A}$ and $J[y,z] > 0$ if y, z is an admissible pair in $\mathcal{A}$ with $y(t) \not\equiv 0$ on $[a+1, b+3]$.

In Exercise 9.2 on page 331 we saw that the self-adjoint difference equation

$$\Delta[P(t)\Delta u(t-1)] + Q(t)u(t) = 0$$

under the change of variables

$$\begin{aligned} y(t) &= u(t-1), \quad t \in [a+1, b+3] \\ z(t) &= P(t)\Delta u(t-1), \quad t \in [a+1, b+2] \end{aligned}$$

is equivalent to a Hamiltonian system (9.1) with

$$A(t) = 0, \qquad B(t) = P^{-1}(t), \qquad C(t) = -Q(t).$$

In this case it is easy to see that

$$\begin{aligned} J[y,z] &= J[u] \\ &= \sum_{t=a+1}^{b+2} \{\Delta u^*(t-1)P(t)\Delta u(t-1) - u^*(t)Q(t)u(t)\}. \end{aligned}$$

Note that this is equivalent to the quadratic form associated with the self-adjoint difference equation of Example 9.2. See, for example, Chapter 8 in [89] or [4, pp. 104–106].

Exercise 9.20 *Show that for the* $2n$*-th order scalar difference equation* (3.21) *on page* 85 (*in Exercise* 9.3 *on page* 333 *we wrote* (3.21) *as an equivalent discrete Hamiltonian system*), *the quadratic form* J *and the set of admissible variations* $\mathcal{A}$ *are equivalent to*

$$J_1[u] = \sum_{t=a+n}^{b+2n} \sum_{i=0}^{n} (-1)^{n+i} r_i(t) |\Delta^i u(t-i)|^2$$

for $u \in \mathcal{A}_1$ *defined by*

$$\mathcal{A}_1 = \{u \ : \ [a, b+2n] \to \mathrm{C} \mid \Delta^i u(a) = 0 = \Delta^i u(b+n+1), \ 0 \leq i \leq n-1\}.$$

Establish that if $y(t)$, $z(t)$ *given in Example* 3.14 *on page* 85 *are related by the equation of motion, then no restriction is placed on* $u(t)$. *Note that if we only consider real* $u(t)$, *then the quadratic form* J_1 *is the quadratic form* Q *introduced in Section* 4.5.

Theorem 9.21 *If the pair $y(t)$, $z(t)$ is in $\mathcal{A}$, then*

$$J[y,z] = -\sum_{t=a+1}^{b+2} y^*(t+1)\left[\Delta z(t) - C(t)y(t+1) + A^*(t)z(t)\right].$$

Proof: Using a summation by parts formula on the first term

$$\begin{aligned}\sum_{t=a+1}^{b+2} & y^*(t+1)[\Delta z(t) - C(t)y(t+1) + A^*(t)z(t)] \\ &= \sum_{t=a+1}^{b+2} \{-(\Delta y^*(t))z(t) - y^*(t+1)C(t)y(t+1) + y^*(t+1)A^*(t)z(t)\} \\ &\quad + [y^*(t)z(t)]_{a+1}^{b+3}.\end{aligned}$$

Since $y(t)$, $z(t)$ is admissible and $y(a+1) = 0 = y(b+3)$ the above quantity is equal to

$$\begin{aligned}\sum_{t=a+1}^{b+2} & \{ - y^*(t+1)A^*(t)z(t) - z^*(t)B(t)z(t) \\ & - y^*(t+1)C(t)y(t+1) + y^*(t+1)A^*(t)z(t)\} = -J[y,z]\end{aligned}$$

which gives the desired result. □

Theorem 9.22 (A Picone Identity) *Assume that the principal solution $Y(t) \equiv Y_1(t, a+1)$, $Z(t) \equiv Z_1(t, a+1)$ at $a+1$ satisfies*

$$\ker Y(t_0+1) \subset \ker Y(t_0)$$

for some $t_0 \in [a+1, b+2]$. Then

$$D(t_0) \equiv Y(t_0)Y^\dagger(t_0+1)F(t_0)$$

is Hermitian. If $y(t)$, $z(t)$ is an admissible pair with $y(t_0) \in Im Y(t_0)$, then $y(t_0+1) \in Im Y(t_0+1)$,

$$y(t_0) = Y(t_0)Y^\dagger(t_0+1)y(t_0+1) - D(t_0)u(t_0) \tag{9.32}$$

where $u(t_0) = z(t_0) - Z(t_0)Y^\dagger(t_0)y(t_0)$ and

$$\begin{aligned}\Delta & \{y^*(t_0)Z(t_0)Y^\dagger(t_0)y(t_0)\} \\ &= y^*(t_0+1)C(t_0)y(t_0+1) + z^*(t_0)B(t_0)z(t_0) \\ &\qquad - u^*(t_0)D(t_0)u(t_0).\end{aligned} \tag{9.33}$$

Note that if Y is nonsingular, this Picone identity (9.33) is closely related to the *Legendre–Clebsch Transformation* given in [4]

Proof: First we prove that $D(t_0)$ is Hermitian. By the first equation in (9.2)

$$\begin{aligned} D(t_0) &= \{[I - A(t_0)]Y(t_0+1) - B(t_0)Z(t_0)\}Y^\dagger(t_0+1)F(t_0) \\ &= \{E^{-1}(t_0)Y(t_0+1) \\ &\quad - B(t_0)E^*(t_0)Z(t_0)\}Y^\dagger(t_0+1)F(t_0). \end{aligned}$$

Using the second equation in (9.2)

$$\begin{aligned} D(t_0) = \{&E^{-1}(t_0)Y(t_0+1) \\ &- B(t_0)E^*(t_0)[Z(t_0+1) - C(t_0)Y(t_0+1)]\}Y^\dagger(t_0+1)F(t_0). \end{aligned}$$

By (9.31) and (9.5)

$$D(t_0) = E^{-1}(t_0)F(t_0) - F^*(t_0)[Z(t_0+1) - C(t_0)Y(t_0+1)]Y^\dagger(t_0+1)F(t_0).$$

Using (9.5) and (9.31) again

$$\begin{aligned} D(t_0) &= \\ &B(t_0) - [Y(t_0+1)Y^\dagger(t_0+1)F(t_0)]^* Z(t_0+1)Y^\dagger(t_0+1)F(t_0) \\ &\qquad + F^*(t_0)C(t_0)F(t_0) \end{aligned}$$

which is Hermitian since $Y(t)$, $Z(t)$ is a prepared solution of (9.2).

Now assume $y(t)$, $z(t)$ is an admissible pair such that $y(t_0)$ is in $\mathrm{Im}Y(t_0)$. Then there is an $\alpha \in C^n$ such that $y(t_0) = Y(t_0)\alpha$. Since $y(t)$, $z(t)$ is an admissible pair

$$\begin{aligned} y(t_0+1) &= E(t_0)y(t_0) + F(t_0)z(t_0) \\ &= E(t_0)Y(t_0)\alpha + F(t_0)z(t_0). \end{aligned}$$

By the first equation in (9.2)

$$y(t_0+1) = Y(t_0+1)\alpha + F(t_0)[z(t_0) - Z(t_0)\alpha].$$

Using (9.29)

$$y(t_0+1) = Y(t_0+1)\{\alpha + Y^{-1}(t_0+1)F(t_0)[z(t_0) - Z(t_0)\alpha]\}$$

which implies that $y(t_0+1) \in \mathrm{Im}Y(t_0+1)$. Next we show that (9.32) holds. To this end, consider

$$\begin{aligned} &Y^\dagger(t_0+1)F(t_0)u(t_0) \\ &\quad = Y^\dagger(t_0+1)F(t_0)[z(t_0) - Z(t_0)Y^\dagger(t_0)y(t_0)] \end{aligned}$$

Since $y(t)$, $z(t)$ is an admissible pair

$$\begin{aligned} &Y^\dagger(t_0+1)F(t_0)u(t_0) \\ &\quad = Y^\dagger(t_0+1)\left[y(t_0+1) - E(t_0)y(t_0) - F(t_0)Z(t_0)Y^\dagger(t_0)y(t_0)\right] \end{aligned}$$

which equals (by the first equation in (9.8))

$$\begin{aligned} &= Y^\dagger(t_0+1)\left[y(t_0+1) - E(t_0)y(t_0) - Y(t_0+1)Y^\dagger(t_0)y(t_0)\right. \\ &\qquad \left. + E(t_0)Y(t_0)Y^\dagger(t_0)y(t_0)\right] \\ &\quad = Y^\dagger(t_0+1)y(t_0+1) - Y^\dagger(t_0+1)E(t_0)y(t_0) \\ &\quad - Y^\dagger(t_0)y(t_0) + Y^\dagger(t_0+1)E(t_0)Y(t_0)Y^\dagger(t_0)Y(t_0)\alpha \end{aligned}$$

by (9.30). Hence

$$Y^\dagger(t_0+1)F(t_0)u(t_0) = Y^\dagger(t_0+1)y(t_0+1) - Y^\dagger(t_0)y(t_0). \tag{9.34}$$

Premultiplying both sides of this by $Y(t_0)$ we get that

$$\begin{aligned} &D(t_0)u(t_0) \\ &\quad = \quad Y(t_0)Y^\dagger(t_0+1)y(t_0+1) - Y(t_0)Y^\dagger(t_0)y(t_0) \\ &\quad = \quad Y(t_0)Y^\dagger(t_0+1)y(t_0+1) - y(t_0) \end{aligned}$$

since $y(t_0) \in \mathrm{Im}Y(t_0)$, which is the desired result (9.32).

Next we prove that (9.33) holds. To this end consider

$$\begin{aligned} &\Delta\left[y^*(t_0)Z(t_0)\right] = y^*(t_0+1)Z(t_0+1) - y^*(t_0)Z(t_0) \\ &\quad = [E(t_0)y(t_0) + F(t_0)z(t_0)]^*[C(t_0)Y(t_0+1) + E^{*-1}(t_0)Z(t_0)] \\ &\qquad - y^*(t_0)Z(t_0) \\ &\quad = [y(t_0) + B(t_0)z(t_0)]^*[E^*(t_0)C(t_0)Y(t_0+1) + Z(t_0)] \\ &\qquad - y^*(t_0)Z(t_0) \end{aligned}$$

from (9.5). Multiply out to obtain that the above is

$$\begin{aligned} &= y^*(t_0)E^*(t_0)C(t_0)Y(t_0+1) + z^*(t_0)B(t_0)E^*(t_0)C(t_0)Y(t_0+1) \\ &\qquad +z^*(t_0)B(t_0)Z(t_0) \\ &= \left[E(t_0)y(t_0) + F(t_0)z(t_0)\right]^*C(t_0)Y(t_0+1) \\ &\qquad + z^*(t_0)B(t_0)Z(t_0) \\ &= y^*(t_0+1)C(t_0)Y(t_0+1) + z^*(t_0)B(t_0)Z(t_0). \end{aligned}$$

Using the first equation in (9.2) this becomes

$$\begin{aligned} &= y^*(t_0+1)C(t_0)Y(t_0+1) + z^*(t_0)[I - A(t_0)]Y(t_0+1) \\ &\quad -z^*(t_0)Y(t_0). \end{aligned}$$

Postmultiply both sides of this sequence of equations by $Y^\dagger(t_0+1)y(t_0+1)$ for

$$\begin{aligned} &\Delta\big[y^*(t_0)Z(t_0)\big]Y^\dagger(t_0+1)y(t_0+1) \\ &\quad = y^*(t_0+1)C(t_0)Y(t_0+1)Y^\dagger(t_0+1)y(t_0+1) \\ &\qquad + z^*(t_0)[I - A(t_0)]Y(t_0+1)Y^\dagger(t_0+1)y(t_0+1) \\ &\qquad - z^*(t_0)Y(t_0)Y^\dagger(t_0+1)y(t_0+1). \end{aligned}$$

Since $y(t_0+1) \in \operatorname{Im} Y(t_0+1)$,

$$\begin{aligned} &\Delta\big[y^*(t_0)Z(t_0)\big]Y^\dagger(t_0+1)y(t_0+1) = y^*(t_0+1)C(t_0)y(t_0+1) \\ &\quad +z^*(t_0)[I - A(t_0)]y(t_0+1) - z^*(t_0)Y(t_0)Y^\dagger(t_0+1)y(t_0+1). \end{aligned}$$

Use the fact that $y(t)$, $z(t)$ is an admissible pair to get that the above is

$$\begin{aligned} &= y^*(t_0+1)C(t_0)y(t_0+1) + z^*(t_0)y(t_0) \\ &\quad +z^*(t_0)B(t_0)z(t_0) - z^*(t_0)Y(t_0)Y^\dagger(t_0+1)y(t_0+1) \\ &= y^*(t_0+1)C(t_0)y(t_0+1) + z^*(t_0)B(t_0)z(t_0) \\ &\quad +z^*(t_0)Y(t_0)Y^\dagger(t_0)y(t_0) - z^*(t_0)Y(t_0)Y^\dagger(t_0+1)y(t_0+1) \end{aligned}$$

since $y(t_0) \in \operatorname{Im} Y(t_0)$. Therefore, this is

$$\begin{aligned} &= y^*(t_0+1)C(t_0)y(t_0+1) + z^*(t_0)B(t_0)z(t_0) \\ &\quad -z^*(t_0)Y(t_0)\Delta\big[Y^\dagger(t_0)y(t_0)\big]. \end{aligned} \tag{9.35}$$

But using the definition of $u(t_0)$

$$\begin{aligned} z^*(t_0)Y(t_0) &= u^*(t_0)Y(t_0) + y^*(t_0)\big[Y^\dagger(t_0)\big]^* Z^*(t_0)Y(t_0) \\ &= u^*(t_0)Y(t_0) + y^*(t_0)\big[Y^\dagger(t_0)\big]^* Y^*(t_0)Z(t_0) \end{aligned}$$

since $Y(t)$, $Z(t)$ is a prepared solution. So

$$\begin{aligned} z^*(t_0)Y(t_0) &= u^*(t_0)Y(t_0) + \big[Y(t_0)Y^\dagger(t_0)y(t_0)\big]^* Z(t_0) \\ &= u^*(t_0)Y(t_0) + y^*(t_0)Z(t_0) \end{aligned}$$

since $y(t_0) \in \operatorname{Im} Y(t_0)$. Hence substituting this last expression in (9.35)

$$\big(\Delta\big[y^*(t_0)Z(t_0)\big]\big)Y^\dagger(t_0+1)y(t_0+1)$$

$$
\begin{aligned}
&= y^*(t_0+1)C(t_0)y(t_0+1) + z^*(t_0)B(t_0)z(t_0) \\
&\quad -\left[u^*(t_0)Y(t_0) + y^*(t_0)Z(t_0)\right]\Delta\left[Y^\dagger(t_0)y(t_0)\right] \\
&= y^*(t_0+1)C(t_0)y(t_0+1) + z^*(t_0)B(t_0)z(t_0) \\
&\quad - u^*(t_0)Y(t_0)\left[Y^\dagger(t_0+1)y(t_0+1) - Y^\dagger(t_0)y(t_0)\right] \\
&\quad - y^*(t_0)Z(t_0)\Delta\left[Y^\dagger(t_0)y(t_0)\right] \\
&= y^*(t_0+1)C(t_0)y(t_0+1) + z^*(t_0)B(t_0)z(t_0) \\
&\quad - u^*(t_0)\left[Y(t_0)Y^\dagger(t_0+1)y(t_0+1) - y(t_0)\right] \\
&\quad - y^*(t_0)Z(t_0)\Delta\left[Y^\dagger(t_0)y(t_0)\right]
\end{aligned}
$$

since $y(t_0) \in \mathrm{Im}Y(t_0)$. Using (9.32)

$$
\begin{aligned}
&\Delta\left[y^*(t_0)Z(t_0)\right]Y^\dagger(t_0+1)y(t_0+1) \\
&\quad = y^*(t_0+1)C(t_0)y(t_0+1) + z^*(t_0)B(t_0)z(t_0) \\
&\quad\quad - u^*(t_0)D(t_0)u(t_0) - y^*(t_0)Z(t_0)\Delta\left[Y^\dagger(t_0)y(t_0)\right].
\end{aligned}
$$

This last equation implies that (9.33) holds and the proof is complete. □

Theorem 9.23 *Assume that there exists a t_0 such that the principal solution $Y(t) = Y_1(t, a+1)$, $Z(t) = Z_1(t, a+1)$ at $t = a+1$ satisfies*

$$\ker Y(t+1) \subset \ker Y(t)$$

for $a+1 \le t < t_0 \le b+2$, but

$$\ker Y(t_0+1) \not\subset \ker Y(t_0).$$

Then if

$$\alpha \in \ker Y(t_0+1) \backslash \ker Y(t_0)$$

and we define

$$
\begin{aligned}
y(t) &= \begin{cases} Y(t)\alpha, & a+1 \le t \le t_0, \\ 0, & t_0+1 \le t \le b+3 \end{cases} \\
z(t) &= \begin{cases} Z(t)\alpha, & a+1 \le t \le t_0, \\ 0, & t_0+1 \le t \le b+3 \end{cases}
\end{aligned}
$$

then the pair $y(t)$, $z(t)$ is in $\mathcal{A}$ with $y(t_0) \neq 0$ and

$$J[y, z] = 0.$$

Proof: Clearly $y(a+1) = y(b+3) = 0$. To see that $y(t)$, $z(t)$ is an admissible pair, note that

$$\begin{aligned} E(t_0)y(t_0) + F(t_0)z(t_0) &= [E(t_0)Y(t_0) + F(t_0)Z(t_0)]\alpha \\ &= Y(t_0+1)\alpha = 0 = y(t_0+1) \end{aligned}$$

since $\alpha \in \ker Y(t_0+1)$. Since $y(t_0) = Y(t_0)\alpha \neq 0$, we know that $y(t)$, $z(t)$ is a nontrivial admissible pair. By the definitions of $y(t)$ and $z(t)$

$$J[y,z] = z^*(t_0)B(t_0)z(t_0) + \sum_{t=a+1}^{t_0-1} \{y^*(t+1)C(t)y(t+1) + z^*(t)B(t)z(t)\}.$$

By Theorem 9.22

$$\begin{aligned} J[y,z] &= \{y^*(t)Z(t)Y^\dagger(t)y(t)\}_{a+1}^{t_0} \\ &\quad + \sum_{t=a+1}^{t_0-1} u^*(t)D(t)u(t) + z^*(t_0)B(t_0)z(t_0) \\ &= y^*(t_0)Z(t_0)Y^\dagger(t_0)y(t_0) + \sum_{t=a+1}^{t_0-1} u^*(t)D(t)u(t) \\ &\quad + z^*(t_0)B(t_0)z(t_0) \end{aligned}$$

since $y(a+1) = 0$. But for $a+1 \leq t \leq t_0 - 1$, we have by Theorem 9.22

$$\begin{aligned} D(t)u(t) &= D^*(t)[z(t) - Z(t)Y^\dagger(t)y(t)] \\ &= \left[Y^\dagger(t+1)F(t)\right]^* Y^*(t)Z(t)\left[\alpha - Y^\dagger(t)Y(t)\alpha\right]. \end{aligned}$$

Since $Y(t)$, $Z(t)$ is a prepared solution

$$D(t)u(t) = \left[Y^\dagger(t+1)F(t)\right]^* Z^*(t)\left[Y(t)\alpha - Y(t)Y^\dagger(t)Y(t)\alpha\right] = 0.$$

Hence

$$\begin{aligned} J[y,z] &= y^*(t_0)Z(t_0)Y^\dagger(t_0)y(t_0) + z^*(t_0)B(t_0)z(t_0) \\ &= \alpha^*Y^*(t_0)Z(t_0)Y^\dagger(t_0)Y(t_0)\alpha + \alpha^*Z^*(t_0)B(t_0)Z(t_0)\alpha \\ &= \alpha^*Z^*(t_0)Y(t_0)Y^\dagger(t_0)Y(t_0)\alpha + \alpha^*Z^*(t_0)B(t_0)Z(t_0)\alpha \end{aligned}$$

since $Y(t)$, $Z(t)$ is a prepared solution. Hence

$$\begin{aligned} J[y,z] &= \alpha^*Z^*(t_0)Y(t_0)\alpha + \alpha^*Z^*(t_0)B(t_0)Z(t_0)\alpha \\ &= \alpha^*Z^*(t_0)E^{-1}(t_0)[E(t_0)Y(t_0) + F(t_0)Z(t_0)]\alpha \end{aligned}$$

using (9.5). By (9.8)

$$J[y,z] = \alpha^* Z^*(t_0)E^{-1}(t_0)Y(t_0+1)\alpha = 0$$

since $\alpha \in \ker Y(t_0+1)$. □

Theorem 9.24 *Assume that the principal solution $Y(t) = Y_1(t,a+1)$, $Z(t) = Z_1(t,a+1)$ at $a+1$ satisfies*

$$\ker Y(t+1) \subset \ker Y(t)$$

for $a+1 \le t \le b+2$. Let t_0 be in $[a+1,b+2]$, and let α be in $\mathbb{C}^n$. Define $y(t)$, $z(t)$ by

$$y(t) = \begin{cases} -Y(t)Y^\dagger(t_0+1)F(t_0)\alpha, & a+1 \le t \le t_0, \\ 0, & t_0+1 \le t \le b+3, \end{cases}$$

$$z(t) = \begin{cases} -Z(t)Y^\dagger(t_0+1)F(t_0)\alpha, & a+1 \le t \le t_0-1, \\ \left[Y(t_0)Y^\dagger(t_0+1)E(t_0)\right]^* \alpha, & t = t_0, \\ 0, & t_0+1 \le t \le b+2. \end{cases}$$

Then the pair $y(t)$, $z(t)$ is in $\mathcal{A}$ and we have

$$J[y,z] = \alpha^* D(t_0)\alpha.$$

Proof: Clearly $y(a+1) = 0 = y(b+3)$. To help see that $y(t)$ $z(t)$ is an admissible pair note that

$$\begin{aligned} &E(t_0)y(t_0) + F(t_0)z(t_0) = \\ &\quad - E(t_0)Y(t_0)Y^\dagger(t_0+1)F(t_0)\alpha + F(t_0)\left[Y(t_0)Y^\dagger(t_0+1)E(t_0)\right]^* \alpha . \end{aligned}$$

Use (9.5) to write this as

$$\begin{aligned} &= -E(t_0)\big(D(t_0)\alpha - B(t_0)\left[Y(t_0)Y^\dagger(t_0+1)E(t_0)\right]^* \alpha\big) \\ &= -E(t_0)[D(t_0)\alpha - D^*(t_0)\alpha] = 0 = y(t_0+1) \end{aligned}$$

since $D(t_0)$ is Hermitian. By the definitions of J, y, and z

$$J[y,z] = \sum_{t=a+1}^{t_0} \{y^*(t+1)C(t)y(t+1) + z^*(t)B(t)z(t)\}.$$

Using (9.33) with t_0 replaced by t

$$\begin{aligned} J[y,z] &= \left[y^*(t)Z(t)Y^\dagger(t)y(t)\right]_{a+1}^{t_0+1} + \sum_{t=a+1}^{t_0} u^*(t)D(t)u(t) \\ &= \sum_{t=a+1}^{t_0} u^*(t)D(t)u(t). \end{aligned}$$

As in the proof of Theorem 9.23

$$D(t)u(t) = 0 \qquad \text{for } a+1 \le t \le t_0 - 1.$$

Hence

$$J[y,z] = u^*(t_0)D(t_0)u(t_0). \tag{9.36}$$

By (9.32)

$$\begin{aligned} D(t_0)u(t_0) &= Y(t_0)Y^\dagger(t_0+1)y(t_0+1) - y(t_0) \\ &= Y(t_0)Y^\dagger(t_0+1)F(t_0)\alpha \qquad (9.37) \\ &= D(t_0)\alpha. \qquad (9.38) \end{aligned}$$

Using (9.36) $D(t_0)$ is Hermitian, and (9.37) gives

$$\begin{aligned} J[y,z] &= u^*(t_0)D(t_0)\alpha = [D(t_0)u(t_0)]^*\alpha \\ &= [D(t_0)\alpha)]^*\alpha = \alpha^* D(t_0)\alpha. \end{aligned}$$

□

We can now prove the following important theorem.

Theorem 9.25 *The quadratic functional J is positive definite on $\mathcal{A}$ iff the principal solution $Y(t) \equiv Y_1(t,a+1)$, $Z(t) \equiv Z_1(t,a+1)$ at $a+1$ satisfies*

$$\ker Y(t+1) \subset \ker Y(t) \tag{9.39}$$

$$D(t) \equiv Y(t)Y^\dagger(t+1)F(t) \ge 0 \tag{9.40}$$

on $[a+1, b+2]$.

Proof: First assume that the principal solution at a

$$Y(t) \equiv Y_1(t,a+1), \qquad Z(t) \equiv Z_1(t,a+1)$$

satisfies (9.39) and (9.40) on the interval $[a+1, b+2]$.

Note that by Theorem 9.22, equation (9.39) implies $D(t)$ is Hermitian on the interval $[a+1, b+2]$.

Let $y(t)$, $z(t)$ be in $\mathcal{A}$. Since $y(a+1) = 0$, $y(a+1) \in \text{Im } Y(a+1)$. By Theorem 9.22 $y(t) \in \text{Im } Y(t)$, for t in $[a+1, b+3]$ and

$$\begin{aligned}\Delta &\left[y^*(t)Z(t)Y^\dagger(t)y(t)\right] \\ &= y^*(t+1)C(t)y(t+1) + z^*(t)B(t)z(t) \\ &\quad -u^*(t)D(t)u(t),\end{aligned} \tag{9.41}$$

where

$$u(t) = z(t) - Z(t)Y^\dagger(t)y(t)$$

for $t \in [a+1, b+2]$. Summing both sides of (9.41) from $a+1$ to $b+2$ we obtain

$$J[y,z] = \left\{y^*(t)Z(t)Y^\dagger(t)y(t)\right\}_{a+1}^{b+3} + \sum_{t=a+1}^{b+2} u^*(t)D(t)u(t).$$

Since $y(a+1) = y(b+3) = 0$,

$$J[y,z] = \sum_{t=a+1}^{b+2} u^*(t)D(t)u(t) \geq 0 \tag{9.42}$$

since condition (9.40) holds on $[a+1, b+2]$.

Next assume y, z is in $\mathcal{A}$ and

$$J[y,z] = 0.$$

From (9.42) we get that

$$D(t)u(t) = 0$$

for $a+1 \leq t \leq b+2$. It follows from (9.32) that

$$y(t) = Y(t)Y^\dagger(t+1)y(t+1)$$

for $t \in [a+1, b+2]$. Since $y(a+1) = y(b+3) = 0$ it follows that $y(t) \equiv 0$ on $[a+1, b+3]$. Hence J is positive definite on $\mathcal{A}$.

Conversely, assume J is positive definite on $\mathcal{A}$. First we show that (9.39) holds for t in $[a+1, b+2]$. Since $Y(a+1) = Y_1(a+1, a+1) = 0$, condition (9.39) holds for $t = a+1$. Assume that there is a t_0 in $[a+2, b+2]$ such that

$$\ker\ Y(t+1) \subset \ker\ Y(t)$$

for t in $[a+1, t_0-1]$ but

$$\ker\ Y(t_0+1) \not\subset \ker\ Y(t_0).$$

By Theorem 9.23 there is a pair $y(t)$, $z(t)$ in $\mathcal{A}$ with $y(t_0) \neq 0$ such that

$$J[y,z] = 0.$$

This is contrary to J being positive definite on $\mathcal{A}$. Hence (9.39) holds on the interval $[a+1, b+2]$.

Next we show that (9.40) holds for t in $[a+1, b+2]$. Fix $\alpha \in \mathbf{C}^n$ and t_0 in $[a+1, b+2]$. Then by Theorem 9.24 there is a pair y, z in $\mathcal{A}$ such that

$$J[y,z] = \alpha^* D(t_0)\alpha \geq 0.$$

This last inequality is true for all $\alpha \in \mathbf{C}^n$ and for all t_0 in $[a+1, b+2]$, because condition (9.40) holds for t in $[a+1, b+2]$. □

Let $y(t)$, $z(t)$ be a prepared solution of (9.1) (or the vector analogue of the symplectic system (9.8)). We say $y(t)$ has a *generalized zero* at the first point in the interval $[a+1, b+3]$ provided $y(a+1) = 0$. We say $y(t)$ has a generalized zero at t_0+1, for $a+2 \leq t_0+1 \leq b+3$ provided

$$y(t_0) \neq 0, \qquad y(t_0+1) \in \text{Im } F(t_0)$$

and

$$y^*(t_0)B^\dagger(t_0)[I - A(t_0)]y(t_0+1) \leq 0.$$

(Note the last inequality would be true if $y(t_0+1) = 0$.) We say that the discrete Hamiltonian system (9.1) is *disconjugate* on $[a+1, b+3]$ provided there is no nontrivial prepared solution matrix with two generalized zeros in $[a+1, b+3]$.

Exercise 9.26 *In Section* 1.2 *we saw that the self-adjoint scalar equation* (1.1) *can be written as an equivalent symplectic system* (1.11) *where*

$$y(t) = u(t-1), \qquad t \in [a+1, b+3]$$
$$a(t) = p(t)\Delta u(t-1), \qquad t \in [a+1, b+2].$$

Show that if $y(t)$ has a generalized zero at t_0+1 as defined above, then $u(t)$ has a generalized zero at t_0 as defined in Section 1.5. In particular, if $p(t) > 0$ on $[a+1, b+2]$, then a generalized zero for a real solution of the scalar equation is the same as a generalized zero as defined by Hartman in [76].

Exercise 9.27 *In Example 3.14 on page 85 we showed that the $2n$–th order scalar equation (3.21) on page 85 can be written as an equivalent symplectic system and in Exercise 9.3 on page 333 we wrote it as an equivalent Hamiltonian system. Show that if a real vector solution $y(t)$ of the Hamiltonian system in Example 3.14 on page 85 has a generalized zero at $t_0 + 1$, then $u(t)$ has a generalized zero at t_0 of order n as defined in Section 4.5. In particluar note that disconjugacy of the Hamiltonian system is equivalent to the (n, n)–disconjugacy of (3.21) as defined in Section 4.6.*

Exercise 9.28 *Using the above definition of a generalized zero of a prepared solution of a Hamiltonian system*

(a) *determine what is meant by a generalized zero for a prepared solution of the discrete Jacobi equation (3.39) in Example 3.17 on page 91;*

(b) *determine what is meant by a generalized zero of a prepared solution of the three term difference equation (3.44) in Example 3.18;*

(c) *determine what is meant by a generalized zero of a prepared solution of the self–adjoint vector equation corresponding to (3.16) in Example 3.8 on page 80.*

We conclude this book with the statement of a general Reid Roundabout Theorem for the discrete Hamiltonian system (9.1). See Bohner [29], Theorem 2, for the proof of this theorem and many related results.

Theorem 9.29 *The following are equivalent:*

(i) *J is positive definite on $\mathcal{A}$.*

(ii) *The Hamiltonian system (9.1) is disconjugate on $[a + 1, b + 3]$.*

(iii) *Every solution $y(t)$, $z(t)$ of (9.1) with $y(a + 1) = 0$ has no generalized zeros in $[a + 2, b + 3]$.*

(iv) *The principal solution $Y(t) = Y_1(t, a + 1)$, $Z(t) = Z_1(t, a + 1)$ at $a + 1$ satisfies (9.39) and (9.40) on $[a + 1, b + 2]$.*

9.4 NOTES

Prior to general use of the **SVD**, the generalized inverse was constructed by bordering A by certain matrices so as to make a larger nonsingular matrix, taking the inverse of this larger matrix, and then using the upper left corner of that inverse for the generalized inverse. See Reid, [134, Appendix B, Section 3, pp. 480-484] for this construction, and [132] for some interesting history of generalized inverses. Additional references are given in Stewart [143, pg. 230].

For the corresponding results of Bohner, our Theorems 9.22 and 9.29 used papers [27, 28, 29, 35].

For generalized zeros for the Sturm–Liouville case, see [20] and [30]. Riccati equations may also be included in the Reid Roundabout Theorem for the setting of this chapter [31]. Disconjugacy of symplectic systems is discussed in [32].

Discrete variational problems are treated by Bohner in [33].

Also, see Bohner's recent JMAA paper [29] in the following connections. Our Theorem 9.5 corresponds to his Remark 2 part (i); Theorem 9.9 corresponds to his Remark 2 part (ii); Theorem 9.17 corresponds to his Remark 2 part (iii); Theorem 9.21 corresponds to his Lemma 1; Theorem 9.22 corresponds to his Theorem 1; Theorem 9.23 corresponds to his Proposition 1; and Theorem 9.25 corresponds to his Theorem 2.

REFERENCES

[1] R. AGARWAL, "*Difference Equations and Inequalities: Theory, Methods, and Applications* ", Marcel Dekker, New York, 1992.

[2] C. D. AHLBRANDT, Disconjugacy criteria for self–adjoint differential systems, *J. Differential Equations* **6** (1969), pp. 271–295.

[3] C. D. AHLBRANDT, The question of equivalence of principal and coprincipal solutions of self–adjoint differential systems, *Illinois J. Math.* **2** (1972), pp. 72–81.

[4] C. D. AHLBRANDT, Discrete variational inequalities, *in* "*General Inequalities 6*", 6th International Conference on General Inequalities, Oberwolfach, Dec. 9-15, 1990, W. Walter, ed., International Series of Numerical Mathematics, Vol 103, Birkhäuser Verlag Basel, 1992, pp. 93–107.

[5] C. D. AHLBRANDT, Continued fraction representations of maximal and minimal solutions of a discrete matrix Riccati equation, *SIAM J. Math. Anal.* **24** (1993), pp. 1597–1621.

[6] C. D. AHLBRANDT, Equivalence of discrete Euler equations and discrete Hamiltonian systems, *J. Math. Anal. Appl.* **180** (1993), pp. 498–517.

[7] C. D. AHLBRANDT, Geometric, analytic, and arithmetic aspects of symplectic continued fractions, *in* "*Analysis, Geometry, and Groups: A Riemann Legacy Volume*", (T. M. Rassias and H. M. Srivastava, Eds.), Hadronic Press, Tarpon Springs, FL, 1993, pp.1–26.

[8] C. D. AHLBRANDT, Dominant and recessive solutions of symmetric three term recurrences, *J. Differential Equations* **107** (1994), pp. 238–258.

[9] C. D. AHLBRANDT, A Pincherle theorem for matrix continued fractions, *J. Approximation Theory* **84** (1996), pp. 188–196.

[10] C. D. AHLBRANDT, C. CHICONE, S. L. CLARK, W. T. PATULA, AND D. STEIGER, Approximate first integrals for discrete Hamiltonian systems, *Dynamics of Continuous, Discrete and Impulsive Systems*, **2**, number 2, (1996), Univ. of Waterloo, pp. 237-264.

[11] C. D. AHLBRANDT, S. L. CLARK, J. W. HOOKER, AND W. T. PATULA, A discrete Interpretation of Reid's roundabout theorem for generalized differential systems, *Computers Math. Applic.* **28** (1994), pp. 11-21.

[12] C. D. AHLBRANDT AND M. HEIFETZ, Discrete Riccati equations of filtering and control, *in "Proceedings of the First International Conference on Difference Equations, Trinity University, San Antonio, Texas, May 25–28, 1994,"* edited by S. N. Elaydi, J. R. Graef, G. Ladas, and A. C. Peterson, Gordon and Breach Publishers, Newark, New Jersey, (1996), pp. 1–16.

[13] C. D. AHLBRANDT, M. HEIFETZ, J. W. HOOKER AND W. T. PATULA, Asymptotics of discrete time Riccati equations, robust control, and discrete linear Hamiltonian systems, *PanAmerican Mathematical Journal* **5** (1969), pp. 1–39.

[14] C. D. AHLBRANDT AND J. W. HOOKER, Riccati transformations and principal solutions of discrete linear systems, *in "Proc.* 1984 *Workshop on Spectral Theory of Sturm-Liouville Differential Operators"*, H. G. Kaper and A. Zettl (eds.), ANL-84-87, Argonne National Lab., Argonne, Illinois, 1984, pp. 1-11.

[15] C. D. AHLBRANDT AND J. W. HOOKER, Disconjugacy criteria for second order linear difference equations, *in "Qualitative Properties of Differential Equations, Proc. of 1984 Edmonton Conference"*, University of Alberta, Edmonton, Alberta, Canada, 1987, pp. 15–26.

[16] C. D. AHLBRANDT AND J. W. HOOKER, A variational view of nonoscillation theory for linear difference equations, *in "Proc. Thirteenth Midwest Differential Equations Conf."*, J.L. Henderson, ed., Institute of Applied Mathematics, University of Missouri-Rolla, Rolla, MO, 1985, pp. 1–21.

[17] C. D. AHLBRANDT AND J. W. HOOKER, Riccati matrix difference equations and disconjugacy of discrete linear systems, *SIAM J. Math. Anal.* **19** (1988), pp. 1183-1197.

[18] C. D. AHLBRANDT AND J. W. HOOKER, Recessive solutions of symmetric three term recurrence relations, *Canadian Mathematical Society, Conference Proceedings* **8** (1987), pp. 3–42.

[19] C. D. AHLBRANDT AND W. T. PATULA, Recessive solutions of block tridiagonal nonhomogeneous systems, *J. Difference Equations and Applications* **1** (1995), pp. 1–15.

[20] C. D. AHLBRANDT AND A. PETERSON, The (n, n)–disconjugacy of a $2n^{th}$ order linear difference equation, *Computers Math. Applic.* **28** (1994), pp. 1–9.

[21] D. ANDERSON, "*Discrete Hamiltonian Systems*", Dissertation, University of Nebraska–Lincoln, 1997.

[22] F. V. ATKINSON, "*Discrete and Continuous Boundary Value Problems*", Academic Press, New York, 1964.

[23] G. BAUR, "*Variationsrechnung und Kontrolltheorie*", Universität Ulm, Ulm, Germany, 1996.

[24] A. BEN–ISRAEL AND T. N. E. GREVILLE, "*Generalized Inverses: Theory and Applications*", John Wiley & Sons, Inc., New York, 1974.

[25] S. BITTNATI, A. J. LAUB, AND J. C. WILLEMS (ED.), "*The Riccati Equation*", Springer Verlag, Berlin, 1991.

[26] G. B. BLISS, "*Lectures on the Calculus of Variations*", Univ. of Chicago Press, Chicago, 1963.

[27] M. BOHNER, Zur Positivität diskreter quadratischer Funktionale, PhD Thesis, Universität Ulm, 1995. English Edition: On positivity of discrete quadratic functionals.

[28] M. BOHNER, Controllability and disconjugacy for linear Hamiltonian difference systems *in* "*Proceedings of the First International Conference on Difference Equations, Trinity University, San Antonio, Texas, May 25–28, 1994,*" edited by S. N. Elaydi, J. R. Graef, G. Ladas, and A. C. Peterson, Gordon and Breach Publishers, Newark, New Jersey, (1996), pp 65–77,

[29] M. BOHNER, Linear Hamiltonian difference systems: disconjugacy and Jacobi–type conditions, *J. Math. Anal. Appl.* **199** (1996), pp. 804–826.

[30] M. BOHNER, On disconjugacy for Sturm–Liouville Difference equations, *J. Difference Equations and Applications* **2** (1996), pp. 227–237.

[31] M. BOHNER, Riccati matrix difference equations and linear Hamiltonian difference systems, *Dynamics of continuous, discrete and impulsive systems*, **2**, number 2, (1996), Univ. of Waterloo, pp.147-160.

[32] M. BOHNER AND O. DOŠLÝ , Disconjugacy and transformations for symplectic systems, *Rocky Mountain J. Math.*, (to appear).

[33] M. BOHNER, Inhomogeneous discrete variational problems, *in* "*Proceed ings of the second International Conference on Difference Equations*", Veszprém, Hungary, 1995, (to appear).

[34] M. BOHNER, Discrete linear Hamiltonian eigenvalue problems, *Compu. Math. Appl.*, 1996 (to appear).

[35] M. BOHNER, Positive definiteness of discrete quadratic functionals, *in* "*Proceedings of the Seventh International Conference on General Inequalities*", Oberwolfach, 1995, (to appear).

[36] O. BOLZA, "*Lectures on the Calculus of Variations*", Chelsea Reprint of 1904 edition, Chelsea Publishing Company, New York.

[37] C. I. BYRNES, A. LINDQUIST, AND T. MCGREGOR, Predictability and unpredictability in Kalman filtering, *IEEE Trans. Automatic Control* **36** (1991), pp. 563-579.

[38] P. E. CAINES AND D. Q. MAYNE, On the discrete–time matrix Riccati equation of optimal control, *Internat. J. Control* **12** (1970), pp. 785–794; also, see "correction" **14** (1971), pp. 205-207.

[39] S. W. CHAN, G. C. GOODWIN, AND K. S. SIN, Convergence properties of the Riccati difference equation in optimal filtering of nonstabilizable systems, *IEEE Trans. Automatic Control* **29** (1984), pp. 110–118.

[40] S. CHEN, Disconjugacy, disfocality, and oscillation of second second order difference equations, *J. Differntial Eqs.* **107** (1994), pp. 383–394.

[41] S. CHEN AND L. ERBE, Oscillation and nonoscillation for systems of self–adjoint second–order difference equations, *SIAM J. Math. Anal.* **20** (1989), pp. 939–949.

[42] S. CHEN AND L. ERBE, Riccati techniques and discrete oscillation, *J. Math. Anal. Appl.* **142** (1989), pp. 468–487.

[43] S. CHEN AND L. ERBE, Oscillation results for second order scalar and matrix difference equations, *Computers. Math. Appl.* **28** (1994), pp. 55–69.

[44] W. A. COPPEL, "*Disconjugacy*", Lecture Notes in Mathematics **220**, Springer–Verlag, New York, 1971.

[45] W. DERRICK AND J. EIDSWICK, Continued fractions, Chebychev polynomials, and chaos, *Amer. Math. Monthly.* **102** (1995), pp. 337–344.

[46] O. DOŠLÝ, Reciprocity principle for Sturm–Liouville difference equations and some of its applications, *in* "*Proceedings of the second International Conference on Difference Equations*", Veszprém, Hungary, 1995, (to appear).

[47] O. DOŠLÝ, Transformations of linear Hamiltonian difference systems and some of their applications, *J. Math. Anal. Appl.* **191** (1995), pp. 250–265.

[48] O. DOŠLÝ, Oscillation criteria for higher order Sturm–Liouville difference equations, (preprint).

[49] O. DOŠLÝ, Factorization of disconjugate higher order Sturm–Liouville difference operators, *Comp. Appl. Math.* (submitted).

[50] J. C. DOYLE, K. GLOVER, P. P. KHARGONEKAR, AND B. A. FRANCIS, State–space solutions to standard H_2 and H_∞ control problems, *IEEE Trans. Automatic Control* **34** (1989), pp. 831–847.

[51] S. ELAYDI, "*An Introduction to Difference Equations*", Springer, New York, 1995.

[52] P. W. ELOE AND J. HENDERSON, Analogues of Fekete and Decartes systems of solutions for difference equations, *J. Approx. Theory* **59** (1989), pp. 38–52.

[53] L. ERBE AND S. HILGER, Sturmian theory on measure chains, *Differential Equations and Dynamical Systems,* **I** (1993), pp. 223–246.

[54] L. H. ERBE AND P. YAN, Disconjugacy for linear Hamiltonian difference systems, *J. Math. Anal. Appl.* **167** (1992), pp. 355–367.

[55] L. H. ERBE AND P. YAN, Qualitative properties of Hamiltonian difference systems, *J. Math. Anal. Appl.* **171** (1992), pp. 334–345.

[56] L. H. ERBE AND P. YAN, Oscillation criteria for Hamiltonian matrix difference systems,*Proc. Amer. Math. Soc.* **119** (1992), pp. 525–533.

[57] L. H. ERBE AND P. YAN, Weighted averaging techniques in oscillation theory for second order difference equations, *Can. Math. Bull.* **35** (1992), pp. 61–69.

[58] L. H. ERBE AND P. YAN, On the discrete Riccati equation and its applications to discrete Hamiltonian systems, *Rocky Mountain J. Math.* **25** (1995), pp. 167–178.

[59] L. H. ERBE AND B. G. ZHANG, Oscillation of second order linear difference equations, *Chinese J. Math.* **16** (1988), pp. 239–252.

[60] L. EULER, Specimen algorithmi singularis, *Novi Commentarii Academiae Scientiarum Imperialis Petropolitanea,* **9** (1762), summary on pp. 10–13; full article on pp. 53–69.

[61] W. N. EVERITT, On the transformation theory of ordinary second–order linear symmetric differential expressions, *Czechoslovak Math. J.* **32 (107)** (1982), pp. 275–306.

[62] W. FAIR, Noncommutative continued fractions, *SIAM J. Math. Anal.* **2** (1971), pp. 226–232.

[63] W. FAIR, A convergence theorem for noncommutative continued fractions, *J. Approximation Theory* **5** (1972), pp. 74–76.

[64] T. FORT, "*Finite Differences and Difference Equations in the Real Domain*", Oxford University Press, London, 1948.

[65] S. FRIEDLANDER, W. STRAUSS AND M. VISHIK, Nonlinear instability in an ideal fluid, *Ann. IHP, J. Nonlinear* (to appear).

[66] É. GALOIS, Démonstration d'un théorème sur les fractions continues périodiques, *Annales de Mathématiques M. Gergonne (Annales de Mathématiques Pures et Appliquées),* **19** (1828–1829) pp. 294–301.

[67] W. GAUTSCHI, Computational aspects of three–term recurrence relations, *SIAM Review* **9** (1967), pp. 24–82.

[68] G. H. GOLUB AND C. F. VAN LOAN, "*Matrix Computations* ," Second Edition, Johns Hopkins University Press, Baltimore, 1989.

[69] A. HALANY AND V. IONESCU, Anticausal stabilizing solution to discrete reverse–time Riccati equation, *Computers Math. Applic.* **28** (1994), pp. 115–126.

[70] A. HALANY AND V. IONESCU, Properties of some global solutions to the discrete–time Riccati equation associated to a contracting input–output operator, *J. Difference Equations* **1** (1995), pp. 61–71.

[71] D. HANKERSON, Right and left disconjugacy in difference equations, *Rocky Mountain J. Math.* **20** (1990), pp. 987–995.

[72] B. HARMSEN, The discrete variational problem with right focal constraints, *PanAmerican Math. J.* **5** (1995), pp. 43–61.

[73] B. HARMSEN, The discrete variational problem: The vector case with right focal constraints, *PanAmerican Math. J.* **6** (1996), pp. 23–37.

[74] B. HARMSEN, The $(2,2)$ focal discrete variational problem, *Communications in Applied Analysis* (to appear).

[75] V. C. HARRIS, "*A system of difference equations and an associated boundary value problem,*" Ph. D. Dissertation, Northwestern University, 1950.

[76] P. HARTMAN, Difference equations: disconjugacy, Green's functions, complete monotonicity, *Trans. Amer. Math. Soc.* **246** (1978), pp. 1–30.

[77] P. HARTMAN, "*Ordinary Differential Equations,*" John Wiley, New York, 1973.

[78] J. J. HENCH AND A. J. LAUB, Numerical solution of the discrete–time Riccati equation, *IEEE Trans. Automatic Control,* **39** (1994), pp. 1197–1210.

[79] D. B. HINTON AND R. T. LEWIS, Spectral analysis of second order difference equations, *J. Math. Anal. Appl.* **63** (1978), pp. 421–438.

[80] J. W. HOOKER, M. K. KWONG, AND W. T. PATULA, Riccati type transformations for second–order linear difference equations II, *J. Math. Anal. Appl.* **107** (1985), pp. 182–196.

[81] J. W. HOOKER, M. K. KWONG, AND W. T. PATULA, Oscillatory second order linear difference equations and Riccati equations, *SIAM. J. Math. Anal.* **18** (1987), pp. 54–63.

[82] J. W. HOOKER AND W. T. PATULA, Riccati type transformations for second–order linear difference equations, *J. Math. Anal. Appl.* **82** (1981), pp. 451–462.

[83] P. IGLESIAS AND K. GLOVER, State space approach to discrete–time H_∞ control, *Int. J. Control* **54** (1991), pp. 1031–1073.

[84] V. IONESCU AND MARTIN WEISS, Two–Riccati formulae for the discrete–time H_∞ control problem, *Int. J. Control* **57** (1993), pp. 141–195.

[85] A. JERRI, "*Linear Difference Equations with Discrete Transform Methods*", Kluwer Academic Publishers, Boston, 1996.

[86] E. A. JONCKHEERE AND L. M. SILVERMAN, Spectral theory of the linear–quadratic optimal control problem: discrete–time single–input case, *IEEE Trans. on Circuits and Systems,* **25** (1978), pp. 810–825.

[87] W. B. JONES AND W. J. THRON, "*Continued Fractions: Analytic Theory and Applications*", Encyclopedia of Math. and its Applications, Volume 11, Addison-Wesley, Reading, MA, 1980.

[88] W. G. KELLEY AND A. C. PETERSON, "*Difference Equations, An Introduction with Applications*", Academic Press, Harcourt Brace Jovanovich, San Diego, California, 1991.

[89] A. N. KHOVANSKII, *The Application of Continued Fractions and Their Generalizations to Problems in Approximation Theory,* translated by P. Wynn, P. Noordhoff, Ltd, Groningen, The Netherlands, 1963.

[90] N. KOMAROFF, Upper bounds for the solution of the discrete Riccati equation, *IEEE Trans. Automatic Control* **9** (1992), pp. 1370–1372.

[91] Q. KONG AND A. ZETTL, Interval oscillation conditions for difference equations, *SIAM J. Math. Anal.* **26** (1995), pp. 1047–1060.

[92] W. KRATZ, "*Quadratic Functionals in Variational Analysis and Control Theory,*" Akademie Verlag, Berlin, 1995.

[93] V. LAKSHMIKANTHAM AND D. TRIGIANTE, "*Theory of Difference Equations: Numerical Methods and Applications,*" Academic Press, New York, 1988.

[94] P. LANCASTER, A. C. M. RAN, AND L. RODMAN, Hermitian solutions of the discrete algebraic Riccati equation, *Int. J. Control* **44** (1986), pp. 777–802.

[95] P. LANCASTER AND L. RODMAN, "*Algebraic Riccati Equations,*" Oxford Science Publications, Clarendon Press, Oxford, 1995.

[96] P. LEVRIE, M. V. BAREL, AND A. BULTHEEL, First–order linear recurrence systems and general N–fractions , *in* "*nonlinear numerical methods and rational approximation II,*" Kluwer Academic Publishers, Boston, 1994, pp. 433–446.

[97] P. LEVRIE AND A. BULTHEEL, First–order linear recurrence systems and matrix continued fractions, *Dept. of Computer Science, K. U. Leuven, Belgium, Report TW 235,* November, 1995.

[98] P. J. MCCARTHY, Note on the oscillation of solutions of second order linear difference equations, *Portugal. Math.* **18** (1959), pp. 203–205.

[99] E. J. MCSHANE AND T. A. BOTTS, "*Real Analysis,*" Van Nostrand, New York, 1959.

[100] F. MERDIVENCI, Green's matrices and positive solutions of a discrete boundary value problem, *PanAmerican Math. J.* **5** (1995), pp. 25–42.

[101] R. E. MICKENS, Construction of finite difference schemes for coupled nonlinear oscillators derived from a discrete energy function, *J. Difference Equations and Appl.* **2** (1996), pp. 185–193.

[102] A. MINGARELLI, "*Volterra–Stieltjes Integral Equations and Generalized Ordinary Differential Expressions,*" *Lecture Notes in Mathematics* **989**, Springer–Verlag, New York, 1983.

[103] T. MORI, N. FUKUMA, AND M. KUWAHARA, On the discrete Riccati equation, *IEEE Trans. Automat. Contr.* **32** (1987), pp. 828–829.

[104] M. MORSE, A generalization of the Sturm separation and comparison theorems in n–space, *Mathematische Annalen* **103** (1930), pp. 52–69.

[105] M. MORSE, "*The Calculus of Variations in the Large,*" AMS Colloquium Publication XVIII, (1934), American Mathematical Society, Providence, R.I.

[106] G. DE NICOLOA, On the time–varying Riccati difference equation of optimal filtering, *SIAM J. Control Optim.* **30** (1992), pp. 1251–1269.

[107] R. NIKOUKHAH, A. S. WILLSKY, AND B. C. LEVY, Kalman filtering and Riccati equations for descriptor systems, *IEEE Trans. Automat. Contr.* **37** (1992), pp. 1325–1342.

[108] F. W. J. OLVER AND D. J. SOOKNE, Note on backward recurrence algorithms, *Math. Comput.* **26** (1972), pp. 941–947.

[109] T. PAPPAS, A. J. LAUB, AND N. R. SANDELL, On the numerical solution of the discrete–time algebraic Riccati equation, *IEEE Trans. Automat. Contr.* **25** (1980), pp. 631–641.

[110] W. PATULA, Growth and oscillation properties of second order linear difference equations, *SIAM J. Math. Anal.* **10** (1979), pp. 55–61.

[111] W. PATULA, Growth, oscillation, and comparison theorems for second order linear difference equations, *SIAM J. Math. Anal.* **10** (1979), pp. 1272–1279.

[112] M. PAVON AND H. K. WIMMER, A comparison theorem for matrix Riccati difference equations, *Systems Control Letters* **19** (1992), pp. 233–239.

[113] T. PEIL AND A. PETERSON, Criteria for C–disfocality of a self–adjoint vector difference equation, *J. Math. Anal. Appl.* **179** (1993), pp. 512–524.

[114] T. PEIL AND A. PETERSON, Asymptotic behavior of solutions of a two term difference equation, *Rocky Mountain J. Math.* **24** (1994), pp. 233–252.

[115] O. PERRON, "*Die Lehre von den Kettenbrüchen,*" Zweite verbesserte Auflage, Chelsea, New York, 1950.

[116] A. PETERSON, Boundary value problems for an nth order linear difference equation, *SIAM J. Math. Anal.* **15** (1984), pp. 124–132.

[117] A. PETERSON, Green's functions and disconjugacy of a vector difference equation, *in "Pitman Research Notes in Mathematics Series"*, J. Wiener and J. K. Hale, Eds. **272**, (1992), pp. 166–180.

[118] A. PETERSON, C-disfocality for linear Hamiltonian difference systems, *J. Differential Equations* **110** (1994), pp. 53–66.

[119] A. PETERSON AND J. RIDENHOUR, Disconjugacy for a second order system of difference equations, *in "Differential Equations: Stability and Control"*, S. Elaydi, Ed., *Lecture Notes in Pure and Applied Math.* **127** (1990), pp. 423–429.

[120] A. PETERSON AND J. RIDENHOUR, Oscillation theorems for second order scalar difference equations, *in"Differential Equations: Stability and Control"*, Marcel Dekker (1990), pp. 417–424.

[121] A. PETERSON AND J. RIDENHOUR, Oscillation of second order linear matrix difference equations, *J. Differential Equations* **89** (1991), pp. 69–88.

[122] A. PETERSON AND J. RIDENHOUR, Atkinson's superlinear oscillation theorem for matrix difference equations, *SIAM J. Math. Anal.* **22** (1991), pp. 774–784.

[123] A. PETERSON AND J. RIDENHOUR, A disconjugacy criterion of W. T. Reid for difference equations, *Proc. Amer. Math. Soc.* **114** (1991), pp. 459–468.

[124] A. PETERSON AND J. RIDENHOUR, A disfocality criterion for an nth order difference equation, *in "Proceedings of the First International Conference on Difference Equations, Trinity University, San Antonio, Texas, May 25–28, 1994,"* edited by S. N. Elaydi, J. R. Graef, G. Ladas, and A. C. Peterson, Gordon and Breach Publishers, Newark, New Jersey, (1996), pp. 411–418.

[125] A. PETERSON AND J. RIDENHOUR, The $(2,2)$–disconjugacy of a fourth order difference equation , *J. Difference Eqs. and Appl.* **1** (1995), pp. 87–93.

[126] P. PFLUGER, "*Matrizenkettenbrüche,*" Dipl. Math. ETH, Zürich, Diss. Nr. 3862, Juris, Zürich, 1966.

[127] S. PINCHERLE, Delle funzioni ipergeometriche e di varie questioni ad esse attinenti, *Giornale di Mathematiche di Battaglini* **32** (1894), especially, Capitolo III, pp. 228–230.

[128] J. Popenda, Oscillation and nonoscillation theorems for second-order difference equations, *J. Math. Anal. Appl.* **123** (1897), pp. 34–38.

[129] A. C. M. Ran and R. Vreugdenhill, Existence and comparison theorems for algebraic Riccati equations for continuous and discrete time systems, *Linear Algebra Appl.* **99** (1988), pp. 63–83.

[130] W. T. Reid, Oscillation criteria for linear differential systems with complex coefficients, *Pacific J. Math.* **6** (1956), pp. 733–751.

[131] W. T. Reid, Principal solutions of non-oscillatory self-adjoint linear differential systems, *Pacific J. Math.* **8** (1958), pp. 147–169.

[132] W. T. Reid, Generalized inverses of differential and integral operators, *in "Proc. of Symposium on Theory and Application of Generalized Inverses of Matrices,"* Texas Technological College, Lubbock, Texas, March, 1968, pp. 1–25.

[133] W. T. Reid, A matrix Liapunov inequality, *J. Math. Anal. Appl.* **32** (1970), pp. 424–434.

[134] W. T. Reid, *"Ordinary Differential Equations,"* John Wiley, New York, 1971.

[135] W. T. Reid, *"Fundamentals of Real Analysis,"* unpublished, Norman Oklahoma, circa 1976.

[136] H.-J. Runckel, Pincherle's theorem for algebraic rings, unpublished notes, Abt. Mathematik, Universität Ulm, November 1995.

[137] J. M. Sanz-Serna, *"Numerical Hamiltonian Problems,"* Applied Mathematics and Mathematical Computation, Vol. 7, Chapman and Hall, New York, 1994.

[138] A. Schelling, *"Matrizenkettenbrüche,"* Dissertation, Universität Ulm, 1993.

[139] H. Schwerdtfeger, Moebius transformations and continued fractions, *Bull. Amer. Math. Soc.* **52** (1946), pp. 307–309.

[140] L. Silverman, Discrete Riccati equations: alternative algorithms asymptotic properties, and system theory interpretations, *Control and Dynamical Systems* **12** (1976), pp. 313–386.

[141] G. F. Simmons, *"Differential Equations,"* McGraw-Hill, New York, 1972.

[142] D. T. SMITH, On the spectral analysis of self adjoint operators generated by second order difference equations, *Proc. Royal Soc. Edinburgh* **118A** (1991), pp. 139–151.

[143] G W. STEWART, "*Introduction to Matrix Computations,*" Academic Press, New York, 1973.

[144] A. A. STOORVOGEL, The discrete–time H_∞ control problem with measurment feedback, *SIAM J. Control Optim.* **30** (1992), pp. 182–202.

[145] A. A. STOORVOGEL AND A. J. T. M. WEEREN, The discrete time Riccati equation related to the H_∞ control problem, "*Proc.* 1992 *American Control Conference,*" Americal Automatic Control Council, Evanston, IL., (1992) pp. 1128–1132.

[146] MI-CHING TSAI, CHIN-SHIONG TSAI, AND YORK-YIH SUN, On discrete-time H_∞ control: A J-lossless coprime factorization approach, *IEEE Trans. Automatic Control* **38** (1993), pp. 1143–1147.

[147] D. R. VAUGHAN, A nonrecursive algebraic solution for the discrete Riccati equation, *IEEE Trans. Automatic Control* **15** (1970), pp. 597–599.

[148] T. VOEPEL, Finite singularities of difference equations, Master's Thesis, University of Missouri, Columbia (1995) *Dynamics of Continuous, Discrete and Impulsive Systems*, Univ. of Waterloo (to appear).

[149] D. J. WALKER, Relationships between three discrete–time H_∞ algebraic Riccati equation solutions, *Int. J. Control* **52** (1990), pp. 801–809.

[150] H. S. WALL, "*Analytic Theory of Continued Fractions*", Van Nostrand, New York, 1948.

[151] A. J. T. M. WEEREN, "*Solving discrete time algebraic Riccati equations,*" Master's thesis, Eindhoven University of Technology, 1991.

[152] H. K. WIMMER, Geometry of the discrete–time algebraic Riccati equation, *J. Math. Syst. Estim. Control* **2** (1992), pp. 123–132.

[153] H. K. WIMMER, Monotonicity and maximality of solutions of discrete–time algebraic Riccati equations, *J. Math. Syst. Estim. Control* **2** (1992), pp. 219–235.

[154] J. WIMP, "*Computation with Recurrence Relations,*" Pitman, Boston, 1984.

[155] L. XIE, C. E. DE SOUZA, AND Y. WANG, Robust control of discrete time uncertain dynamical systems, *Automatica* **29** (1993), pp. 1133–1137.

[156] I. YAESH AND U. SHAKED, Minimum H_∞–norm regulation of linear discrete-time systems and its relation to linear quadratic discrete games, *IEEE Trans. Automatic Control* **35** (1990), pp. 1061–1064.

[157] I. YAESH AND U. SHAKED, A transfer function approach to the problems of discrete–time systems: H_∞–optimal linear control and filtering, *IEEE Trans. Automatic Control* **36** (1991), pp. 1246–1271.

[158] I. YAESH AND U. SHAKED, H_∞–optimal one–step–ahead output feedback control of discrete–time systems, *IEEE Trans. Automatic Control* **37** (1992), pp. 1245–1250.

[159] I. YAESH AND U. SHAKED, Game theory approach to state estimation of linear discrete–time processes and its relation to H_∞ optimal estimation, *Int. J. Control* **55** (1992), pp. 1443–1452.

[160] B. G. ZHANG, Oscillation and asymptotic behavior of second order difference equations, *J. Math. Anal. Appl.* **173** (1993), pp. 58–68.

[161] M. ZNOJIL, The generalized continued fractions and potentials of the Lennard-Jones type, *J. Math. Phys.* **31** (1990), pp. 1955–1961.

Index

Kluwer Texts in the Mathematical Sciences

1. A.A. Harms and D.R. Wyman: *Mathematics and Physics of Neutron Radiography.* 1986 ISBN 90-277-2191-2
2. H.A. Mavromatis: *Exercises in Quantum Mechanics.* A Collection of Illustrative Problems and Their Solutions. 1987 ISBN 90-277-2288-9
3. V.I. Kukulin, V.M. Krasnopol'sky and J. Horácek: *Theory of Resonances.* Principles and Applications. 1989 ISBN 90-277-2364-8
4. M. Anderson and Todd Feil: *Lattice-Ordered Groups.* An Introduction. 1988 ISBN 90-277-2643-4
5. J. Avery: *Hyperspherical Harmonics.* Applications in Quantum Theory. 1989 ISBN 0-7923-0165-X
6. H.A. Mavromatis: *Exercises in Quantum Mechanics.* A Collection of Illustrative Problems and Their Solutions. Second Revised Edition. 1992 ISBN 0-7923-1557-X
7. G. Micula and P. Pavel: *Differential and Integral Equations through Practical Problems and Exercises.* 1992 ISBN 0-7923-1890-0
8. W.S. Anglin: *The Queen of Mathematics.* An Introduction to Number Theory. 1995 ISBN 0-7923-3287-3
9. Y.G. Borisovich, N.M. Bliznyakov, T.N. Fomenko and Y.A. Izrailevich: *Introduction to Differential and Algebraic Topology.* 1995 ISBN 0-7923-3499-X
10. J. Schmeelk, D. Takači and A. Takači: *Elementary Analysis through Examples and Exercises.* 1995 ISBN 0-7923-3597-X
11. J.S. Golan: *Foundations of Linear Algebra.* 1995 ISBN 0-7923-3614-3
12. S.S. Kutateladze: *Fundamentals of Functional Analysis.* 1996 ISBN 0-7923-3898-7
13. R. Lavendhomme: *Basic Concepts of Synthetic Differential Geometry.* 1996 ISBN 0-7923-3941-X
14. G.P. Gavrilov and A.A. Sapozhenko: *Problems and Exercises in Discrete Mathematics.* 1996 ISBN 0-7923-4036-1
15. R. Singh and N. Singh Mangat: *Elements of Survey Sampling.* 1996. ISBN 0-7923-4045-0
16. C.D. Ahlbrandt and A.C. Peterson: *Discrete Hamiltonian Systems.* Difference Equations, Continued Fractions, and Riccati Equations. 1996 ISBN 0-7923-4277-1

KLUWER ACADEMIC PUBLISHERS – DORDRECHT / BOSTON / LONDON

GPSR Compliance
The European Union's (EU) General Product Safety Regulation (GPSR) is a set of rules that requires consumer products to be safe and our obligations to ensure this.

If you have any concerns about our products, you can contact us on

ProductSafety@springernature.com

In case Publisher is established outside the EU, the EU authorized representative is:

Springer Nature Customer Service Center GmbH
Europaplatz 3
69115 Heidelberg, Germany

www.ingramcontent.com/pod-product-compliance
Ingram Content Group UK Ltd.
Pitfield, Milton Keynes, MK11 3LW, UK
UKHW022327190726
13856UKWH00001B/252

* 9 7 8 1 4 7 5 7 2 4 6 8 4 *